LIBRARY
Institute of Cancer Research
15 Cotswold Road
Sutton
SM2 5NG

Institute of Cancer Research Library

Ribozymes and RNA Catalysis

RSC Biomolecular Sciences

Editorial Board:

Professor Stephen Neidle (Chairman), *The School of Pharmacy, University of London, London, UK*
Dr Simon F Campbell CBE, FRS
Dr Marius Clore, *National Institutes of Health, USA*
Professor David M J Lilley FRS, *University of Dundee, UK*

This Series is devoted to coverage of the interface between the chemical and biological sciences, especially structural biology, chemical biology, bio- and chemo-informatics, drug discovery and development, chemical enzymology and biophysical chemistry. Ideal as reference and state-of-the-art guides at the graduate and post-graduate level.

Titles in the Series:

Biophysical and Structural Aspects of Bioenergetics
Edited by Mårten Wikström, *University of Helsinki, Finland*
Computational and Structural Approaches to Drug Discovery: Ligand-Protein Interactions
Edited by Robert M Stroud and Janet Finer-Moore, *University of California in San Francisco, San Francisco, CA, USA*
Exploiting Chemical Diversity for Drug Discovery
Edited by Paul A. Bartlett, *Department of Chemistry, University of California, Berkeley* and Michael Entzeroth, *S*Bio Pte Ltd, Singapore*
Metabolomics, Metabonomics and Metabolite Profiling
Edited by William J. Griffiths, *University of London, The School of Pharmacy, University of London, London, UK*
Protein–Carbohydrate Interactions in Infectious Disease
Edited by Carole A. Bewley, *National Institutes of Health, Bethesda, Maryland, USA*
Quadruplex Nucleic Acids
Edited by Stephen Neidle, *The School of Pharmacy, University of London, London, UK* and Shankar Balasubramanian, *Department of Chemistry, University of Cambridge, Cambridge, UK*
Ribozymes and RNA Catalysis
Edited by David MJ Lilley FRS, *University of Dundee, Dundee, UK* and Fritz Eckstein, *Max-Planck-Institut for Experimental Medicine, Goettingen, Germany*
Sequence-specific DNA Binding Agents
Edited by Michael Waring, *Department of Pharmacology, University of Cambridge, Cambridge, UK*
Structural Biology of Membrane Proteins
Edited by Reinhard Grisshammer and Susan K. Buchanan, *Laboratory of Molecular Biology, National Institutes of Health, Bethesda, Maryland, USA*
Structure-based Drug Discovery: An Overview
Edited by Roderick E. Hubbard, *University of York, UK and Vernalis (R&D) Ltd, Cambridge, UK*

Visit our website on www.rsc.org/biomolecularsciences

For further information please contact:
Sales and Customer Care, Royal Society of Chemistry, Thomas Graham House, Science Park, Milton Road, Cambridge, CB4 0WF, UK
Telephone: +44 (0)1223 432360, Fax: +44 (0)1223 426017, Email: sales@rsc.org

Ribozymes and RNA Catalysis

Edited by

David M.J. Lilley FRS
University of Dundee, Dundee, UK

Fritz Eckstein
Max-Planck-Institut for Experimental Medicine, Göttingen, Germany

RSCPublishing

ISBN: 978-0-85404-253-1

A catalogue record for this book is available from the British Library

© The Royal Society of Chemistry 2008

All rights reserved

Apart from fair dealing for the purposes of research for non-commercial purposes or for private study, criticism or review, as permitted under the Copyright, Designs and Patents Act 1988 and the Copyright and Related Rights Regulations 2003, this publication may not be reproduced, stored or transmitted, in any form or by any means, without the prior permission in writing of The Royal Society of Chemistry, or in the case of reproduction in accordance with the terms of licences issued by the Copyright Licensing Agency in the UK, or in accordance with the terms of the licences issued by the appropriate Reproduction Rights Organization outside the UK. Enquiries concerning reproduction outside the terms stated here should be sent to The Royal Society of Chemistry at the address printed on this page.

Published by The Royal Society of Chemistry,
Thomas Graham House, Science Park, Milton Road,
Cambridge CB4 0WF, UK

Registered Charity Number 207890

For further information see our web site at www.rsc.org

Preface

Until 25 years ago it was thought that proteins were the only macromolecular catalysts in the cell. The discovery that RNA could also act in this role was a great paradigm shift in molecular biology. This had important implications for thinking about how life emerged on the planet, supplying one of the prerequisites of the hypothetical RNA world. And while it initially seemed that ribozymes were confined to some relatively obscure corners of biology, we now know that this was not true at all. Some extremely important cellular reactions such as the condensation of amino acids to form polypeptides and probably the splicing of mRNA are catalysed by RNA molecules. So far from being confined to a few reactions in some slimy pond-dwellers and the like, catalytic RNA is very important in biology generally.

RNA catalysis also threw down a fascinating challenge to the biological chemist. Proteins have all the advantages as catalysts – decorated with chemically contrasting side chains that make them Nature's chemistry set. By comparison, RNA is chemically much less adventurous, yet ribozymes perform some impressive feats of catalysis. So just what are the mechanistic origins of chemical catalysis in ribozymes? That is the essence of this volume.

It is ten years since we edited a volume on essentially the same topic. In that time huge advances have been made. The atomic structures of most ribozymes have been determined over this period, and we know much more about the folding of these species. But at the same time our mechanistic understanding of ribozymes has advanced greatly, providing insight into the origins of catalysis. This can never reach a final conclusion. Probably no protein enzyme can be said to be totally understood, and chemists have been studying these for much longer. But some consensus on general mechanisms of RNA catalysis is now developing, and thus it is a good moment to review the area again.

To do this we have assembled a list of authors who are unquestionably the world leaders in each topic. They have provided a comprehensive view of ribozyme mechanism at this time. In each case these present the personal viewpoint of the authors. Of course the subject will develop further, and some things will change. Some aspects may prove to be wrong in time. But we believe that most of the general principles will survive. We are indebted to our authors who have written a wonderful set of reviews, both clear and authoritative. Finally, we are very pleased that Tom Cech has written the Foreword to this volume, providing some historical perspective – so appropriately since he started us on this journey 25 years ago.

We hope this volume will serve as a resource for the RNA community, and an inspiration for the next generation of scientists who will go on to take this fascinating field further.

David M.J. Lilley, Dundee
Fritz Eckstein, Göttingen
May 2007

FOREWORD
Twenty-Five Years of Ribozymes

Twenty-five years have passed since the discovery of the first ribozyme! It hardly seems possible. In 1982, when we announced that the pre-ribosomal RNA intron from the ciliated protozoan *Tetrahymena* was a self-splicing RNA with enzyme-like properties, it required a bit of daring to speak of catalytic RNA. And we really stuck our necks out when we coined the word "ribozyme" to describe the universe of ribonucleic acid molecules with enzyme-like activities, because at that time this universe consisted of just the one example.[1] Now, the field is so well-established that several books, including this one, have been dedicated to the topic. And high school and college textbooks have included sections on ribozymes for so long that the students think they've been known "forever".

Of course, had there been just the one example, the ribozyme would have remained a curiosity, not a field. The first addition to the list was RNase P, an endonuclease that produces the mature 5′ end of transfer RNAs in all organisms.[2,3] This was a key addition, especially because the RNA acted naturally as a true enzyme, processing a limitless number of substrate RNAs without being altered in the process. It later turned out that ribozymes including the self-splicing *Tetrahymena* intron, self-splicing group II introns, and the self-cleaving hammerhead and hairpin ribozymes could be easily engineered to be trans-acting, *i.e.*, to cleave exogenous substrates with multiple turnover.[4-8] Thus, the distinction between a self-acting ribozyme and a true enzyme remains biologically significant, but the distinction is not so important when considering the chemical and mechanistic features of transition state stabilization in the active site.

At the same time that these boutique catalytic RNAs were being studied, Harry Noller was steadfastly claiming that the central ribonucleoprotein particle of life, the ribosome, was an RNA machine.[9,10] His elegant biochemical experiments and the compelling evolutionary argument for an ancestral RNA-based protein synthetic apparatus[11-13] convinced many of us, but it was nevertheless thrilling to see directly the RNA forming the peptidyltransferase center[14] and binding a transition-state analog.[15] Not only is the ribosome the most important ribozyme, but it also serves as a rich model to understand the contributions of protein components of RNP enzymes.

Many details of the chemistry, biology, and structure of ribozymes have by now been revealed, thanks to the intense international effort documented in this volume. Thus, it is with some apprehension that I look back on what we initially predicted about RNA catalysis in 1982. One prediction of general interest concerned nuclear pre-mRNA splicing, which was still two years away from being recapitulated *in vitro*.[16,17] We suggested that "while mRNA precursors are unlikely to be self-splicing, it remains possible that they undergo such a reaction when complexed with small nuclear RNPs. If the RNA moiety of the small nuclear RNP bound a nucleotide cofactor or participated in catalysis in any way, it would be a ribozyme."[1] The finding that group II introns underwent self-splicing by the same lariat-formation reaction as pre-mRNA introns provided support for this contention,[18–20] which is now widely accepted but not yet definitive.[21]

Another conjecture was that "the autoexcision of the IVS may be reversible," the reverse reaction consisting of "integration of the IVS into another RNA molecule" that might then provide a retrotransposition pathway "for introduction of an IVS into a gene that was formerly contiguous."[1] [IVS, intervening sequence, is synonymous with "intron."] It was later possible to demonstrate such activity for group I and group II introns.[22,23] Even more exciting was the finding that group II introns directly insert themselves into double-stranded DNA, a reaction that may provide a new tool for *in vivo* mutagenesis and potentially for gene therapy.[24,25]

The last general prediction was that "single-stranded DNA might, under some conditions, be self-splicing. Such an event might take place when the DNA strands were separated for replication, and, if it could be regulated, might provide a mechanism for DNA rearrangements during cellular differentiation."[1] In fact, *in vitro* selection was later used to identify catalytic single-stranded DNAs with RNA-cleavage activity, dubbed "DNAzymes" or "deoxyribozymes".[26] However, the speculation about a biological role for DNA catalysis has yet to be substantiated.

Finally, we made specific proposals that the IVS RNA would have several enzyme-like properties.[1]

- "It lowers the activation energy for specific bond cleavage and formation events." This was a safe prediction, because it was just restating our observations in chemical terms! However, in this first paper we completely missed the role of metal ion catalysis, which subsequently became a significant research area.[27–30]
- "Its activity depends on a precise structure." After an immense effort, the crystal structure of a domain of the *Tetrahymena* ribozyme was solved,[31] and then an entire active ribozyme at modest resolution,[32] and finally three different group I introns at higher resolution.[33–35]
- "It has a specific binding site for the guanosine cofactor." When we provided indirect evidence for direct ligand-binding by the *Tetrahymena* ribozyme;[36] the idea was met with disbelief by those who thought that only proteins could have such ability. Yet the site was eventually

pin-pointed within the RNA[37] and visualized in the various crystal structures, and, with the discovery of riboswitches, the concept has been found to be central to a long-undiscovered mode of regulation of gene expression.[38]

- "The RNA forms an active site cleft or hole that can exclude water, thereby preventing hydrolysis after each cleavage step." A concave active site in fact existed,[32] and the exclusion of water by RNA active sites has been best described for the case of the ribosome.[39]

In retrospect, it is amusing to see how these proposals, which seemed rather bold in 1982, seem so obvious today.

On a more personal note, the study of RNA catalysis has been great fun because it has brought me into contact with such great people. The list of contributors to this volume, for example, includes several former postdoctoral fellows and collaborators, and the rest are all "science friends". Students who are still deciding on a career may not yet appreciate how the life of a scientist is not at all a solitary pursuit, but involves entry into a community of scholars who share common goals, collaborate, critique each others' work, and generally enjoy each others' company.

Thomas R. Cech
Department of Chemistry and Biochemistry, University of Colorado
and Howard Hughes Medical Institute
Chevy Chase
Maryland
USA

References

1. K. Kruger et al., Self-splicing RNA: Autoexcision and autocyclization of the ribosomal RNA intervening sequence of *Tetrahymena*, *Cell*, 1982, **31**, 147–157.
2. C. Guerrier-Takada et al., The RNA moiety of ribonuclease P is the catalytic subunit of the enzyme, *Cell*, 1983, **35**, 849–857.
3. C. Guerrier-Takada and S. Altman, Catalytic activity of an RNA molecule prepared by transcription in vitro, *Science*, 1984, **223**, 286–186.
4. A.J. Zaug and T.R. Cech, The intervening sequence RNA of *Tetrahymena* is an enzyme, *Science*, 1986, **231**, 470–475.
5. A. Jacquier and M. Rosbash, Efficient trans-splicing of a yeast mitochondrial RNA group II intron implicates a strong 5' exon-intron interaction, *Science*, 1986, **234**, 1099–1104.
6. O.C. Uhlenbeck, A small catalytic oligoribonucleotide, *Nature*, 1987, **328**, 596–600.
7. J. Haseloff and W.L. Gerlach, Simple RNA enzymes with new and highly specific endoribonuclease activities, *Nature*, 1988, **334**, 585–591.

8. J. Hampel and R. Tritz, RNA catalytic properties of the minimum (-)sTRSV sequence, *Biochemistry*, 1989, **28**, 4929–4933.
9. H.F. Noller *et al.*, Secondary structure model for 23S ribosomal RNA, *Nucleic Acids Res.*, 1981, **9**, 6167–6189.
10. H.F. Noller, V. Hoffarth and L. Zimniak, Unusual resistance of peptidyl transferase to protein extraction methods, *Science*, 1992, **256**, 1416–1419.
11. F.H.C. Crick, The origin of the genetic code, *J. Mol. Biol.*, 1968, **38**, 367–379.
12. L.E. Orgel, Evolution of the genetic apparatus, *J. Mol. Biol.*, 1968, **38**, 381–393.
13. D.R. Woese, Just so stories and Rube Goldberg machines: Speculations on the origin of the protein synthetic machinery. In *Ribosomes: Structure, Function and Genetics*, ed. G. Chambliss *et al.*, University Park Press, Baltimore, 1980, pp. 357–373.
14. N. Ban *et al.*, The complete atomic structure of the large ribosomal subunit at 2.4A resolution, *Science*, 2000, **289**, 905–920.
15. M. Welch, J. Chastang and M. Yarus, An inhibitor of ribosomal peptidyl transferase using transition-state analogy, *Biochemistry*, 1995, **34**, 385–390.
16. P.J. Grabowski, R.A. Padgett and P.A. Sharp, Messenger RNA splicing in vitro: An excised intervening sequence and a potential intermediate, *Cell*, 1984, **37**, 415–427.
17. A.R. Krainer *et al.*, Normal and mutant human beta-globin pre-mRNAs are faithfully and efficiently spliced in vitro, *Cell*, 1984, **36**, 993–1005.
18. C.L. Peebles *et al.*, A self-splicing RNA excises an intron lariat, *Cell*, 1986, **44**, 213–223.
19. C. Schmelzer and R.J. Schweyen, Self-splicing of group II introns in vitro: Mapping of the branch point and mutational inhibition of lariat formation, *Cell*, 1986, **46**, 557–565.
20. R. van der Veen *et al.*, Excised group II introns in yeast mitochondria are lariats and can be formed by self-splicing in vitro, *Cell*, 1986, **44**, 225–234.
21. T. Villa, J.A. Pleiss and C. Guthrie, Spliceosomal snRNAs: Mg(2+)-dependent chemistry at the catalytic core? *Cell*, 2002, **109**, 149–152.
22. S.A. Woodson and T.R. Cech, Reverse self-splicing of the Tetrahymena group I intron: Implication for the directionality of splicing and for intron transposition, *Cell*, 1989, **57**, 335–345.
23. S. Augustin, M.W. Muller and R.J. Schweyen, Reverse self-splicing of group II intro RNAs in vitro, *Nature*, 1990, **343**, 383–386.
24. J. Yang, S. Zimmerly, P.S. Perlman and A.M. Lambowitz, Efficient integration of an intron RNA into double-stranded DNA by reverse splicing, *Nature*, 1996, **381**, 332–335.
25. B. Cousineau *et al.*, Retrohoming of a bacterial group II intron: Mobility via complete reverse splicing, independent of homologous DNA recombination, *Cell*, 1998, **94**, 451–462.
26. R.R. Breaker and G.F. Joyce, A DNA enzyme that cleaves RNA, *Chem. Biol.*, 1994, **1**, 223–229.

27. C. Guerrier-Takada, K. Haydock, L. Allen and S. Altman, Metal ion requirements and other aspects of the reaction catalyzed by M1 RNA, the RNA subunit of ribonuclease P from Escherichia coli, *Biochemistry*, 1986, **25**, 1509–1515.
28. J.A. Piccirilli *et al.*, Metal ion catalysis in the Tetrahymena ribozyme reaction, *Nature*, 1993, **361**, 85–88.
29. S. Shan *et al.*, Defining the catalytic metal ion interactions in the Tetrahymena ribozyme reaction, *Biochemistry*, 2001, **40**, 5161–5171.
30. M.R. Stahley and S.A. Strobel, Structural evidence for a two-metal-ion mechanism of group I intron splicing, *Science*, 2005, **309**, 1587–1590.
31. J.H. Cate *et al.*, Crystal structure of a group I ribozyme domain: Principles of RNA packing, *Science*, 1996, **273**, 1678–1685.
32. G.L. Golden *et al.*, A preorganized active site in the crystal structure of the Tetrahymena ribozyme, *Science*, 1998, **282**, 259–264.
33. P.L. Adams *et al.*, Crystal structure of a self-splicing group I intron with both exons, *Nature*, 2004, **430**, 45–50.
34. B.L. Golden, H. Kim and E. Chase, Crystal structure of a phage Twort group I ribozyme-product complex, *Nat. Struct. Mol. Biol.*, 2005, **12**, 82–89.
35. F. Guo, A.R. Gooding and T.R. Cech, Structure of the *Tetrahymena* ribozyme: Base triple sandwich and metal ion at the active site, *Mol. Cell*, 2004, **16**, 351–362.
36. B.L. Bass and T.R. Cech, Specific interaction between the self-splicing RNA of *Tetrahymena* and its guanosine substrate: Implications for biological catalysis by RNA, *Nature*, 1984, **308**, 820–826.
37. F. Michel *et al.*, The guanosine binding site of the Tetrahymena ribozyme, *Nature*, 1989, **342**, 391–395.
38. J.E. Barrick and R.R. Breaker, The power of riboswitches, *Sci. Am.*, 2007, **296**, 50–57.
39. T.M. Schmeing *et al.*, An induced-fit mechanism to promote peptide bond formation and exclude hydrolysis of peptidyl-tRNA, *Nature*, 2005, **438**, 520–524.

Contents

Chapter 1 Ribozymes and RNA Catalysis: Introduction and Primer
David M.J. Lilley and Fritz Eckstein

1.1	What are Ribozymes?	1
1.2	What is the Role of Ribozymes in Cells?	1
1.3	Ribozymes Bring about Significant Rate Enhancements	4
1.4	Why Study Ribozymes?	4
1.5	Folding RNA into the Active Conformation	5
1.6	The Catalytic Resources of RNA – Making a Lot of a Little	6
1.7	Mechanisms and Catalytic Strategies of Ribozymes	7
1.8	Impact of New Methodologies to Study Ribozymes	8
1.9	Finally	8
References		8

Chapter 2 Proton Transfer in Ribozyme Catalysis
Philip C. Bevilacqua

2.1	Scope of Chapter and Rationale	11
2.2	Overview of Proton Transfer Chemistry	12
2.3	General Considerations for Proton Transfer in RNA Enzymes	17
	2.3.1 Classes of Protonation Sites in RNA	17
	2.3.2 Driving Forces for pK_a Shifting in RNA	18
	2.3.3 Quantitative Contributions of Proton Transfer to RNA Catalysis	19
2.4	Proton Transfer in Small Ribozymes: Five Case Studies	20
	2.4.1 Why Small Ribozymes?	20
	2.4.2 Proton Transfer in the Hepatitis Delta Virus Ribozyme	22
	2.4.3 Proton Transfer in the Hairpin Ribozyme	27
	2.4.4 Proton Transfer in the Hammerhead Ribozyme	28

		2.4.5	Proton Transfer in the VS Ribozyme	29
		2.4.6	Proton Transfer in the *glmS* Ribozyme	30
	2.5	Conclusion and Perspectives		31
	Acknowledgement			32
	References			32

Chapter 3 Finding the Hammerhead Ribozyme Active Site
Dominic Lambert and John M. Burke

	3.1	Introduction		37
	3.2	Background		38
	3.3	Experimental Data		40
		3.3.1	Mechanistic Hypothesis Leads to Identification and Functional Test of Active Site Components	40
		3.3.2	Structural Hypothesis – Large-scale Conformational Changes are Required for Catalysis	41
		3.3.3	Molecular Modeling of a Hammerhead Active Fold that Satisfies Structural and Biochemical Constraints	43
	3.4	Current Status and Future Prospects		46
	Acknowledgements			46
	References			46

Chapter 4 Hammerhead Ribozyme Crystal Structures and Catalysis
William G. Scott

	4.1	Introduction	48
	4.2	A Catalytic RNA Prototype	49
	4.3	A Small Ribozyme	49
	4.4	Chemistry of Phosphodiester Bond Isomerization	50
	4.5	Hammerhead Ribozyme Structure	51
	4.6	Catalysis in the Crystal	53
	4.7	Making Movies from Crystallographic Snapshots	53
	4.8	An Ever-growing List of Concerns	55
	4.9	Occam's Razor Can Slit Your Throat	56
	4.10	Structure of a Full-length Hammerhead Ribozyme	57
	4.11	Do the Minimal and Full-length Hammerhead Crystal Structures have Anything in Common?	59
	4.12	How Does the Minimal Hammerhead Work?	60
	4.13	A Movie Sequel with a Happy Ending	61
	4.14	Concluding Remarks	62
	Acknowledgements		62
	References		62

Chapter 5 **The Hairpin and Varkud Satellite Ribozymes**
David M.J. Lilley

	5.1	Nucleolytic Ribozymes	66
	5.2	Hairpin Ribozyme	66
		5.2.1 Structure of the Hairpin Ribozyme	67
		5.2.2 Metal Ion-dependent Folding of the Hairpin Ribozyme	69
		5.2.3 Observing the Cleavage and Ligation Activities of the Hairpin Ribozyme	71
		5.2.4 Mechanism of the Hairpin Ribozyme	73
	5.3	VS Ribozyme	76
		5.3.1 Structure of the VS Ribozyme	77
		5.3.2 Structure of the Substrate	80
		5.3.3 Location of the Substrate	80
		5.3.4 Active Site of the VS Ribozyme	82
		5.3.5 Candidate Catalytic Nucleobases	82
		5.3.6 Mechanism of the VS Ribozyme	84
	5.4	Some Striking Similarities between the Hairpin and VS Ribozymes	88
	Acknowledgements		88
	References		88

Chapter 6 **Catalytic Mechanism of the HDV Ribozyme**
Selene Koo, Thaddeus Novak and Joseph A. Piccirilli

	6.1	Introduction	92
		6.1.1 Hepatitis Delta Virus Biology	92
		6.1.2 Cleavage Reactions of Small Ribozymes	93
	6.2	HDV Ribozyme Structure	95
		6.2.1 Determination of Crystal Structures	95
		6.2.2 Structure Overview	97
		6.2.3 Active Site	97
	6.3	Catalytic Strategies for RNA Cleavage	99
	6.4	The Active Site Nucleobase: C75	100
		6.4.1 Exogenous Base Rescue Reactions	101
		6.4.2 Role of C75 in HDV Catalysis	103
		6.4.3 Resolving the Kinetic Ambiguity	105
		6.4.3.1 Reaction in the Absence of Divalent Cations	105
		6.4.3.2 Sulfur Substitution of the Leaving Group	106
	6.5	Metal Ions in the HDV Ribozyme	108
		6.5.1 Structural Metal Ions	108
		6.5.2 Catalytic Metal Ions	111

	6.6 Contributions of Non-active-site Structures to Catalysis	112
	6.7 Dynamics in HDV Function	113
	6.8 Varieties of Experimental Systems	115
	6.9 Models for HDV Catalysis	117
	6.10 Conclusion	119
	Acknowledgements	120
	References	120

Chapter 7 Mammalian Self-Cleaving Ribozymes
Andrej Lupták and Jack W. Szostak

	7.1 Introduction	123
	7.2 General Features of Small Self-cleaving Sequences	124
	7.3 Genome-wide Selection of Self-cleaving Ribozymes	124
	7.4 *CPEB3* Ribozyme	125
	7.4.1 Expression of the *CPEB3* Ribozyme	126
	7.4.2 Structural Features of the *CPEB3* and HDV Ribozymes	127
	7.4.3 Linkage of HDV to the Human Transcriptome	129
	7.5 Possible Biological Roles of Self-cleaving Ribozymes	130
	7.6 Closing Remarks	131
	References	131

Chapter 8 The Structure and Action of *glmS* Ribozymes
Kristian H. Link and Ronald R. Breaker

	8.1 Introduction	134
	8.2 Biochemical Characteristics of *glmS* Ribozymes	136
	8.2.1 Divalent Metal Ions Support Structure and Not Chemistry	136
	8.2.2 Ligand Specificity of *glmS* Ribozymes	137
	8.2.3 Evidence for a Coenzyme Role for GlcN6P	139
	8.3 Atomic-resolution Structure of *glmS* Ribozymes	141
	8.3.1 Secondary and Tertiary Structures of *glmS* Ribozymes	141
	8.3.2 Metabolite Recognition by *glmS* Ribozymes	143
	8.4 Mechanism of *glmS* Ribozyme Self-cleavage	145
	8.5 Can *glmS* Ribozymes be Drug Targets?	148
	8.6 Conclusions	149
	References	150

Chapter 9   A Structural Analysis of Ribonuclease P
Steven M. Marquez, Donald Evans, Alexei V. Kazantsev and Norman R. Pace

9.1	Introduction	153
9.2	Chemistry of RNase P RNA	155
	9.2.1 Universal	155
	9.2.2 S_N2-type Reaction	155
	9.2.3 pH-Dependence of the Reaction: Hydroxide Ion as the Nucleophile	157
	9.2.4 Metal Ions in Catalysis	157
9.3	Phylogenetic Variation and Structure of RNase P RNA	158
9.4	Early Studies of the RNase P RNA Structure	159
9.5	Crystallographic Studies of Bacterial RNase P RNAs	160
9.6	Modeling an RNase P RNA:tRNA Complex	162
9.7	Modeling the Bacterial RNase P Holoenzyme	163
9.8	Substrate Recognition	165
9.9	Archaeal and Eucaryal Holoenzymes – More Proteins	166
9.10	Concluding Remarks	170
Acknowledgements		171
References		171

Chapter 10   Group I Introns: Biochemical and Crystallographic Characterization of the Active Site Structure
Barbara L. Golden

10.1	Group I Intron Origins	178
10.2	Group I Intron Self-splicing	178
10.3	What has Changed in Group I Intron Knowledge in the Last Decade	181
10.4	Structure of Group I Introns	181
10.5	Crystallography of Group I Introns	182
	10.5.1 *Tetrahymena LSU* P4-P6 Domain	182
	10.5.2 *Tetrahymena* Intron Catalytic Core	183
	10.5.3 Twort orf142-I2 Ribozyme	183
	10.5.4 *Azoarcus* sp. BBH72 tRNA^{Ile} Intron	184
10.6	Structural Basis for Group I Intron Self-splicing	184
	10.6.1 Recognition of the 5′-Splice Site	185
	10.6.2 Does the Ribozyme Undergo Conformational Changes upon P1 Docking?	186
	10.6.3 A Binding Pocket for Guanosine	187
	10.6.4 Packed Stacks	189

	10.7	Biochemical Characterization of the Structure	191
		10.7.1 Metal Ion Binding and Specificity Switches	191
		10.7.2 Identification of Ligands to the Catalytic Metal Ions	192
		10.7.3 Correlation with Metal Ion Binding Sites within the Crystal Structures	193
		10.7.4 Nucleotide Analog Interference Techniques	194
	10.8	What Makes a Catalytic Site?	196
	10.9	Back to the Origins	197
	References		198

Chapter 11 Group II Introns: Catalysts for Splicing, Genomic Change and Evolution
Anna Marie Pyle

	11.1	Introduction: The Place of Group II Introns Among the Family of Ribozymes	201
	11.2	The Basic Reactions of Group II Introns	201
	11.3	The Biological Significance of Group II Introns	204
		11.3.1 Evolutionary Significance	204
		11.3.2 Significance and Prevalence in Modern Genomes	204
		11.3.3 The Potential Utility of Group II Introns	204
	11.4	Domains and Parts: The Anatomy of a Group II Intron	205
		11.4.1 Domain 1	206
		11.4.2 Domain 2	206
		11.4.3 Domain 3	206
		11.4.4 Domain 4	206
		11.4.5 Domain 5	206
		11.4.6 Domain 6	207
		11.4.7 Other Domains and Insertions	207
		11.4.8 Alternative Structural Organization and Split Introns	208
	11.5	A Big, Complicated Family: The Diversity of Group II Introns	208
	11.6	Group II Intron Tertiary Structure	209
	11.7	Group II Intron Folding Mechanisms	211
		11.7.1 A Slow, Direct Path to the Native State	211
		11.7.2 A Folding Control Element in the Center of D1	212
		11.7.3 Proteins and Group II Intron Folding	212
	11.8	Setting the Stage for Catalysis: Proximity of the Splice Sites and Branch-site	213

		11.8.1	Recognition of Exons and Ribozyme Substrates	213
		11.8.2	Branch-site Recognition and the Coordination Loop	213
	11.9	A Single Active-site for Group II Intron Catalysis		215
	11.10	The Group II Intron Active-site: What are the Players?		216
		11.10.1	Active-site Players in D1 and Surrounding Linker Regions	217
		11.10.2	Domain 3 and the J2/3 Linker	217
		11.10.3	Domain 5: Structural and Catalytic Regions	218
	11.11	The Chemical Mechanism of Group II Intron Catalysis		219
	11.12	Proteins and Group II Intron Function		221
		11.12.1	Maturases	221
		11.12.2	CRM-domain Plant Proteins	221
		11.12.3	ATPase Proteins	221
	11.13	Group II Introns and Their Many Hypothetical Relatives		222
	11.14	Group II Introns: RNA Processing Enzymes, Transposons, or Tiny Living Things?		223
	References			223

Chapter 12   The GIR1 Branching Ribozyme
Henrik Nielsen, Bertrand Beckert, Benoit Masquida and Steinar D. Johansen

	12.1	Introduction		229
	12.2	Distribution and Structural Organization of Twin-ribozyme Introns		231
	12.3	Biological Context		234
		12.3.1	Three Processing Pathways of a Twin-ribozyme Intron	234
		12.3.2	Processing of the I-*Dir*I mRNA	235
		12.3.3	Conformational Switching in GIR1	236
	12.4	Biochemical Characterization		238
		12.4.1	GIR1 Catalyzes Three Different Reactions	239
		12.4.2	Characterization of the Branching Reaction	240
		12.4.3	Biochemistry of GIR1	240
	12.5	Modelling the Structure of GIR1		241
		12.5.1	Overall Structure	242
		12.5.2	Coaxially Stacked Helices	242
		12.5.3	Junctions and Tertiary Interactions Involving Peripheral Elements	245

		12.5.4 The Active Site	245
	12.6	Phylogenetic Considerations	247
	12.7	Concluding Remarks	248
	References		249

Chapter 13 Is the Spliceosome a Ribozyme?
Dipali G. Sashital and Samuel E. Butcher

13.1	Introduction	253
13.2	Similarity to Group II Self-splicing Introns	253
13.3	Role of snRNA in the Spliceosome Active Site	255
13.4	Conformation of the U2-U6 Complex and Parallels to Group II Intron Structures	260
13.5	RNA-mediated Regulation in the Spliceosome	262
References		266

Chapter 14 Peptidyl Transferase Mechanism: The Ribosome as a Ribozyme
Marina V. Rodnina

14.1	Introduction: Historical Background	270
14.2	The Ribosome	271
14.3	Peptidyl Transfer Reaction	272
	14.3.1 Characteristics of the Reaction off the Ribosome	273
	14.3.2 Enzymology of the Peptidyl Transfer Reaction	274
	14.3.2.1 Potential Mechanisms of Rate Acceleration by the Ribosome	274
	14.3.2.2 Experimental Approaches to Reaction on the Ribosome	275
	14.3.2.3 pH–Rate Profiles	277
	14.3.2.4 Activation Parameters	278
14.4	The Active Site	279
	14.4.1 Structures of the Reaction Intermediates	281
	14.4.2 Conformational Rearrangements of the Active Site	282
	14.4.2.1 Induced Fit	282
	14.4.2.2 Role of the P-site Substrate	283
	14.4.2.3 Conformational Flexibility of the Active Site	284
	14.4.3 Probing the Catalytic Mechanism: Effects of Base Substitutions	285
	14.4.4 Importance of the 2′-OH of A76 of the P-site tRNA	286
14.5	Conclusions and Evolutionary Considerations	287
References		288

Chapter 15 Folding Mechanisms of Group I Ribozymes
Sarah A. Woodson and Prashanth Rangan

15.1	Introduction	295
15.2	Multi-domain Architecture of Group I Ribozymes	296
15.3	RNA Folding Problem	297
	15.3.1 Hierarchical Folding of tRNA	297
	15.3.2 Coupling of Secondary and Tertiary Structure	298
15.4	Late Events: Formation of Tertiary Domains in the *Tetrahymena* Ribozyme	298
	15.4.1 Time-resolved Footprinting of Intermediates	298
	15.4.2 Misfolding of the Intron Core	300
	15.4.3 Peripheral Stability Elements	300
15.5	Kinetic Partitioning among Parallel Folding Pathways	301
	15.5.1 Theory and Experiment	301
	15.5.2 Single Molecule Folding Studies	301
	15.5.3 Estimating the Flux through Footprinting Intermediates	302
	15.5.4 Kinetic Partitioning *In Vivo*	302
15.6	Early Events: Counterion-dependent RNA Collapse	302
	15.6.1 Compact Non-native Form of bI5 Ribozyme	303
	15.6.2 Small Angle X-ray Scattering of *Tetrahymena* Ribozyme	303
	15.6.3 Native-like Folding Intermediates in the *Azoarcus* Ribozyme	304
	15.6.4 Early Folding Intermediates of the P4-P6 RNA	305
15.7	Counterions and Folding of Group I Ribozymes	305
	15.7.1 Metal Ions and RNA Folding	305
	15.7.2 Valence and Size of Counterions Matter	306
	15.7.3 Specific Metal Ion Coordination and Folding	307
15.8	Protein-dependent Folding of Group I Ribozymes	307
	15.8.1 Stabilization of RNA Tertiary Structure	308
	15.8.2 Stimulation of Refolding by RNA Chaperones	308
15.9	Conclusion	309
	References	309

Subject Index 315

CHAPTER 1
Ribozymes and RNA Catalysis: Introduction and Primer

DAVID M.J. LILLEY[a] AND FRITZ ECKSTEIN[b]

[a] Cancer Research UK Nucleic Acid Structure Research Group, MSI/WTB Complex, The University of Dundee, Dow Street, Dundee DD1 5EH, UK;
[b] Max-Planck-Institut Für Experimentelle Medizin, Hermann-Rein-Str. 3, Göttingen D-37075, Germany

1.1 What are Ribozymes?

Ribozymes are RNA molecules that act as chemical catalysts, a shortening of *ribo*nucleic acid en*zymes*. In the contemporary biosphere, the known ribozymes carry out a relatively limited range of reactions (Figure 1.1), mostly involving phosphoryl transfer, notably transesterification (the large majority) and hydrolysis reactions. However, the discovery that peptidyl transferase is catalysed by the rRNA component of the large ribosomal subunit significantly extends the range of chemistry to include the condensation of an amine with an sp^2-hybridized carbonyl centre. A significantly greater range of chemical reactions may be catalysed by RNA species selected for the purpose, so that ribozymes catalyzing a broader set of reactions may have existed in the past.

1.2 What is the Role of Ribozymes in Cells?

Contemporary ribozymes (Table 1.1) are used for various biological purposes. The nucleolytic ribozymes bring about the site-specific cleavage (or the reverse ligation process) of RNA by attack of a 2′-hydroxyl group on the adjacent 3′-phosphorus (Figure 1.2A) (Chapters 2–8). This reaction is exploited for the processing of replication intermediates, and in the control of gene expression by metabolite-induced cleavage of mRNA. Ribonuclease P carries out the processing of tRNA in all kingdoms of life, using a hydrolytic reaction (Chapter 9). Several introns are spliced out autocatalytically by ribozyme action, initiated either by the attack of a 2′-hydroxyl group located remotely within the intron

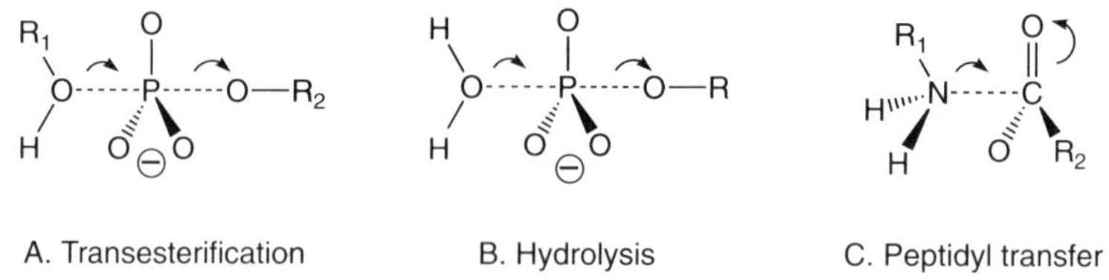

A. Transesterification   B. Hydrolysis   C. Peptidyl transfer

Figure 1.1 Reactions catalysed by the known natural ribozymes observed in biology. The biggest number carry out transesterification reaction (*A*), notably the nucleolytic ribozymes and the self-splicing introns. RNaseP carries out a hydrolytic reaction to process tRNA molecules (*B*). In the peptidyl transfer reaction (*C*) the amine of the A-site aminoacyl-tRNA attacks the carbonyl group of the peptidyl-tRNA (or initiator aminoacyl t-RNA) held in the P site of the large ribosomal subunit.

Table 1.1 Classes of natural ribozymes.

Ribozyme	Mechanism	Reference
Hammerhead	Transesterification 2'-O	Chapters 3, 4[a]
Hairpin	Transesterification 2'-O	Chapter 5
Hepatitis Delta Virus	Transesterification 2'-O	Chapters 6, 7[b]
Varkud Satellite	Transesterification 2'-O	Chapter 5
glmS	Transesterification 2'-O	Chapter 8[c]
Group I Intron	Nucleotidyl transfer	Chapter 10
Group I-like	Nucleotidyl transfer	Chapter 12
Group II Intron	Nucleotidyl transfer	Chapter 11
RNase P	Hydrolysis	Chapter 9
Ribosome	Peptidyl transfer	Chapter 14

[a] This is a large group of ribozymes. Although many are found in plant pathogens, hammerhead sequences have been found in amphibia (Newt) and plants (*Arabidopsis thaliana*).
[b] A closely related sequence has been identified in the genomes of mammals, in gene CPEB3.
[c] The *glmS* ribozyme is the only nucleolytic ribozyme that uses a small molecule cofactor in its catalytic chemistry.

(group II introns, Figure 1.2B) (Chapter 11), or by an exogenous guanosine molecule (group I introns, Fig 1.2C) (Chapter 10). While "smoking gun" evidence has not yet been found, the similarity of the chemistry of mRNA splicing in the spliceosome to that of the group II introns makes it very likely that this too is RNA catalysed, with the snU4/U6 RNA as the ribozyme (Chapter 13). Lastly, the peptidyl transferase activity of the ribosome catalyses what is arguably the most important reaction of the cell, the condensation of amino acids into polypeptides (Chapter 14). Ribozymes are widespread in nature, from bacteria and their phages, archaea, yeasts and fungi and higher eukaryotes. They are also present in clinically-significant human pathogens such as the hepatitis D virus (Chapter 6). New ribozymes are still being found, both by biochemical approaches and by the bioinformatic analysis of genome sequencing data.

Ribozymes and RNA Catalysis: Introduction and Primer

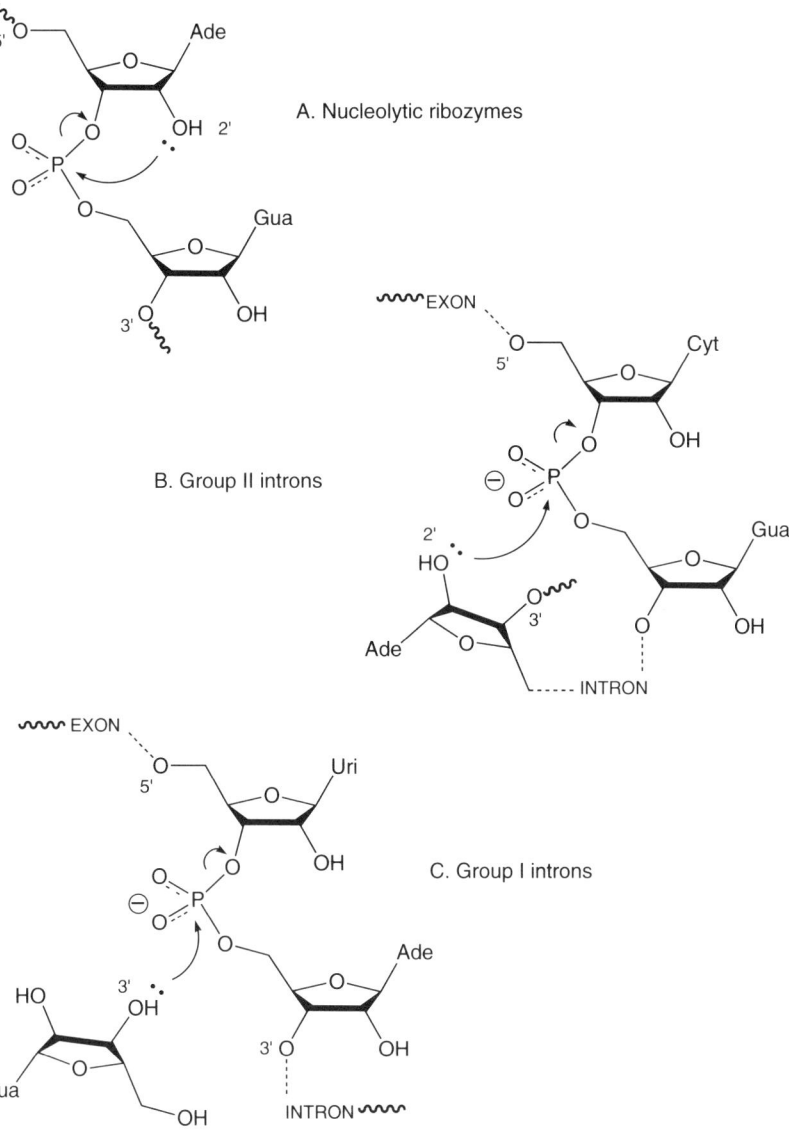

Figure 1.2 Three types of transesterification reactions catalysed by natural ribozymes. The cleavage reaction of the nucleolytic ribozymes (*A*) involves attack of the 2'-O on the adjacent 3'-P, with a 5'-oxyanion leaving group. In the reverse ligation reaction the 5'-O attacks the P of the cyclic phosphate. A similar reaction occurs in the group II intron (*B*), except that the 2'-O nucleophile is provided by a nucleotide located elsewhere in the intron. The nucleophile of the first reaction of the group I intron (*C*) is the 3'-O of an exogenous guanosine.

1.3 Ribozymes Bring about Significant Rate Enhancements

Protein enzymes can achieve some extraordinary catalytic rate enhancements. Values of almost 10^{18}-fold are possible, although many generate much smaller accelerations. RNA catalysts tend to produce more modest rate enhancements. For example, the nucleolytic ribozymes typically accelerate their transesterification reactions by around a million-fold relative to the uncatalysed reaction in a dinucleotide, with rates of around $1\,\text{min}^{-1}$. For those ribozymes this may be as fast as it needs to be, since a given site needs to be cut just once. However, while this rate was previously discussed as some kind of speed limit,[1] it appears that this is not an intrinsic limitation, and some redesign of some ribozymes has resulted in very respectable catalytic rates $\geq 10\,\text{s}^{-1}$.[2,3]

1.4 Why Study Ribozymes?

There are several reasons for studying ribozymes. First, they are active in contemporary living cells, carrying out reactions that are critical for cell viability in some cases; they are therefore legitimate subjects of interest in the complete description of cellular metabolism.

Second, they may have had a key role in the evolution of life on the planet.[4,5] There is clearly a rather severe "chicken and egg" problem involved in the origins of proteins and translation systems, both of which seem to require the prior existence of the other. Yet in principle a biosphere in which RNA was simultaneously the informational and catalytic macromolecule provides a temporary solution to that problem. Such an RNA world might have existed around 3.5 billion years ago, yet would have been relatively short lived in geological terms, being swiftly replaced by polypeptides that it would have produced. Some of the ribozymes that currently exist, most notably the ribosome perhaps, may be molecular fossils from that time, and therefore their study may offer a partial glimpse of that early metabolism. Although contemporary ribozymes carry out a very limited range of chemistries, selected ribozymes provide an indication of what is achievable by RNA catalysts,[6–12] and potentially offer a kind of proof-of-principle of an RNA world.

A third reason for studying ribozymes is that they are rather basic biocatalysts, providing a simplified and contrasting perspective on macromolecular catalytic mechanisms compared with enzymes. The last few years have seen significant advances in our understanding of the chemical origins of ribozyme catalysis, and this may cast light on protein-based catalysis in turn.

Lastly, there has been some effort to exploit the potential specificity of ribozymes as therapeutic agents. In principle, the great selectivity of ribozyme-induced cleavage of a chosen sequence could provide an opportunity to interfere with gene expression if targeted to a specific mRNA; this should ideally be the basis for their development into therapeutic drugs. However, this requires that many more problems be overcome, including stability in serum, delivery to the required location of the chosen cell and correct folding into the active conformation in competition with the native folding of the target RNA *in vivo*. So far only two ribozymes have found their way into clinical trials. One is a chemically synthesized and modified hammerhead ribozyme targeting the vascular endothelial growth factor receptor-1 (VEGFR) mRNA.[13] In preclinical trials it has exhibited antitumor and antimetastatic activity by interfering with VEGF-dependent angiogenesis. Angiogenesis inhibition is important in patients with refractory solid tumours. The other example involved the use of a hammerhead ribozyme as part of a vector to combat HIV-1.[14] The ribozyme directs the cleavage of the transcript of the chemokine receptor CCR5 that is essential for HIV-1 infection. To optimize efficiency the vector contains in addition a TAR decoy and a short hairpin RNA targeting the *rev* and *tat* mRNA of HIV-1. Potent inhibition of HIV-1 replication was achieved with this construct in a human T cell line.

1.5 Folding RNA into the Active Conformation

Just as protein enzymes must be correctly folded into the conformation required for catalytic activity, so must RNA. Moreover, it is clear that the folding processes of ribozymes is intimately associated with their function in many cases. Marked differences between the chemical nature of RNA and proteins results in very different folding processes. In general the precise nature of Watson–Crick basepairing leads to the relatively easy formation of secondary structure, although a requirement to "un-do" unfavourable pairings can provide significant barriers to correct folding. But most of the attention in RNA folding is focussed on the formation of the tertiary structure. The polyelectrolyte character of RNA results in a strong electrostatic contribution to this process, and thus a dependence on the presence of metal ions. The resulting folded RNA structure can bind metal ions, either site-specifically or diffusely, and these bound ions can play a direct role in catalysis. Site-bound ions are inner-sphere complexes where one of more water molecules in the first coordination sphere have exchanged with ligands provided by the RNA; such ions exchange slowly with bulk solvent. Diffusely bound ions do not undergo ligand substitution, and consequently exchange with solvent much more rapidly. They can nevertheless exhibit high occupancy in sites of strong electrostatic potential.

The smaller ribozymes, notably the nucleolytic ribozymes, exhibit some common structural themes, and their architectures are based around either helical junctions (hammerhead, hairpin and VS) or pseudoknots (HDV, *glmS*); evidently these motifs are efficient ways to construct small, autonomously folding species. Furthermore, some of these ribozymes contain additional elements that are not strictly essential for catalytic activity, yet result in marked enhancement of folding, such as the interacting loops of the hammerhead[2,15,16] and the four-way junction of the hairpin ribozyme.[17]

Most studies of RNA folding *in vitro* have therefore focussed on ion-induced folding. The small nucleolytic ribozymes generally exhibit relatively simple folding, typically two or three-state processes. However, larger ribozymes like the group I introns undergo complex folding pathways, beset with kinetic traps (Chapter 15). In the cell, proteins may assist the folding processes.

1.6 The Catalytic Resources of RNA – Making a Lot of a Little

The chemical nature of proteins has evolved to provide a highly adaptable catalytic framework with a broad repertoire of functional groups. It is based on an electrically neutral backbone, with sidechains that introduce a wide variety of chemistries, including carboxylic acids, amines, hydroxyl and thiol groups as well as hydrophobic side chains that may be either aliphatic or aromatic. By contrast, RNA consists of just four nucleotide bases of rather similar chemical nature, connected by an electrically-charged ribose-phosphate backbone.

So what resources are available to RNA that can be used to build a catalyst? Firstly, there are the nucleobases. These have hydrogen bond donors and acceptors that can be used to bind the substrate, and potentially to stabilize a transition state. In principle they could also act as general acids and bases. However, first they must overcome the problem of their pK_a values, which are unsuitable for general acid–base catalysis at neutral pH. Adenine N1 and cytosine N3 have low pK_a values, while those of guanine N1 and uracil N3 are relatively high. For example, a cytosine with a pK_a of 4 is a relatively strong acid, but only one molecule in 1000 is protonated at neutral pH. Thus most ribozymes will be in the wrong form to carry out a protonation. The great majority of molecules are in the deprotonated form and able to act as a general base, but the conjugate base of a strong acid is weak, so it is rather unreactive. However, the situation can be improved because nearby anionic phosphate groups may raise the pK_a significantly, and values of 5.5–6.5 are quite possible,[18,19] making the nucleobase more available as an acid. Similarly, the pK_a of guanine might be reduced if it is located close to a bound metal ion, thereby making it basic at a lower pH.

The second potential players are metal ions, and their associated water molecules. The folding of a ribozyme may create specific ion binding sites, or pockets in which there is high occupancy of more weakly bound ions. Metal ions can act as Lewis acids, or provide electrostatic stabilization of negative charge such as a dianionic phosphorane transition state. Water molecules contained within the inner sphere of coordination may participate in general acid–base catalysis, as exemplified by the HDV ribozyme.[20,21]

In addition to chemical participants, RNA can also potentially exploit its structure to contribute to catalysis. Substrate binding can result in acceleration of reaction velocity due to proximity and orientation, together with structural stabilization of the transition state.

1.7 Mechanisms and Catalytic Strategies of Ribozymes

Given the relative paucity of potential catalytic groups present in RNA molecules, ribozymes achieve some impressive rate accelerations. How they achieve this is a major topic of this volume, but we consider this briefly here. We will take the nucleolytic ribozymes as an example – most ribozymes carry out phosphoryl transfer reactions of various kinds, so that similar considerations will apply.

The chemical mechanism is shown in Figure 1.3. The cleavage reaction involves a nucleophilic attack of the 2′-oxygen on the adjacent 3′-phosphorus, with departure of the 5′-oxygen to create a cyclic 2′,3′-phosphate product. The chirality of the phosphorus becomes inverted during the reaction,[22–24] indicating concerted bond formation and breaking to some degree, and passage through a phosphorane transition state (or possibly high energy intermediate). This requires an in-line attack by the nucleophile, and thus a degree of rate enhancement can arise from prealigning the reactants into this geometry. The phosphorane transition state might be stabilized relative to the ground state by specific hydrogen bonding, or electrostatically. The latter might be achieved by juxtaposition of a metal ion, or perhaps a protonated nucleobase.

A hydroxyl group is a relatively weak nucleophile. Removal of its proton by a base would create a much more reactive alkoxide ion.[25] The reaction would

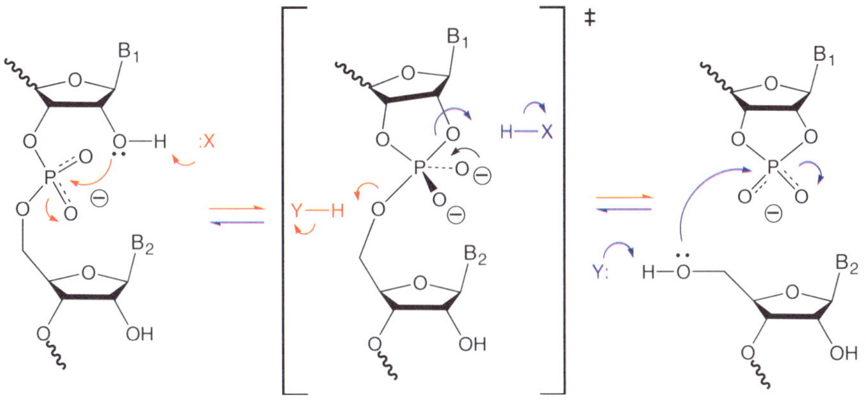

Figure 1.3 Mechanism of the nucleolytic ribozymes, together with a suggested role of general acid–base catalysis. Cleavage is rightward (red arrows) while ligation is leftward (blue arrows) in this scheme. The trigonal bipyramidal phosphorane is indicated as the transition state, although it could be a high energy intermediate flanked by a less symmetrical transition state. The cleavage reaction could be catalysed by a general base (X) to remove the proton from the 2′-O nucleophile and a general acid (YH) to protonate the 5′-oxyanion leaving group (both indicated red). By the principle of microscopic reversibility, the same groups will play converse roles in the ligation reaction, i.e. XH is the general acid that protonates the 2′-oxyanion leaving group, and Y is the base removing the proton from the 5′-O nucleophile (both shown red).

also be assisted by protonation of the 5′-oxyanion leaving group. Thus it would be expected that the reaction would be subject to general acid–base catalysis, and considerable evidence has been collected that this is generally the case in the nucleolytic ribozymes (Chapter 2). To date the active participants have included the nucleobases adenine, cytosine and guanine, and hydrated metal ions. Note that, the groups that act as acid and base in the cleavage reaction will reverse roles in the ligation reaction by the principle of microscopic reversibility.

1.8 Impact of New Methodologies to Study Ribozymes

While a lot of mechanistic insight into ribozyme action has come from the application of more or less conventional enzymological approaches, structural and biophysical methods have played key roles. Atomic resolution X-ray crystallographic structures have been determined for all the nucleolytic ribozymes except the VS ribozyme, and multiple forms have been determined in general. Crystal structures have also been solved for some of the larger ribozymes, including several examples of the group I ribozyme[26–28] (Chapter 10), RNaseP[29–31] (Chapter 9) and of course the peptidyl transferase centre within the 50S ribosomal subunit[32] (Chapter 14). All these studies have provided a wealth of structural data that then feeds back into mechanistic studies in an iterative process.

Single-molecule methods have also had a significant impact in the study of ribozyme folding[17,33–35] and activity.[36] These can provide a different perspective upon kinetic processes, free of the averaging that occurs with the ensemble and opening up the study of processes that cannot be synchronized. Both fluorescence and force spectroscopy have been applied to ribozymes. Other biophysical methods are also providing valuable information on RNA folding processes, such as small-angle X-ray scattering[37–39] and the combination of chemical footprinting and rapid reaction techniques.[40,41]

1.9 Finally

The field of RNA catalysis provides great challenges, and tremendous excitement. It has seen enormous development over the last 20 or so years, and it continues to spring surprises on a regular basis. This chapter provides an introduction for the reader who might not be directly involved in ribozyme research, Much more detail is provided in the following chapters. So please read on.

References

1. R.R. Breaker, G.M. Emilsson, D. Lazarev, S. Nakamura, I.J. Puskarz, A. Roth and N. Sudarsan, *RNA*, 2003, **9**, 949–957.

2. M.D. Canny, F.M. Jucker, E. Kellogg, A. Khvorova, S.D. Jayasena and A. Pardi, *J. Am. Chem. Soc.*, 2004, **126**, 10848–10849.
3. R. Zamel, A. Poon, D. Jaikaran, A. Andersen, J. Olive, D. De Abreu and R.A. Collins, *Proc. Natl. Acad. Sci. U.S.A.*, 2004, **101**, 1467–1472.
4. C. Woese, in *The Genetic Code*, Harper & Row, New York, 1967, pp. 179–195.
5. F.H.C. Crick, *J. Mol. Biol.*, 1968, **38**, 367–379.
6. E.H. Ekland, J.W. Szostak and D.P. Bartel, *Science*, 1995, **269**, 364–370.
7. T.M. Tarasow, S.L. Tarasow and B.E. Eaton, *Nature*, 1997, **389**, 54–57.
8. H. Suga, P.A. Lohse and J.W. Szostak, *J. Am. Chem. Soc.*, 1998, **120**, 1151–1156.
9. P.J. Unrau and D.P. Bartel, *Nature*, 1998, **395**, 260–263.
10. W.K. Johnston, P.J. Unrau, M.S. Lawrence, M.E. Glasner and D.P. Bartel, *Science*, 2001, **292**, 1319–1325.
11. G. Sengle, R.A. Eisenfuh, P.S. Arora, J.S. Nowick and M. Famulok, *Chem. Biol.*, 2001, **8**, 459–473.
12. G.F. Joyce, *Nature*, 2002, **418**, 214–221.
13. D.E. Weng, P.A. Masci, S.F. Radka, T.E. Jackson, P.A. Weiss, R. Ganapathi, P.J. Elson, W.B. Capra, V.P. Parker, J.A. Lockridge, J.W. Cowens, N. Usman and E.C. Borden, *Mol. Cancer Therap.*, 2005, **4**, 948–955.
14. M.J. Li, J. Kim, S. Li, J. Zaia, J.K. Yee, J. Anderson, R. Akkina and J. Rossi, *J. Mol. Therap.*, 2005, **12**, 900–909.
15. A. Khvorova, A. Lescoute, E. Westhof and S.D. Jayasena, *Nat. Struct. Biol.*, 2003, **10**, 1–5.
16. J.C. Penedo, T.J. Wilson, S.D. Jayasena, A. Khvorova and D.M.J. Lilley, *RNA*, 2004, **10**, 880–888.
17. E. Tan, T.J. Wilson, M.K. Nahas, R.M. Clegg, D.M.J. Lilley and T. Ha, *Proc. Natl. Acad. Sci. U.S.A.*, 2003, **100**, 9308–9313.
18. P. Legault and A. Pardi, *J. Am. Chem. Soc.*, 1997, **119**, 6621–6628.
19. S. Ravindranathan, S.E. Butcher and J. Feigon, *Biochemistry*, 2000, **39**, 16026–16032.
20. S. Nakano, D.M. Chadalavada and P.C. Bevilacqua, *Science*, 2000, **287**, 1493–1497.
21. A. Ke, K. Zhou, F. Ding, J.H. Cate and J.A. Doudna, *Nature*, 2004, **429**, 201–205.
22. H. van Tol, J.M. Buzayan, P.A. Feldstein, F. Eckstein and G. Bruening, *Nucleic Acids Res.*, 1990, **18**, 1971–1975.
23. M. Koizumi and E. Ohtsuka, *Biochemistry*, 1991, **30**, 5145–5150.
24. G. Slim and M.J. Gait, *Nucleic Acids Res.*, 1991, **19**, 1183–1188.
25. M. Oivanen, S. Kuusela and H. Lonnberg, *Chem. Rev.*, 1998, **98**, 961–990.
26. P.L. Adams, M.R. Stahley, J. Wang and S.A. Strobel, *Nature*, 2004, **430**, 45–50.
27. F. Guo, A.R. Gooding and T.R. Cech, *Mol. Cell*, 2004, **16**, 351–362.
28. B.L. Golden, H.D. Kim and E. Chase, *Nat. Struct. Mol. Biol.*, 2005, **12**, 82–89.
29. A.S. Krasilnikov, X. Yang, T. Pan and A. Mondragon, *Nature*, 2003, **421**, 760–764.

30. A.S. Krasilnikov, Y. Xiao, T. Pan and A. Mondragon, *Science*, 2004, **306**, 104–107.
31. A.V. Kazantsev, A.A. Krivenko, D.J. Harrington, S.R. Holbrook, P.D. Adams and N.R. Pace, *Proc. Natl. Acad. Sci. U.S.A.*, 2005, **102**, 13392–13397.
32. P. Nissen, J. Hansen, N. Ban, P.B. Moore and T.A. Steitz, *Science*, 2000, **289**, 920–930.
33. X. Zhuang, L.E. Bartley, H.P. Babcock, R. Russel, T. Ha, D. Herschlag and S. Chu, *Science*, 2000, **288**, 2048–2051.
34. X.W. Zhuang, H.D. Kim, M.J.B. Pereira, H.P. Babcock, N.G. Walter and S. Chu, *Science*, 2002, **296**, 1473–1476.
35. G. Bokinsky, D. Rueda, V.K. Misra, M.M. Rhodes, A. Gordus, H.P. Babcock, N.G. Walter and X. Zhuang, *Proc. Natl. Acad. Sci. U.S.A.*, 2003, **100**, 9302–9307.
36. M.K. Nahas, T.J. Wilson, S. Hohng, K. Jarvie, D.M.J. Lilley and T. Ha, *Nat. Struct. Mol. Biol.*, 2004, **11**, 1107–1113.
37. R. Russell, I.S. Millett, S. Doniach and D. Herschlag, *Nat. Struct. Biol.*, 2000, **7**, 367–370.
38. R. Das, L.W. Kwok, I.S. Millett, Y. Bai, T.T. Mills, J. Jacob, G.S. Maskel, S. Seifert, S.G. Mochrie, P. Thiyagarajan, S. Doniach, L. Pollack and D. Herschlag, *J. Mol. Biol.*, 2003, **332**, 311–319.
39. G. Caliskan, C. Hyeon, U. Perez-Salas, R.M. Briber, S.A. Woodson and D. Thirumalai, *Phys. Rev. Lett.*, 2005, **95**, 268–303.
40. M.L. Deras, M. Brenowitz, C.Y. Ralston, M.R. Chance and S.A. Woodson, *Biochemistry*, 2000, **39**, 10975–10985.
41. T. Adilakshmi, P. Ramaswamy and S.A. Woodson, *J. Mol. Biol.*, 2005, **351**, 508–519.

CHAPTER 2
Proton Transfer in Ribozyme Catalysis

PHILIP C. BEVILACQUA

The Pennsylvania State University, Department of Chemistry, University Park, Pennsylvania 16802, USA

2.1 Scope of Chapter and Rationale

This chapter provides an overview of recent progress in understanding proton transfer in RNA enzymes. It was over 10 years ago that the previous book by the same editors was published. In that volume, there was no chapter on proton transfer. Indeed, it was unclear whether proton transfer made important contributions to RNA catalysis. Many things have changed in the last decade. Most importantly, RNA structural biology has flourished.[1] There are now high-resolution structures of many RNA enzymes, both large and small, and many features of these structures have been tested and confirmed in biochemical experiments. Relevant to this chapter, ribozyme structures have provided inspiration for probing and, in some instances, demonstrating roles for proton transfer in RNA catalysis.

In this chapter, I provide an overview of the evolution of proton transfer research in RNA enzymes, largely from an historical perspective. The chapter begins with a definition of proton transfer and consideration of why proton transfer is important to the mechanisms of most enzyme-catalyzed reactions. A brief overview of proton transfer in protein reactions is provided, including the usage of amino acids other than histidine, and the importance of pK_a shifting. Next, the chapter focuses on RNA enzymes. The extent to which the four natural nucleobases[†] can directly partake in proton transfer is assessed. The benefit that large and small ribozymes might derive from proton transfer is considered from the point of view of the uncatalyzed reaction, as are possible quantitative contributions to catalysis. The chapter concludes with case studies

---
[†] The term "nucleobase" is preferred over "base" in referring to the four sidechains in RNA and DNA, to avoid confusing with general base or specific base.

of evidence for proton transfer in five small ribozymes: the hepatitis delta virus (HDV), hairpin, hammerhead, Varkud Satellite (VS), and *glmS* ribozymes.

The involvement of proton transfer in ribozyme mechanism is an area that has moved rapidly in the last few years. In some instances, mechanistic hypotheses flowed from high-resolution structural data; in others, functionally relevant high-resolution data followed experimental findings, while in other instances only secondary structures have provided the inspiration for experiments. In each case, discussion is limited to evidence for or against proton transfer, as full chapters are presented on the structures and mechanisms of these ribozymes elsewhere in this book. The chapter concludes with a look forward into outstanding questions and possible directions of the field.

Lastly, a comment about what is not covered is in order. This chapter focuses on proton transfer chemistry. Obviously, this is not the only catalytic device that nucleobases can offer to rate acceleration. For example, the nucleobases can provide a cationic charge, acting as oxyanion holes to stabilize an anionic transition state and provide electrostatic catalysis. Alternatively, the bases can hydrogen bond to the transition state to stabilize it and drive catalysis. In at least some instances, it is likely that two or more of these catalytic devices operate simultaneously, making it essential to consider cooperativity in parsing the contributions to catalysis.[2] The interested reader is referred to several reviews that discuss the many roles nucleobases can have in catalysis.[3–6] Although the focus of this chapter is on proton transfer, other catalytic devices are often intertwined and will be addressed as necessary.

2.2 Overview of Proton Transfer Chemistry

During chemical changes in biology, multiple covalent bonds are made and broken. In the phosphodiester self-cleavage reaction common to small ribozymes, a bond is made between the O2′ and the phosphorus ("m2" in Figure 2.1), and a bond is broken between the phosphorus and the bridging 5′-oxygen ("b2" in Figure 2.1).

While we often focus on non-hydrogen atom bond making and breaking events, it is important to also consider the concomitant making and breaking of covalent bonds with hydrogen atoms. These latter reactions are grouped under the designation "proton transfer" or "acid–base chemistry" because the most common isotope of hydrogen, protium, has just 1 proton, 1 electron, and 0 neutrons, and it is the proton, without its electron, that is transferred in this reaction (see electron pushing in Figure 2.1). Proton transfer to and from biopolymer sidechains is designated "general acid–base catalysis", while proton transfer to and from water, or hydroxide and hydronium ions is designated "specific acid–base catalysis".[7–8] Indeed, the reaction depicted in Figure 2.1 involves the making and breaking of six bonds overall, with two of the making (m1 and m3) and two of the breaking (b1 and b3) events being proton transfers. Initially, it might seem surprising that transfer of a few protons would be important to the mechanism of ribozymes, typically having thousands of

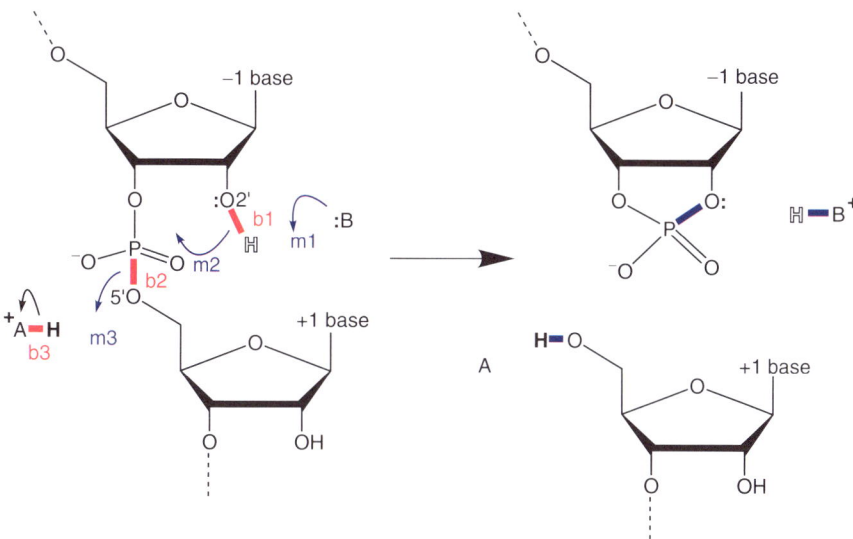

Figure 2.1 Self-cleavage reaction of small ribozymes. Reaction involves attack by the 2′-hydroxyl of the −1 nucleotide on the phosphorus center of the +1 nucleotide. During the reaction, three bonds are broken (bold, red in the reactant state and designated "b1", "b2", and "b3"), and three bonds are made (bold, blue in the product state; positions of bonds to be made are designated "m1", "m2", and "m3" in the reactant state). Of these six bond making and breaking events, two of the making events ("m1" and "m3") and two of the breaking events ("b1" and "b3") involve proton transfers.

atoms. However, there is ample evidence for the critical importance of such transfers in protein enzymes and, increasingly so, in RNA enzymes.

For biochemical reactions, proton transfers such as those shown in Figure 2.1 are ubiquitous. A survey of Silverman's text, *The Organic Chemistry of Enzyme-Catalyzed Reactions* reveals that virtually every enzyme-catalyzed chemical reaction (radical reactions excepted) involves some degree of general acid–base chemistry.[8] How important are such transfers to rate acceleration? Jencks pointed out that to the organic chemist such proton transfers are of no great concern because acid or base solutions can be added to the flask to drive the reaction.[7] However, to the cell these proton transfers present a significant hurdle because the pH is buffered near neutrality. Jencks stated the importance of proton transfer reactions to protein enzymes as, "Facilitation of these proton transfers is the most important mechanism by which the rate of such a reaction may be increased in the chemistry laboratory and is almost certainly a major contributing mechanism in enzyme catalysis."[7] Contribution of proton transfer to ribozyme reactions from the perspective of charge accumulation is considered in this section, with quantitative estimates to ribozyme reactions provided in Section 2.3.

For ribozyme reactions, proton transfers are important because they prevent the accumulation of unfavorable intermediates that bear charge on bridging oxygen atoms. This principle is illustrated in the transition states drawn in Figure 2.2.

14   Chapter 2

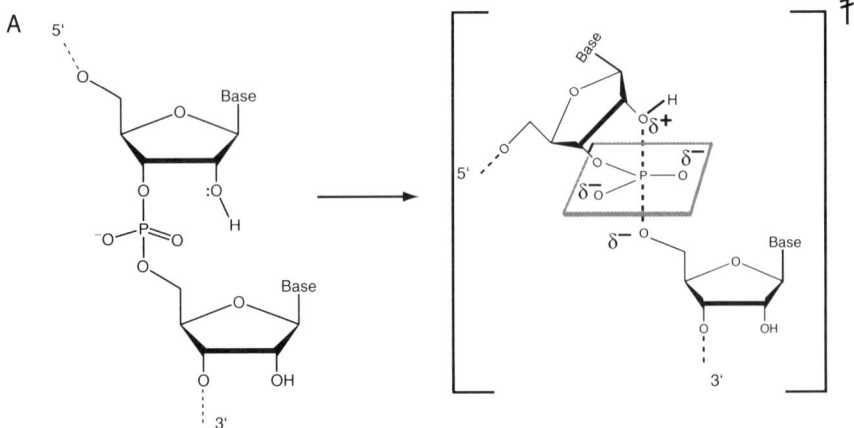

No general acid-base catalysis

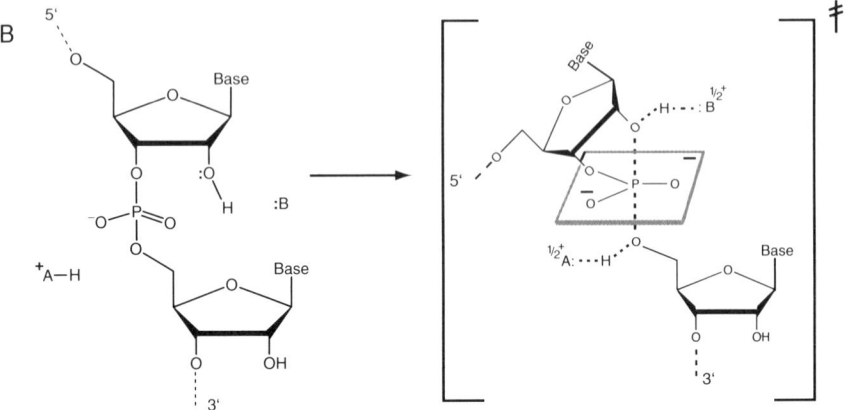

General acid-base catalysis

Figure 2.2 Stabilization of charge development by general acid–base catalysis. Both panels (*A*) and (*B*) depict the mechanism for self-cleavage reaction of a small ribozyme, similar to that in Figure 2.1. The transition state is assumed to be trigonal bipyramidal. (*A*) Reaction in the absence of general acid–base catalysis. Assuming that the transition state is synchronous, as described elsewhere,[9] the sum of the bond order on the phosphorus attached to the O2′ nucleophile and O5′ leaving group remains relatively constant near 1 throughout the reaction. This causes positive charge to accumulate on the O2′ and negative charge on the O5′. (*B*) Reaction in the presence of general acid–base catalysis. It is assumed that the acid is cationic in the reactant state and that the base is neutral, as in Figure 2.1. Here, charge accumulates on the general acid and general base rather than the O2′ and O5′ atoms.

In this illustration, it is assumed that the reaction is synchronous, as previously demonstrated for both uncatalyzed and alkaline phosphatase-catalyzed phosphate diester hydrolysis reactions.[9,‡] In a ribozyme self-cleavage reaction *without* general acid–base catalysis, positive charge accumulates on the nucleophilic 2′-oxygen and negative charge on the bridging 5′-oxygen (Figure 2.2A). In a ribozyme self-cleavage reaction *with* general acid–base catalysis, charge does not accumulate on the 2′- or 5′-oxygens, but instead on the general acid–base species (Figure 2.2B). Observation of Brønsted α and β values of ~ 0.5 in the HDV ribozyme (discussed below) is consistent with the synchronous transition state in Figure 2.2(B). The reaction depicted in Figure 2.2(B) is much more favorable energetically than that in Figure 2.2(A). This is because in the product state (Figure 2.1) the 5′-oxygen has a pK_a for protonation near 15, and in the reactant state this oxygen has a $pK_a < 0$ for protonation.[10] As such, positive or negative charge development on an oxygen atom, such as in the transition state without general acid–base catalysis (Figure 2.2A), makes for a highly unfavorable reaction at neutral pH. Indeed, formation of the conjugate acids and bases of the reactants, such as these, is highly unfavorable for most enzymes.[7]

It has been argued that even though the favorable reaction trajectory depicted in Figure 2.2(B) involves proton transfer in the transition state, it does not preclude the need for pK_a shifting in the reactant state.[3] This is because in order for the acid and base to donate and accept protons synchronously throughout bond cleaving and making[9] they must be in their functional protonated and deprotonated forms in the reactant state. As such, for *optimal* general acid–base catalysis, pK_a values need to be shifted towards *neutrality* in the *ground state*, which is not to say that pK_a shifting does not occur on reacting atoms as a reaction progresses from the reactant state to the product state.

Indeed, it is possible to express pH-dependent populating of the functional form of a ribozyme using a partition function and ground state acid and base pK_a values, and to express the effects of acid and base pK_as on the transition state in a pH-independent intrinsic rate for proton transfers and bond cleavage.[11] This approach separates effects of pK_as on population and chemistry, and allows a graphical approach to analyzing the pH dependence of reaction kinetics (see below). This approach has led to remarkably similar estimates for the intrinsic rate of cleavage for RNA enzymes and protein enzymes.[11–13]

Any properly positioned functional group with a pK_a intermediate between the reactant and product state pK_as can and will act as a general acid–base catalyst – a principle Jencks colorfully referred to as the "libido rule".[14] Considering that the pK_a on the bridging oxygen cycles between <0 and 15 during the course of the reaction (see above), many of the functional groups on the bases, which fall in this range (Figure 2.3),[3,15,16] could be acid–base catalysts.

However, atoms with pK_as far from neutrality are poor general acid–bases. Both A and C have imino nitrogens with pK_a values well below neutrality, at 3.5 and 4.2, respectively, and G and U have imino nitrogens with pK_as above

---

‡ By "synchronous", it is meant that the sum of the axial bond orders is near one.

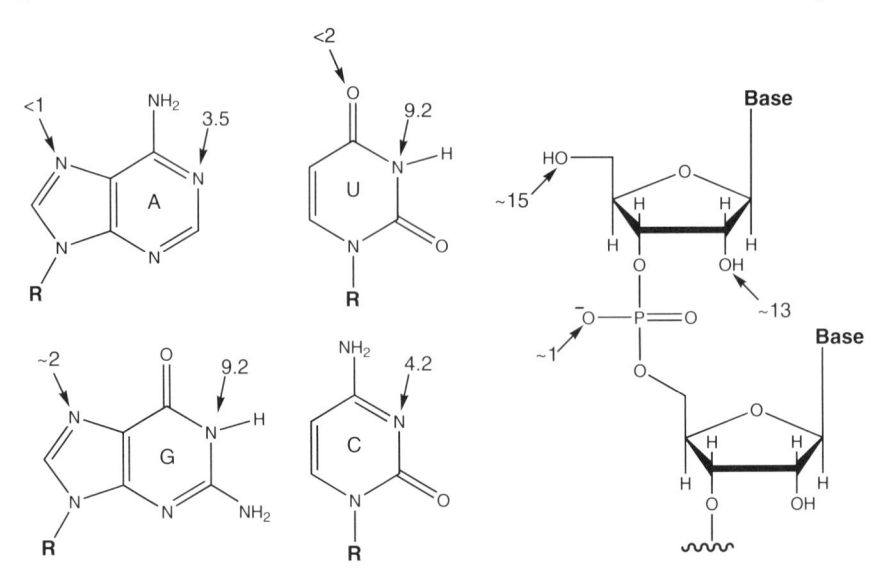

Figure 2.3 Unperturbed pK_a values for RNA. The four nucleobases are shown in their typical tautomeric forms, and pK_as for various atoms are designated. Also shown is the phosphodiester backbone with pK_a values. Note that the pK_as closest to neutrality, and therefore most likely to function in general acid–base catalysis, are the imino nitrogens that participate in Watson–Crick base pairing. The pK_a of the 2′-hydroxyl has been estimated to be between 12.5 and 14.9.[56,128,129] (Figure is used with permission from *Biopolymers*[3] Copyright 2004, John Wiley & Sons, Inc.)

neutrality, near 9.2;[15] moreover, these pK_as shift even further from neutrality upon hydrogen binding in Watson–Crick base pairs.[17,18] Poor general acid–base character for pK_as removed from neutrality is due to one of two factors. *First* is the aforementioned need for the functional group to be in the proper functional state to carry out the transfer. For example, the N3 of cytidine has an unperturbed pK_a near 4.2. Although its pK_a is intermediate between 0 and 15, this group is ill-suited to act as a general acid without a pK_a shift. This is because, at neutrality, only ~0.1% of the cytidine is protonated, although the actual penalty is somewhat less than this because the 0.1% in the functional form is a better acid than a pK_a 7 group, with the effect dependent on the Brønsted value for the transfer.[11] *Second* is that this group is suboptimal for acting as a general base because, although 99.9% in the functional form, having a pK_a of 4.2 makes it a poor proton acceptor. Thus, the compromise is to have a pK_a near neutrality. The pK_as closest to neutrality, and therefore most likely to function in general acid–base chemistry, are the imino nitrogens that participate in Watson–Crick base pairing (Figure 2.3).

In proteins, histidine has an unperturbed pK_a of ~6, close to physiological pH.[19] Evolution thus appears to have endowed proteins with an ideal species for general acid–base chemistry – which is logical given that the vast majority of enzymes in nature are proteins. Indeed, histidine plays important roles in general acid–base chemistry in many proteins, *e.g.*, RNase A.[20] Interestingly, despite having the choice of histidine, proteins often use amino acids other than histidine for proton transfer. In many instances, the pK_a of these groups are shifted towards neutrality, making them more efficient acids and bases. For example, Glu-35 of hen egg white lysozyme has a pK_a shifted upwards from 4.5 to 6.5–8.2 (with the actual value depending on the presence of an inhibitor), allowing it to act as a general acid to protonate an alkoxide leaving group.[21] In addition, upward shifting of pK_as plays an important role in acid–base chemistry in aspartyl proteases such as HIV-1 protease.[22] In other cases, *e.g.*, 5-aminolevulinate synthase and 6-phosphogluconate dehydrogenase, lysine has been shown to act as a general base, with a pK_a shifted downwards from ~10.5 to ~8.0–8.5.[23,24]

2.3 General Considerations for Proton Transfer in RNA Enzymes

2.3.1 Classes of Protonation Sites in RNA

Given that proteins use pK_a shifting to optimize an amino acid for general acid–base catalysis, it seems logical that RNA too might employ pK_a shifting to optimize general acid–base chemistry. Two classes of shifted pK_as have been suggested for nucleobases.[3,18] In Class I, the loaded proton is sequestered in hydrogen bonding, and in Class II this proton is exposed. Figure 2.4 provides representative examples of each class.

Class I sites include **1**, the A$^+$·C wobble pair that is common in RNA and DNA helices,[25–27] and near the cleavage site of the hairpin ribozyme[28,29] and the lead-dependent ribozyme;[17,30] **2**, protonated adenine in a Hoogsteen interaction with guanine in a DNA duplex;[31–33] **3**, protonated cytosine forming a wobble base pair with guanine in the *Tetrahymena* ribozyme;[34] **4**, a protonated-cytosine–cytosine base pair;[35,36] and **5**, protonated cytosine forming a Hoogsteen pair with guanine found in DNA in a complex with an antibiotic.[37] Class II sites include **6**, a model for the transition state of the HDV ribozyme self-cleavage reaction;[12] **7**, Hoogsteen–Hoogsteen interactions for two protonated adenines found in polyadenylic acid;[38] **8**, adenine and protonated adenine in a DNA duplex;[39,40] and **9**, guanine and protonated adenine in a Hoogsteen pairing.[40]

One reason for separating ionized base interactions into two classes is that they could have distinct functions.[3,18] In Class II sites, the loaded proton is free to transfer from the nucleobase to the reaction leaving group and so, in principle, the nucleobase could act as a general acid; likewise, the proton is free to transfer from the nucleophile to the nucleobase and so the nucleobase could act as a general base. In Class I sites, the loaded proton is sequestered in hydrogen bonding, suggesting that, while having potential importance in electrostatic catalysis as an oxyanion hole or as a pH-conformational switch,

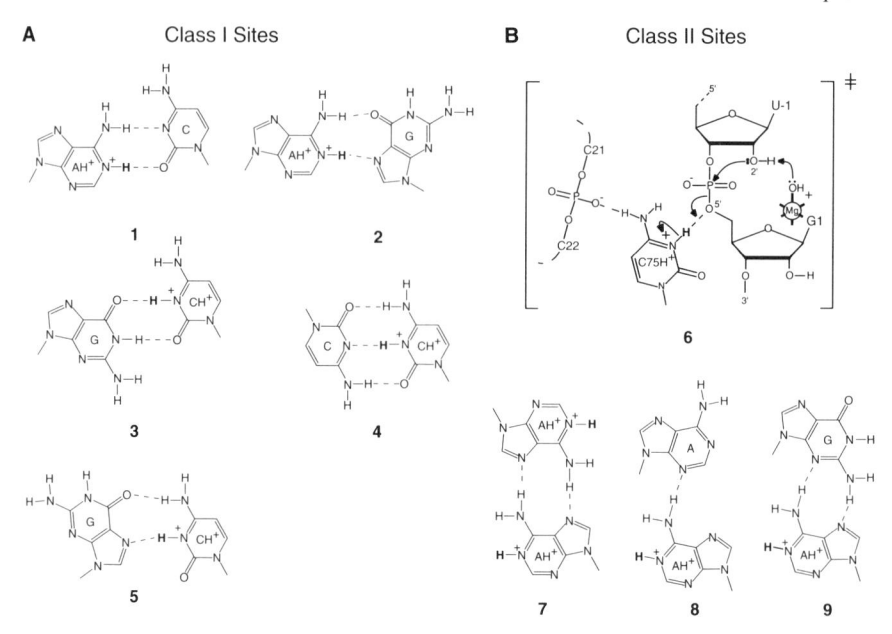

Figure 2.4 Class I and II sites for nucleobase protonation in RNA. Protons can be loaded onto an RNA at sites that involve sequestering of the proton within base pairing (Class I, panel A), or at sites that do not involve sequestering (Class II, panel B). References for species shown are as follows: Structure **1**,[17,25–30] **2**,[31–33] **3**,[34] **4**,[35,36] **5**,[37] **6**,[12] **7**,[38] **8**,[39,40] and **9**.[40] (Figure is used with permission from *Biopolymers*[3] copyright 2004, John Wiley & Sons, Inc.)

this proton may not be able to participate in proton transfer.[3,18] However, interestingly, protonation of one atom on a nucleobase typically perturbs the pK_a of other atoms. For instance, protonation of cytidine on the N3 shifts the pK_a of the N4 downward by ~9 units, from 18 to 9.[41–43] This exceptionally large shift can be understood on the basis of electron donation from N4 into the ring system, which delocalizes positive charge onto N4. Given such pK_a perturbation of the ring, it seems possible that Class I sites *could* be participants in general acid–base chemistry, albeit where the labile proton transferred is not the loaded and sequestered proton. Such a scenario would greatly expand the repertoire of general acid–base-capable motifs in RNA, as Class I sites appear to be more common than Class II sites.

2.3.2 Driving Forces for pK_a Shifting in RNA

Several studies have been published describing driving forces for pK_a shifting in RNA. A statistical thermodynamic formalism based on a linkage model has provided insight into the coupling between folding and pK_a shifting.[44,45] In this case, the unfolded state was treated as an ensemble of microscopic pK_as, and the few protonations in the folded state were treated explicitly. This led to a set of equations for extracting pK_a values from pH-dependent melts, and identified cooperativity in RNA folding as a major driving force in pK_a shifting.[45] In essence, the more interactions that form after loading of a folded-state proton, the bigger the pK_a shift. Cooperative folding has been shown to be operative in small and large RNA and DNA motifs, suggesting the potential for significant pK_a shifts.[2,46–48] The linkage model has provided a good description of protonation behavior in pseudoknotted RNAs with a protonated cytosine motif, in which pK_a shifts from 4.2 to 7.7–9.6 were determined.[47] The linkage model led to the description of two driving forces for pK_a shifting: (1) Those involving cooperative formation of hydrogen bonding and stacking interactions, and (2) those involving close positioning of positive and negative charges, especially a "−+−" sandwich of charges.[49,50] The first driving force is an essential feature of Class I sites, while the second driving force is essential to Class II sites; however, both driving forces could be operative for a given site. With regards to the second driving force, nonlinear Poisson–Boltzmann (NLPB) calculations have been used to identify regions of extreme negative potential in RNAs, as they may serve to favor nucleobase protonation and pK_a shifting.[3,51–53]

Recently, pK_a calculations for RNA were carried out on the basis of solutions to the Poisson–Boltzmann equation.[53] According to their findings, the authors developed a few simple rules for pK_a shifting: compact motifs with conserved electrostatic arrangements lead to large pK_a shifts, as do phosphates positioned to focus electrostatic potential on certain atoms. In particular, they calculated significant pK_a shifts in the HDV and hairpin ribozymes.[53] In the HDV ribozyme, C75 was predicted to have a pK_a of ~9.6 in the product structure, which is just slightly higher than the range of 6–8 range measured experimentally,[12,54] while in the hairpin ribozyme A10, A22, and A38 were predicted to have elevated pK_as of 6.6, 7.2, and 5.9, respectively. The pK_a shift for A38 is intriguing as this base has been suggested to provide electrostatic stabilization in the transition state and is positioned to serve as a general acid.[11,55] Overall, it is encouraging that general methods to predict nucleobase pK_a values are being developed, as they may be helpful for identifying shifted pK_a values in other RNA molecules.

2.3.3 Quantitative Contributions of Proton Transfer to RNA Catalysis

For proton transfer to merit special attention it should make significant quantitative contributions to chemical catalysis. Quantitating the contribution of any particular catalytic device to overall catalysis is problematic given that the contributions of various devices to rate acceleration are often non-additive.[2,3] For example, mutation of a nucleobase could change not only the rate of proton transfer but also positioning of atoms for in-line attack. One way to limit non-additivity is to study rate contributions on very simple model systems. In particular, Li and Breaker studied specific base-catalyzed cleavage of the phosphodiester backbone of a DNA-RNA chimeric strand and found that general base catalysis contributes $\sim 10^5$ to rate acceleration under physiological conditions.[56] Mutation of each of the histidines to alanine in the protein enzyme RNase A revealed k_{cat}/K_m losses of 10^3- to 10^5-fold per mutation,[57] similar to those in the chimeric model phosphodiester system. Thus, a reasonable upper limit to the contribution of one proton transfer is 10^5-fold.

Quantitative contributions of proton transfer to overall catalysis have been estimated for the HDV ribozyme (described more fully in Section 2.4.2). Mutation of the catalytically important C75 to uracil, in C75U, led to losses in rate of greater than 7×10^5- and 2×10^6-fold for the antigenomic and genomic ribozymes, respectively.[3,58] These effects are larger than the limit of 10^5 from the model phosphodiester system,[56] suggesting the presence of non-additivity.[2,3] One interpretation is that general acid–base chemistry[§] contributes near its maximum of $\sim 10^4$–10^5-fold,[56,57] and that the additional catalytic effect originates from positioning defects from loss of the hydrogen bonding of the N4 of C75 as seen in the crystal structure of the product form of the ribozyme.[59] In summary, general acid–base chemistry contributes substantially to rate acceleration in proteins and in nucleic acid model systems and enzymes.

2.4 Proton Transfer in Small Ribozymes: Five Case Studies

2.4.1 Why Small Ribozymes?

In this section, five case studies on proton transfer in ribozymes are presented. Various experimental approaches are discussed, including crystal structures, kinetics experiments, solvent isotope effects, proton inventories, Brønsted studies, and theoretical treatments. All five case studies involve proton transfer to and from nucleobases in the *smaller* ribozymes. This focus is because at present there is no evidence that larger ribozymes use nucleobases in proton transfer.

As shown in Figure 2.5, the nature of the cleavage reaction for large and small ribozymes is quite different. For the larger ribozyme, the nucleophile (Nu:) is the 3'-hydroxyl and the leaving group is the bridging 3'-oxygen. Moreover, the bridging 3'-oxygen is also the leaving group for the RNA-catalyzed peptidyl transferase reaction of the ribosome. One of the defining characteristics of a 3'-oxygen leaving group is the presence of a nearby source of labile protons for transfer to the leaving group in the form of the vicinal 2'-hydroxyl. Indeed, the 2'-hydroxyl vicinal to the 3'-oxygen leaving group has been shown to play a critical role in the reaction mechanism of group I introns[60] and the ribosome.[61]

For the group I intron, detailed mechanistic studies are consistent with an intramolecular hydrogen bond from the 2'-hydroxyl to the developing negative charge on the bridging 3'-oxygen leaving group of the substrate.[60,62]

[§]Evidence that C75 is involved in general acid–base chemistry and not just folding or electrostatic catalysis is described in Section 2.4.2.

In addition, biochemical[63,64] and recent crystallographic studies[65] implicate a metal ion(s) in coordinating to the 2′-hydroxyl and bridging 3′-oxygen leaving group of the ωG at the 3′-splice site, albeit with these studies arriving at differences in the number of metal ions making these contributions.

For the ribosome, the 2′-hydroxyl of the P-site tRNA substrate contributes substantially to peptide bond formation.[61] In this case, experiments and calculations support the notion that the 2′-hydroxyl participates in a proton shuttle, acting as both a general acid and a general base, either in a concerted or stepwise fashion to transfer a proton from the amine to the bridging 3′-oxygen (Figure 2.5A).[66–68] Recent experiments on a slow-reacting A-site tRNA substrate, for which chemistry is rate-limiting, revealed absence of pH-dependence for peptide bond formation, which is most consistent with the absence of involvement of ribosomal nucleobases in general acid–base chemistry.[67] It is

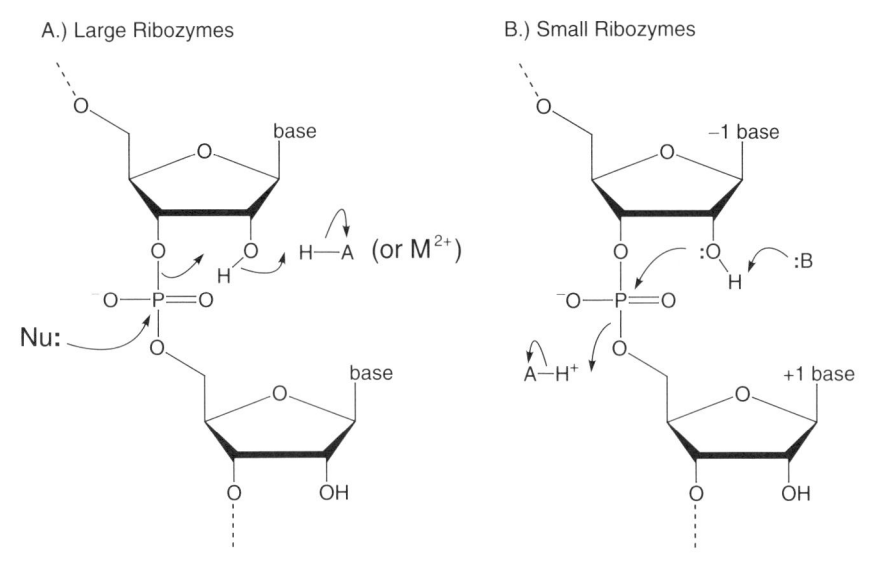

Figure 2.5 Comparison of reactions for large and small ribozymes. (*A*) Larger ribozymes such as the group I intron or the ribosome. With the group I intron, an exogenous nucleophile (Nu:) participates in the reaction and yields termini with a 2′,3′-cis-diol and 5′-phosphate. General acid–base catalysis can be mediated, directly or indirectly, by the vicinal 2′-hydroxyls. Proton shuttles such as the one depicted here seem to provide simple solutions to proton transfer. For the ribosome, the nucleophile is the amino nitrogen of the A-site tRNA. (*B*) Smaller ribozymes such as the HDV, hairpin, hammerhead, VS, and *glmS* ribozymes. A vicinal 2′-OH participates in the reaction as the nucleophile and yields termini with a 2′,3′-cyclic phosphate and 5′-hydroxyl. General acid catalysis is mediated by a new species, shown in this depiction as A–H$^+$, which could be a protonated nucleobase. The scissile phosphate is positioned between nucleotides numbered −1 and +1 by convention. (Reprinted from *Curr. Opin. Chem. Biol.*,[6] Copyright 2006, with permission from Elsevier.)

astonishing that a vicinal 2′-hydroxyl, whose identity changes with each step in protein synthesis, would trump the entire ribosome in terms of acid–base chemistry. This elegant finding does not lessen the importance of proton transfer to the overall reaction. It does, however, suggest that RNA enzymes, even very complex ones like the group I intron and the ribosome, will follow the simplest route to catalysis.

As depicted in Figure 2.5(B), the nucleophile in small ribozyme cleavage reactions is the 2′-hydroxyl near the reactive phosphorus. Unlike the larger ribozymes, the small ribozymes do not have the convenience of labile protons vicinal to the leaving group. This is problematic given that protonation of this group is particularly important to rate acceleration, as revealed by physical organic studies.[69,70] As detailed below, proton transfer appears to come instead from close positioning of a general acid nucleobase.

Evidence for proton transfer in five different small ribozymes will be the focus of the remainder of this chapter. Since a full chapter is written on each of these ribozymes, I provide only the background needed to discuss proton transfer. The reader is referred to the later chapters for further background.

2.4.2 Proton Transfer in the Hepatitis Delta Virus Ribozyme

The hepatitis delta virus ribozyme (Chapter 6) was the first catalytic RNA to have proton transfer from a nucleobase implicated in its mechanism and occurs as closely related genomic and antigenomic versions. Crystal structures from the Doudna laboratory have been instrumental in developing models for proton transfer in the HDV ribozyme.[59,71] The idea that a nucleobase could participate in proton transfer followed from inspection of a high-resolution crystal structure of the self-cleaved product form of the ribozyme.[59] Close positioning of the 5′-hydroxyl of G1 and the N3 of C75 (2.7 Å) led to the notion that C75/C76 could act as a general acid–base where C76 is the antigenomic counterpart of C75.

Studies from Been and co-workers showed that self-cleavage of inactive C75/76U or C75/76Δ (where "Δ" represents a deletion) versions of the ribozyme could be rescued by addition of imidazole to the reaction media.[58,72,73] This result is striking given that imidazole is the side chain for histidine, which is optimal for general acid–base chemistry without pK_a shifting. These rescue experiments gave similar results on the genomic and antigenomic ribozymes.[12,58,72,73]

Kinetic ambiguity prevented assignment of C75 as a general acid or base. However, the crystal structure of the product form of the ribozyme[59] suggested that C75 might act as a general acid in the direction of cleavage. This is because, according to the principle of microscopic reversibility, accepting a proton by C75 from the 5′-hydroxyl in the product state would translate into donating a proton to the bridging 5′-oxygen in the reactant state. Kinetic experiments on the genomic form of the ribozyme were consistent with this notion, and led to a model for cleavage in which protonated C75 serves as the general acid and hydrated magnesium hydroxide, with a pK_a of 11.4, acts as the general base.[12,51]

Evidence that C75 is involved in proton transfer comes directly from Brønsted plots. The dependence of log k_{obs} on pK_a of the rescuing small molecule was linear with a slope of ~0.5 for both the antigenomic and genomic ribozymes.[72,73] These data do not rule out an electrostatic role for C75, especially given that it is cationic when protonated, but if C75 were acting only as an oxyanion hole then the rate of the reaction would not depend on the basicity of the rescuing small molecule.[72]

Key biochemical experiments that can implicate proton transfer in ribozyme mechanism are the pH-dependence of the rate, kinetic solvent isotope effects (KSIE), and proton inventories. The pH dependence of the self-cleavage rate for the HDV ribozyme gives pK_a values near neutrality that couple anticooperatively with binding of a hydrated Mg^{2+} ion.[12] Fitting to a model encompassing all four Mg^{2+}- and H^+-bound states gave intrinsic pK_a values for C75 of 7.25 and 5.9 for Mg^{2+}-free and -bound states, respectively.[54] Identification of a pK_a at biological pH is congruent with the ability of a ribozyme to tune a pK_a for proton transfer.

The pH dependence of the reaction can be deconvoluted graphically to extract the contributions of populations of states and intrinsic rates of the reaction.[11] As shown in Figure 2.6, a typical rate–pH profile for the HDV ribozyme,¶ in which rate increases log–linearly with pH and then levels off at high pH with a pK_a of 7, can be accounted for by two models.

In the first model, referred to as "GAB Model 1", a general base has a pK_a of 7 and is accompanied by an acid with a high pK_a, such as hexahydrated magnesium, which has a pK_a of 11.4 (Figure 2.6A). In the other model, referred to as "GAB Model 2", a general acid has a pK_a of 7 and is accompanied by a general base with a high pK_a, such as pentahydrated magnesium hydroxide (Figure 2.6B). The ability of these two models to provide indistinguishable data is a manifestation of kinetic ambiguity.

Another way to consider GAB Models 1 and 2 is in simple kinetic schemes (Schemes 2.1 and 2.2).[54] The two schemes have identical upper banks of states and differ only in which of the two states is chemically competent. Scheme 2.1 is the reaction model in which C75 acts as the general base and hexahydrated magnesium acts as the general acid, while Scheme 2.2 is the reaction model in which protonated C75 acts as the general acid and pentahydrated magnesium hydroxide acts as the general base. Indeed, the rate law for each scheme has an identical dependence on pH,[54] as expected from the graphical consideration in Figure 2.6. In either case, observation of a pK_a near 7 supports proton transfer from a nucleobase. When adenosine was substituted at residue 75 in C75A (or C76A), a pK_a difference similar to that between CMP and AMP was observed in both the genomic and antigenomic ribozymes, consistent with assignment of the pK_a to residue 75.[12,58]

¶ Notably rate–pH profiles for some HDV ribozyme constructs show a second pK_a at higher pH, which results in a bell-shaped curve. The origin of this pK_a is presently unclear and will not be discussed here.

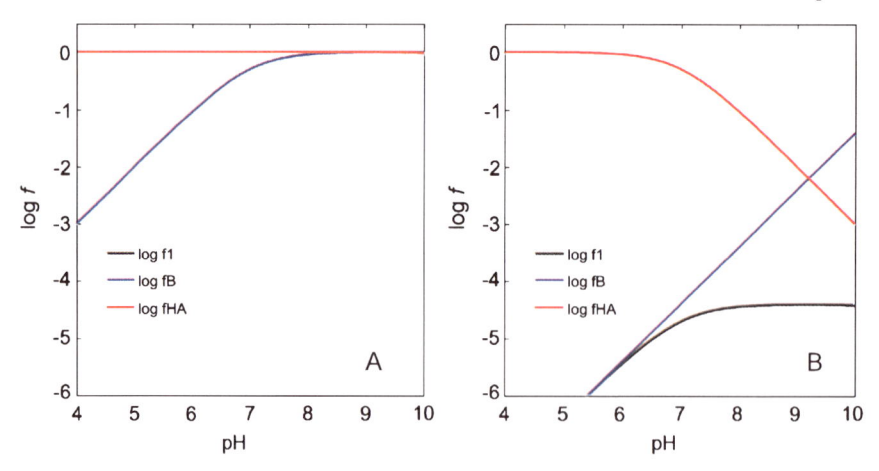

Figure 2.6 Simulation of pH–species plots according to the kinetic models in Schemes 2.1 and 2.2. Simulations were carried out according to a partition function approach.[11] In each case the log of the functional form of the general acid, log f_{HA}, is given in red, the log of the functional form of the general base, log f_B, is given in blue, and the log of the fraction of the ribozyme in the functional form, log f_1, is given in black. Note that log f_1 = log f_{HA} + log f_B, which can be seen graphically.[11] (Notation of "HA" for the functional form of the general acid and "B" for the functional form of the general base do not specify overall charge of these species.) (*A*) GAB Model 1. Simulation for Scheme 2.1 for the HDV ribozyme in which C75 is the general base with a pK_a of 7 and hexahydrated magnesium is the general acid (pK_a of 11.4). In this instance, the log f_1 curve is hidden under the log f_B curve. (*B*) GAB Model 2. Simulation for Scheme 2.2 for the HDV ribozyme in which protonated C75 is the general acid with a pK_a of 7 and pentahydrated magnesium hydroxide is the general base (pK_a of 11.4). Note that the simulations in panels (*A*) and (*B*) both afford log f_1 curves of identical shape.

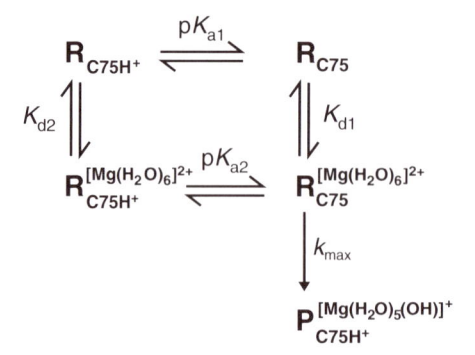

Scheme 2.1 GAB Model 1. General acid–base model for the HDV ribozyme in which C75 is the general base and hexahydrated magnesium is the general acid.

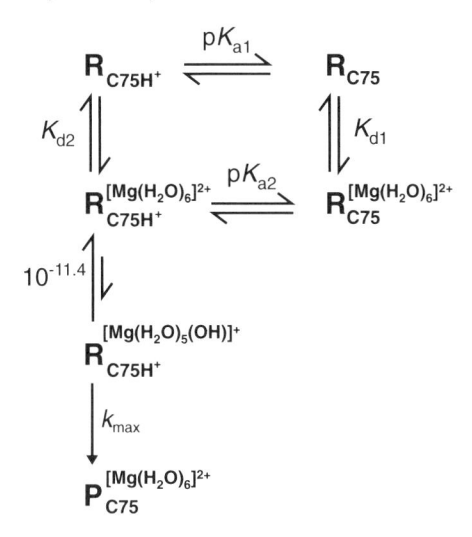

$$R_{C75H^+} \underset{}{\overset{pK_{a1}}{\rightleftharpoons}} R_{C75}$$

$$K_{d2} \updownarrow \qquad \qquad \updownarrow K_{d1}$$

$$R^{[Mg(H_2O)_6]^{2+}}_{C75H^+} \underset{}{\overset{pK_{a2}}{\rightleftharpoons}} R^{[Mg(H_2O)_6]^{2+}}_{C75}$$

$$10^{-11.4} \downarrow$$

$$R^{[Mg(H_2O)_5(OH)]^+}_{C75H^+}$$

$$\downarrow k_{max}$$

$$P^{[Mg(H_2O)_6]^{2+}}_{C75}$$

Scheme 2.2 GAB Model 2. General acid–base model for the HDV ribozyme in which protonated C75 is the general acid and pentahydrated magnesium hydroxide is the general base.

Rate–pH profiles of enzymes are, notably, fraught with difficulties in interpretation. These include the possibility that a pK_a could be due only to a change in a rate-limiting step rather than ionization of a chemically essential residue,[74,75] as well as the contribution of acid- and base-denaturation phenomena.[44,45] Evidence against the pK_a from the low pH-arm arising from a change in the rate-limiting step is the presence of a thio effect,[||] albeit with unusual properties,[76,77] and a KSIE in the plateau region.[12,73,78] With regards to the thio effect, the Sp phosphorothioate diastereomer was originally reported to react at the same rate as the unsubstituted phosphodiester (i.e., no thio effect), while the Rp was reported to have very little reactivity but with the portion that did react doing so at the same rate as the Sp.[76] Recent results on HDV cleavage in monovalent cations on unseparated phosphorothioate diastereomers revealed thio effects on the reactive half of the substrate that were attributed to the Sp diastereomer reacting slower than the unsubstituted phosphodiester.[77] Overall, these thio effects can be taken as support that chemistry is rate limiting in the HDV ribozyme both in the presence and absence of divalent metal ions.

Kinetic solvent isotope experiments have been conducted on both the genomic and antigenomic versions of the ribozyme. In each case, KSIEs in the plateau region of the plot were ~3–4-fold,[12,73,78] and the pK_a in D_2O, after correction for the pH meter effect, was 0.4–0.6 units higher, as expected for proton transfers from similar functionalities.[79,80]

[||] A "thio effect" refers to the slowing of the reaction rate when a phosphorothioate is substituted at the reactive phosphorus.

Proton inventory studies have been carried out on Mg^{2+}-containing reactions on both the genomic and antigenomic HDV ribozymes.[73,78] These experiments, which describe the dependence of observed cleavage rates on the percent D_2O composition of the solvent, have the potential to give information on the number of protons "in-flight" during the rate-limiting step. Inventories of 1 and 2 were found for the antigenomic[73] and genomic[78] ribozymes, respectively. The mechanistic basis for these differences is unclear; however, it can be noted that the steepness of the bowl-shape for the genomic ribozyme was dependent on the concentration of Mg^{2+} in the reaction, and tended toward an inventory of 1 at higher concentrations of Mg^{2+}.[78] In addition, inventories of 1 can always be fit with a greater number of proton transfers.[80] An inventory of 1 is generally consistent with a stepwise proton-transfer mechanism in which one of the steps is rate limiting.[3] An inventory of 2 is consistent with two proton transfers in the rate limiting step, such as in a concerted proton transfer mechanism; proton inventories of 2 were found for RNase A.[81] In any case, the data support proton transfer being involved in the mechanism.

Proton inventories and KSIEs must be interpreted with caution, however. Walter and co-workers showed that conformational changes of the HDV ribozyme, which have no pH-dependence, can give rise to kinetic solvent isotope effects and bowl-shaped proton inventories, although they generally do not fit simple Gross–Butler equations for 1 or 2 proton transfers.[82] This study indicated that KSIEs and proton inventories alone do not prove that chemistry is rate-limiting.

Returning to the kinetic ambiguity issue, Das and Piccirilli carried out an incisive experiment for determining which of the reactant states is catalytically active.[83] They eliminated the need for general acid–base chemistry by replacing the bridging 5'-oxygen with a good leaving group, a bridging sulfur. This substitution rescued the C75U mutant, associating residue 75 with facilitating the leaving group in the reaction (Scheme 2.2). This experiment is the ribozyme counterpart of similar experiments that have been conducted on protein enzymes.[57] The reaction model in Scheme 2.2 is also supported by inversion of the rate–pH profiles in the absence of high pK_a divalent ions and at high ionic strength, which support RNA folding,[12,51] although this data can be interpreted two ways.[54,84] Notably, the HDV ribozyme is active in the absence of divalent ions at low pH, as this supports a limited role of divalent ions in catalysis.[12,51,84] Indeed quantitative considerations suggest that Mg^{2+} ions contribute a modest 25-fold to the reaction rate.[51]

The cleavage mechanism of the HDV ribozyme is not settled. Recent structures of the pre-cleaved HDV ribozyme, inactivated by a C75U mutation or omission of Mg^{2+}, show Mg^{2+} positioned to act as the general acid in cleavage, and C75 can be modeled as the general base.[71] Ultimately, resolution of the cleavage mechanism will require accord between structure and function. This topic is all the more interesting given that a ribozyme nearly equivalent to the HDV ribozyme has been recently discovered in the human genome (Chapter 7) and seems to operate by nucleobase catalysis.[85]

Recently, theoretical studies have been conducted on the HDV ribozyme, which have arrived at mixed conclusions.[86–89] In one study, the authors used quantum mechanics (QM)-molecular mechanics (MM) techniques that led them to favor a mechanism in which C75 acts as a general acid and hydrated magnesium hydroxide acts as a general base (Scheme 2.2).[86] The ribozyme was suggested to carry out two proton transfers, separated by a high energy trigonal bipyramidal intermediate. In another study, density functional theory was used to investigate the cleavage mechanism.[87] The authors reached a similar conclusion as the QM-MM study, favoring a two-step mechanism with C75 acting as the general acid (Scheme 2.2). These authors additionally implicated a five-coordinate phosphorane as a high energy intermediate and N4-imino tautomerization of C75 as a way to stabilize and then deprotonate the phosphorane.

However, other computational studies, including molecular dynamics (MD) studies beginning from the pre-cleaved state of the ribozyme, suggest that C75 likely acts as the general base in the reaction (Scheme 2.1).[88,89] Simulations were carried out with either unprotonated or protonated C75. Unprotonated C75 allows for a productive general base conformation, while protonated C75 allows for a productive general acid conformation. Although these various calculations reach different mechanistic conclusions, there is optimism that theory can contribute to our ability to understand proton transfer in ribozymes. Ultimate success will likely require a close connection between calculations and experiments.

2.4.3 Proton Transfer in the Hairpin Ribozyme

The final sections of this chapter consider data on proton transfer in the mechanism of four other small ribozymes. As with the HDV ribozyme, the role of the nucleobases in catalysis in the hairpin ribozyme (Chapter 5) was inspired by high-resolution crystal structures. However, landmark biochemical experiments preceded the structures, namely experiments showing that the hairpin, hammerhead, and VS ribozymes are active in the absence of divalent ions.[90–93] These experiments provided strong, albeit indirect, evidence that nucleobases may be active participants in chemistry. Ferre-D'Amare and co-workers have solved multiple structures of the hairpin ribozyme, including reactant, product, and vanadate-bound transition state complexes.[94,95] These structures revealed the N1 of G8 positioned near the nucleophilic 2′-hydroxyl and the N1 of A38 near 5′-bridging oxygen leaving group.[94,95] More recent crystal structures from the Wedekind laboratory, solved in the absence of the U1A protein used in the prior studies, revealed four ordered water molecules that might be involved in chemistry,[96] and showed altered positioning of the bases.[97]

The hairpin ribozyme has also been subjected to rate–pH profiles, KSIEs and proton inventory. Burke and co-workers showed that one of the pK_as in the rate–pH profile was sensitive to the pK_a of the base at position 8.[98] The resultant rate–pH profiles supported the importance of the ionization of G8 to cleavage, although kinetic ambiguity precluded assignment of the neutral or anionic form as being functional. In addition, a KSIE of 3–4 was observed, as was a proton inventory of 2.[98] These findings parallel those on the HDV ribozymes that supported proton transfer.

As stated, two of the residues implicated in catalysis by the hairpin ribozyme are G8 and A38.[94,95] However, their roles in proton transfer in the mechanism of the hairpin ribozyme are uncertain. Small molecule rescue experiments on G8 and A38 have been interpreted to favor a role for the nucleobases in making electrostatic contributions to catalysis;[55,99] in this scenario, water molecules, perhaps those seen in recent crystal structures,[96] serve roles as specific acids and bases.[5] In addition, recent MD calculations on the transition state structure led to the suggestion that water might be a specific base in the reaction mechanism,[100] although G8 and A38 maintained general base and acid positionings, respectively, from the crystal structure. However, rate–pH profiles and positioning of these atoms in crystal structures are also consistent with general acid–base chemistry, in which G8 acts as the general base for cleavage and A38 acts as the general acid.[11] As stated for the HDV ribozyme, it is not impossible that the nucleobases serve both roles in catalysis. Recent calculations of pK_a values are consistent with several adenosines in the hairpin having elevated pK_as, including A38,[53] which is supportive of either electrostatic or proton transfer roles.

Lilley and co-workers have conducted two series of experiments that implicate nucleobases in hairpin ribozyme catalysis. In one instance, they incorporated imidazole covalently into the backbone of the hairpin ribozyme in place of G8 using phosphoramidite chemistry;[101] they developed this technique because the hairpin ribozyme is not rescuable by imidazole addition.[102,103] Not only was the modified ribozyme reactive, it gave identical rate–pH profiles for cleavage and ligation reactions, which is interpretable as having either a general acid–base or an electrostatic role in the reaction.[11,101] The Lilley laboratory has also conducted single molecule experiments on the hairpin ribozyme as a function of pH. Importantly, these experiments were done on a natural four-way junction construct that allowed separation of folding and chemistry steps.[104] As with the imidazole nucleoside experiments, these studies gave identical rate–pH profiles for cleavage and ligation, showing again the possibility for general acid–base roles for nucleosides in the mechanism.

In summary, nucleobases clearly play essential roles in the mechanism of the hairpin ribozyme. In addition, several experiments support G8 and A38 serving as the general acid and base in the reaction, although recent crystallographic, small molecule rescue, and computational studies support *specific* acid–base chemistry in the mechanism. As mentioned for larger ribozymes, proton shuttles involving water are formally possible, as is a combination of proton transfer and electrostatic roles by the bases themselves.

2.4.4 Proton Transfer in the Hammerhead Ribozyme

The HDV and hairpin ribozymes followed courses of investigation that were not dissimilar. In each case, initial biochemical studies established reliable secondary structures. High-resolution crystal structures were then solved, which dominated the design and interpretation of biochemical experiments that were for the most part congruent with the structural data. The hammerhead ribozyme, however, has followed a more complex course of development (Chapters 3 and 4).

Several high-resolution structures were solved for this ribozyme from Scott and co-workers.[105–108] However, findings from biochemical experiments suggested that the positioning of some of the atoms seen in the early crystal structures were not catalytically relevant.[109] In particular, metal ion rescue experiments and mutants required close positioning of A9 and the scissile phosphate, which were 20 Å apart in the crystal structure.[110] Indeed, quite recent results of the pH-dependence of G8 and G12 substitutions,[111] and crosslinking/reactivity studies[112,113] supported the notion that the early structures were not of the catalytic fold.

Recently, Martick and Scott solved the crystal structure of a hammerhead ribozyme with peripheral domains that aid folding and discovered a very different catalytic core.[114] Strikingly, many of the earlier structure–function discrepancies were resolved by the new structure. The structure has suggestive roles for nucleobases in catalysis. In particular, G8 and G12, implicated above in solution biochemical studies,[111–113] are positioned to participate in catalysis.[114] The N1 of G12 potentially serves as a general base to deprotonate the 2′-hydroxyl nucleophile, while the 2′-hydroxyl of G8 potentially serves as a general acid to protonate the bridging 5′-hydroxyl. The hammerhead ribozyme, like the HDV and hairpin ribozyme, is now poised for in-depth mechanistic studies to help determine if the bases themselves are carrying out proton transfer, as suggested by the recent structure.

2.4.5 Proton Transfer in the VS Ribozyme

Unlike the HDV, hairpin, and hammerhead ribozymes, there is presently not a crystal structure of the entire VS ribozyme (Chapter 5), although NMR structures have been reported for various domains, including the cleavage site.[115–117] Despite the absence of a crystal structure, there is an amazing amount of recent biochemical work that has implicated nucleobases in catalysis, potentially as general acid–bases. In particular, work from Lilley and co-workers with the imidazole phosphoramidites, discussed above for the hairpin ribozyme, has shown A756 to imidazole as reactive, albeit at a reduced rate.[103] Residue 756 was implicated in catalysis on the basis of modeling from FRET and electrophoresis data,[118] as well as the kinetics of various mutants,[119,120] which was the basis for targeting it with imidazole.

Most recently, these authors went in search for another nucleobase that might be active in catalysis.[121] They made numerous modifications in the substrate stem-loop and identified G638 as important to catalysis. Rate–pH profiles of substitutions at this position showed that this residue is responsible for one of the pK_as in the rate–pH profile. The mechanism of the VS ribozyme that follows from these studies is consistent with general acid–base chemistry by G638 and A756, and is remarkably similar to the hairpin ribozyme, suggestive of a convergent evolution in mechanism.

Recently, Smith and Collins reported rate–pH profiles, KSIEs, and proton inventories for the VS ribozyme that strongly support the presence of proton transfer in the mechanism.[13] Using a VS ribozyme construct that they developed that reacts exceptionally quickly,[122] they performed a battery of mechanistic tests. They found evidence in the rate–pH profile for nucleobase pK_as with values of 5.8 and 8.3. Solvent isotope effect experiments led to a KSIE of 2, pK_a shifts in D_2O of 0.6 to 0.7 units, and proton inventories of 1 (but also consistent with 2). These observations are remarkably similar to those that support proton transfer in the HDV and hairpin ribozymes. Moreover, the two pK_a values are supportive of the model from Lilley and co-workers in which an A and G participate in general acid–base[121] since these bases have unshifted pK_a values relatively close to the observed pK_as (Figure 2.3). One other notable feature of the Smith and Collins study is the testing of the effects of viscosity on the rate.[13] In particular, D_2O is 1.2-fold more viscous than H_2O at room temperature. The authors show that the effect of D_2O on the reaction is not a viscosity effect. Moreover, this study shows that caution must also be exercised in evaluating the effects of D_2O on reactions that are limited by conformational changes.[82]

The course of research development on the VS ribozyme is interesting from a historical perspective. While high-resolution studies contributed profoundly to mechanistic development on the HDV, hairpin, and hammerhead ribozymes, the VS ribozyme's mechanism has developed without recourse to a structure of the intact ribozyme. Despite this limitation, there is now strong evidence to support proton transfer by nucleobases.

2.4.6 Proton Transfer in the *glmS* Ribozyme

The *glmS* ribozyme (Chapter 8), which consists of the expression domain of the glucosamine-6-phosphate (GlcN6P) riboswitch, is the most recently discovered small ribozymes.[123] Unusually, it requires a small molecule, GlcN6P, for activity. As with the HDV, hairpin, hammerhead, and VS ribozymes, it has been shown that divalent ions play only structural roles in the *glmS* ribozyme mechanism, again suggestive of a direct role for nucleobases in the mechanism.[124] In addition, rate–pH profiles give a pK_a of neutrality – consistent with pK_a shifting and optimization for general acid–base chemistry.[124] Intriguingly, recent studies support the direct participation of the GlcN6P cofactor itself in chemistry.

Remarkably, only two years after its discovery, two crystal structures of the *glmS* ribozyme bound to cofactor were solved.[125,126] One crystal structure is of the *glmS* ribozyme from *T. tengcongenis* bound to the competitive inhibitor, Glc6P,[125] while the other structure is of the *glmS* ribozyme from *B. anthracis* bound to the GlcN6P cofactor.[126] Both structures are notable in that they implicate the cofactor in proton transfer. In both structures, the amine of the cofactor is 2.7 Å from the bridging 5′-oxygen, identical to the positioning of the N3 of C75 to its bridging 5′-oxygen in the HDV ribozyme product structure.[127] In addition, both *glmS* structures revealed a guanosine well-positioned to use its N1 as a general base in deprotonating the 2′-hydroxyl nucleophile.

2.5 Conclusion and Perspectives

Proton transfer is an essential catalytic device in promoting virtually all chemical reactions. Enzymes stabilize the transition state at a constant biological pH of neutrality by proton transfer. In large ribozymes and the ribosome, proton shuttles may be key in the transfers. In small ribozymes, nucleobase catalysis seems to be important, especially near the bridging 5′-oxygen where there is a dearth of labile protons. In these cases, evidence for perturbations of nucleobase pK_as towards neutrality have been provided. Such pK_a shifts allow the nucleobases to function at the molecular level much like histidine and lysine in proteins. High-resolution crystal structures have been invaluable in developing models of proton transfer, although remarkable progress has been made recently on the mechanism of the VS ribozyme where a crystal structure of the intact ribozyme is not yet available.

Recently, accord between structure and function has improved for several ribozymes. Nonetheless, there is still much to learn about proton transfer. The new crystal structures on the hammerhead and *glmS* ribozymes will undoubtedly drive new biochemical experiments and theoretical calculations to probe mechanism. Additional crystal structures may shed light on small ribozyme mechanism. A full understanding of the role of proton transfer in mechanism will undoubtedly require structures of reactant, product, and transition state models for each ribozyme, as accomplished only for the hairpin ribozyme to date. In addition, structures of catalytically important ribozyme mutants will be required to achieve a full and deep understanding of proton transfer. Theory is also likely to play an increased role in understanding of proton transfer, especially given that structure and function is reaching a stage where questions of proton transfer are difficult to address experimentally.

Certain themes are developing in ribozyme proton transfer. Curiously, guanosine may hold a special role in deprotonating the 2′-hydroxyl for cleavage (and protonating it for ligation). In particular, the hammerhead, hairpin, and *glmS* crystal structures all reveal a guanosine in the position for general base catalysis for cleavage; even the VS ribozyme, for which a full-length structure in not yet available, has a guanosine implicated as the general base or acid. Of the small ribozymes, only the HDV ribozyme seems to not use guanosine as its general base, which may explain why only the HDV ribozyme requires low pH to operate proficiently in the absence of divalent ions.[12,93] It is somewhat surprising that guanosine would act as a general base, given that it would presumably be anionic in its functional form. However, the pK_a of the N1 of guanosine is only 2 units removed from neutrality (Figure 2.3), and its anionic form could be stabilized by cooperativity arising from through-space electrostatic interactions with cationic species at the general acid position such as a protonated adenosine or GlcN6P.

Acknowledgement

I would like to thank the National Science Foundation for support through grant MCB-0527102.

References

1. S.R. Holbrook, *Curr. Opin. Struct. Biol.*, 2005, **15**, 302.
2. D.A. Kraut, K.S. Carroll and D. Herschlag, *Annu. Rev. Biochem.*, 2003, **72**, 517.
3. P.C. Bevilacqua, T.S. Brown, S. Nakano and R. Yajima, *Biopolymers*, 2004, **73**, 90.
4. J.A. Doudna and J.R. Lorsch, *Nat. Struct. Mol. Biol.*, 2005, **12**, 395.
5. M.J. Fedor and J.R. Williamson, *Nat. Rev. Mol. Cell. Biol.*, 2005, **6**, 399.
6. P.C. Bevilacqua and R. Yajima, *Curr. Opin. Chem. Biol.*, 2006, **10**, 455.
7. W.P. Jencks, in *Catalysis in Chemistry and Enzymology*, Dover Publications Inc., New York, 1969.
8. R.B. Silverman, *The Organic Chemistry of Enzyme-Catalyzed Reactions*, Academic Press, San Diego, CA, 2000.
9. J.G. Zalatan and D. Herschlag, *J. Am. Chem. Soc.*, 2006, **128**, 1293.
10. W.P. Jencks and J. Regenstein, in *Handbook of Biochemistry: Selected data for molecular biology*, 2nd edition (Ed. H.A. Soper) J150–189, 1970.
11. P.C. Bevilacqua, *Biochemistry*, 2003, **42**, 2259.
12. S. Nakano, D.M. Chadalavada and P.C. Bevilacqua, *Science*, 2000, **287**, 1493.
13. M.D. Smith and R.A. Collins, *Proc. Natl. Acad. Sci. U.S.A.*, 2007, **104**, 4818.
14. W.P. Jencks, *J. Am. Chem. Soc.*, 1972, **94**, 4731.
15. R.M. Izatt, J.J. Christensen and J.H. Rytting, *Chem. Rev.*, 1971, **71** 439.
16. W. Saenger, *Principles of Nucleic Acid Structure*, Springer-Verlag, New York, 1984.
17. P. Legault and A. Pardi, *J. Am. Chem. Soc.*, 1997, **119**, 6621.
18. G.J. Narlikar and D. Herschlag, *Annu. Rev. Biochem.*, 1997, **66**, 19.
19. R.M.C. Dawson, D.C. Elliot, W.H. Elliot and K.M. Jones, *Data for Biochemical Research*, Clarendon Press, Oxford, UK, 1969.
20. R.T. Raines, *Chem. Rev.*, 1998, **98**, 1045.
21. S.M. Parsons and M.A. Raftery, *Biochemistry*, 1972, **11**, 1633.
22. J. Trylska *et al.*, *Protein Sci.*, 1999, **8**, 180.
23. G.A. Hunter and G.C. Ferreira, *Biochemistry*, 1999, **38**, 3711.
24. L. Zhang, L. Chooback and P.F. Cook, *Biochemistry*, 1999, **38**, 11231.

25. C. Wang, H. Gao, B.L. Gaffney and R.A. Jones, *J. Am. Chem. Soc.*, 1991, **113**, 5486.
26. H.T. Allawi and J. SantaLucia, Jr., *Biochemistry*, 1998, **37**, 9435.
27. B. Pan, S.N. Mitra and M. Sundaralingam, *J. Mol. Biol.*, 1998, **283**, 977.
28. Z. Cai and I. Tinoco, Jr., *Biochemistry*, 1996, **35**, 6026.
29. S.E. Butcher, F.H. Allain and J. Feigon, *Nat. Struct. Biol.*, 1999, **6**, 212.
30. P. Legault and A. Pardi, *J. Am. Chem. Soc.*, 1994, **116**, 8390.
31. X. Gao and D.J. Patel, *J. Am. Chem. Soc.*, 1988, **110**, 5178.
32. C. Carbonnaux, G.A. van der Marel, J.H. van Boom, W. Guschlbauer and G.V. Fazakerley, *Biochemistry*, 1991, **30**, 5449.
33. G.A. Leonard, E.D. Booth and T. Brown, *Nucleic Acids Res.*, 1990, **18**, 5617.
34. D.S. Knitt, G.J. Narlikar and D. Herschlag, *Biochemistry*, 1994, **33**, 13864.
35. D.S. Pilch and R.H. Shafer, *J. Am. Chem. Soc.*, 1993, **115**, 2565.
36. B. Borah and J.L. Wood, *J. Mol. Struct.*, 1976, **30**, 13.
37. G.J. Quigley *et al.*, *Science*, 1986, **232**, 1255.
38. A. Rich, D.R. Davies, F.H. Crick and J.D. Watson, *J. Mol. Biol.*, 1961 **3**, 71.
39. S.-H. Chou, J.-W. Cheng, O.Y. Fedoroff, V.P. Chuprina and B.R. Reid, *J. Am. Chem. Soc.*, 1992, **114**, 3114.
40. K. Maskos, B.M. Gunn, D.A. LeBlanc and K.M. Morden, *Biochemistry*, 1993, **32**, 3583.
41. B. McConnell, *Biochemistry*, 1978, **17**, 3168.
42. B. McConnell and D. Politowski, *Biophys. Chem.*, 1984, **20**, 135.
43. M. Gueron and J.L. Leroy, *Methods Enzymol.*, 1995, **261**, 383.
44. D.S. Knitt and D. Herschlag, *Biochemistry*, 1996, **35**, 1560.
45. E.M. Moody, J.T. Lecomte and P.C. Bevilacqua, *RNA*, 2005, **11**, 157.
46. E.M. Moody and P.C. Bevilacqua, *J. Am. Chem. Soc.*, 2003, **125**, 16285.
47. P.V. Cornish and D.P. Giedroc, *Biochemistry*, 2006, **45**, 11162.
48. N.A. Siegfried, S.L. Metzger and P.C. Bevilacqua, *Biochemistry*, 2007 **46**, 172.
49. A. Warshel, *Proc. Natl. Acad. Sci. U.S.A.*, 1978, **75**, 5250.
50. A. Fersht, *Enzyme Structure and Mechanism*, Freeman, New York, 1985.
51. S. Nakano, D.J. Proctor and P.C. Bevilacqua, *Biochemistry*, 2001 **40**, 12022.
52. P.C. Bevilacqua *et al.*, *Biochem. Soc. Trans.*, 2005, **33**, 466.
53. C.L. Tang, E. Alexov, A.M. Pyle and B. Honig, *J. Mol. Biol.* 2007, **366**, 1475.
54. S. Nakano and P.C. Bevilacqua, *Biochemistry*, 2007, **46**, 3001.
55. Y.I. Kuzmin, C.P. Da Costa, J.W. Cottrell and M.J. Fedor, *J. Mol. Biol.*, 2005, **349**, 989.
56. Y. Li and R.R. Breaker, *J. Am. Chem. Soc.*, 1999, **121**, 5364.
57. J.E. Thompson and R.T. Raines, *J. Am. Chem. Soc.*, 1994, **116**, 5467.
58. A.T. Perrotta, I. Shih and M.D. Been, *Science*, 1999, **286**, 123.

59. A.R. Ferre-D'Amare, K. Zhou and J.A. Doudna, *Nature*, 1998 **395**, 567.
60. D. Herschlag, F. Eckstein and T.R. Cech, *Biochemistry*, 1993, **32**, 8312.
61. J.S. Weinger, K.M. Parnell, S. Dorner, R. Green and S.A. Strobel, *Nat. Struct. Mol. Biol.*, 2004, **11**, 1101.
62. J.L. Hougland, J.A. Piccirilli, M. Forconi, J. Lee and D. Herschlag, in *RNA World*, 3rd edn., ed. R.F. Gesteland, T.R. Cech and J.F. Atkins, Cold Spring Harbor Press, Cold Spring Harbor, New York, 2006, p. 768.
63. S. Shan, A. Yoshida, S. Sun, J.A. Piccirilli and D. Herschlag, *Proc. Natl. Acad. Sci. U.S.A.*, 1999, **96**, 12299.
64. S. Shan, A.V. Kravchuk, J.A. Piccirilli and D. Herschlag, *Biochemistry*, 2001, **40**, 5161.
65. M.R. Stahley and S.A. Strobel, *Science*, 2005, **309**, 1587.
66. T.M. Schmeing, K.S. Huang, D.E. Kitchen, S.A. Strobel and T.A. Steitz, *Mol. Cell*, 2005, **20**, 437.
67. P. Bieling, M. Beringer, S. Adio and M.V. Rodnina, *Nat. Struct. Mol. Biol.*, 2006, **13**, 423.
68. S. Trobro and J. Aqvist, *Proc. Natl. Acad. Sci. U.S.A.*, 2005, **102**, 12395.
69. A.J. Kirby and M. Younas, *J. Chem. Soc. B*, 1970, 1165.
70. M. Oivanen, S. Kuusela and H. Lonnberg, *Chem. Rev.*, 1998, **98**, 961.
71. A. Ke, K. Zhou, F. Ding, J.H. Cate and J.A. Doudna, *Nature*, 2004 **429**, 201.
72. A.T. Perrotta, T.S. Wadkins and M.D. Been, *RNA*, 2006, **12**, 1282.
73. I.H. Shih and M.D. Been, *Proc. Natl. Acad. Sci. U.S.A.*, 2001, **98**, 1489.
74. D. Herschlag and M. Khosla, *Biochemistry*, 1994, **33**, 5291.
75. P.C. Bevilacqua, T.S. Brown, D. Chadalavada, A.D. Parente and R. Yajima, *Kinetic Analysis of Ribozyme Cleavage*, Oxford University Press, Oxford, 2003.
76. H. Fauzi, J. Kawakami, F. Nishikawa and S. Nishikawa, *Nucleic Acids Res.*, 1997, **25**, 3124.
77. A.T. Perrotta and M.D. Been, *Biochemistry*, 2006, **45**, 11357.
78. S. Nakano and P.C. Bevilacqua, *J. Am. Chem. Soc.*, 2001, **123**, 11333.
79. J.H. Shim and S.J. Benkovic, *Biochemistry*, 1999, **38**, 10024.
80. K.B. Schowen and R.L. Schowen, *Methods Enzymol.*, 1982, **87**, 551.
81. M.S. Matta and D.T. Vo, *J. Am. Chem. Soc.*, 1986, **108**, 5316.
82. R.A. Tinsley, D.A. Harris and N.G. Walter, *J. Am. Chem. Soc.*, 2003, **125**, 13972.
83. S.R. Das and J.A. Piccirilli, *Nat. Chem. Biol.*, 2005, **1**, 1.
84. T.S. Wadkins, I. Shih, A.T. Perrotta and M.D. Been, *J. Mol. Biol.*, 2001, **305**, 1045.
85. K. Salehi-Ashtiani, A. Luptak, A. Litovchick and J.W. Szostak, *Science*, 2006, **313**, 1788.
86. K. Wei, L. Liu, Y.H. Cheng, Y. Fu and Q.X. Guo, *J. Phys. Chem. B*, 2007, **111**, 1514.

87. H. Liu, J.J. Robinet, S. Ananvoranich and J.W. Gauld, *J. Phys. Chem. B*, 2007, **111**, 439.
88. M.V. Krasovska, J. Sefcikova, N. Spackova, J. Sponer and N.G. Walter, *J. Mol. Biol.*, 2005, **351**, 731.
89. M.V. Krasovska *et al.*, *Biophys. J.*, 2006, **91**, 626.
90. A. Hampel and J.A. Cowan, *Chem. Biol.*, 1997, **4**, 513.
91. S. Nesbitt, L.A. Hegg and M.J. Fedor, *Chem. Biol.*, 1997, **4**, 619.
92. K.J. Young, F. Gill and J.A. Grasby, *Nucleic Acids Res.*, 1997, **25**, 3760.
93. J.B. Murray, A.A. Seyhan, N.G. Walter, J.M. Burke and W.G. Scott, *Chem. Biol.*, 1998, **5**, 587.
94. P.B. Rupert and A.R. Ferre-D'Amare, *Nature*, 2001, **410**, 780.
95. P.B. Rupert, A.P. Massey, S.T. Sigurdsson and A.R. Ferre-D'Amare, *Science*, 2002, **298**, 1421.
96. J. Salter, J. Krucinska, S. Alam, V. Grum-Tokars and J.E. Wedekind, *Biochemistry*, 2006, **45**, 686.
97. S. Alam, V. Grum-Tokars, J. Krucinska, M.L. Kundracik and J.E. Wedekind, *Biochemistry*, 2005, **44**, 14396.
98. R. Pinard *et al.*, *EMBO J.*, 2001, **20**, 6434.
99. Y.I. Kuzmin, C.P. Da Costa and M.J. Fedor, *J. Mol. Biol.*, 2004 **340**, 233.
100. H. Park and S. Lee, *J. Chem. Theory Comput.*, 2006, **2**, 858.
101. T.J. Wilson *et al.*, *RNA*, 2006, **12**, 980.
102. L. Araki *et al.*, *Tetrahedron Lett.*, 2004, **45**, 2657–2661.
103. Z.Y. Zhao *et al.*, *J. Am. Chem. Soc.*, 2005, **127**, 5026.
104. M.K. Nahas *et al.*, *Nat. Struct. Mol. Biol.*, 2004, **11**, 1107.
105. W.G. Scott, J.T. Finch and A. Klug, *Cell*, 1995, **81**, 991.
106. W.G. Scott, J.B. Murray, J.R. Arnold, B.L. Stoddard and A. Klug, *Science*, 1996, **274**, 2065.
107. J.B. Murray *et al.*, *Cell*, 1998, **92**, 665.
108. J.B. Murray, H. Szoke, A. Szoke and W.G. Scott, *Mol. Cell*, 2000, **5**, 279.
109. K.F. Blount and O.C. Uhlenbeck, *Annu. Rev. Biophys. Biomol. Struct.*, 2005, **34**, 415.
110. S. Wang, K. Karbstein, A. Peracchi, L. Beigelman and D. Herschlag, *Biochemistry*, 1999, **38**, 14363.
111. J. Han and J.M. Burke, *Biochemistry*, 2005, **44**, 7864.
112. J.E. Heckman, D. Lambert and J.M. Burke, *Biochemistry*, 2005 **44**, 4148.
113. D. Lambert, J.E. Heckman and J.M. Burke, *Biochemistry*, 2006, **45**, 7140.
114. M. Martick and W.G. Scott, *Cell*, 2006, **126**, 309.
115. P.J. Michiels, C.H. Schouten, C.W. Hilbers and H.A. Heus, *RNA*, 2000, **6**, 1821.
116. J. Flinders and T. Dieckmann, *J. Mol. Biol.*, 2001, **308**, 665.
117. B. Hoffmann *et al.*, *Proc. Natl. Acad. Sci. U.S.A.*, 2003, **100**, 7003.
118. D.A. Lafontaine, D.G. Norman and D.M. Lilley, *EMBO J.*, 2002 **21**, 2461.

119. D.A. Lafontaine, T.J. Wilson, D.G. Norman and D.M. Lilley, *J. Mol. Biol.*, 2001, **312**, 663.
120. V.D. Sood and R.A. Collins, *J. Mol. Biol.*, 2002, **320**, 443.
121. T.J. Wilson, A. McLeod and D.M. Lilley, *EMBO J.* 2007, **26**, 2489.
122. R. Zamel *et al.*, *Proc. Natl. Acad. Sci. U.S.A.*, 2004, **101**, 1467.
123. W.C. Winkler, A. Nahvi, A. Roth, J.A. Collins and R.R. Breaker, *Nature*, 2004, **428**, 281.
124. A. Roth, A. Nahvi, M. Lee, I. Jona and R.R. Breaker, *RNA*, 2006 **12**, 607.
125. D.J. Klein and A.R. Ferre-D'Amare, *Science*, 2006, **313**, 1752.
126. J.C. Cochrane, S.V. Lipchock and S.A. Strobel, *Chem. Biol.*, 2007, **14**, 97.
127. M.D. Been, *Science*, 2006, **313**, 1745.
128. P.D. Lyne and M. Karplus, *J. Am. Chem. Soc.*, 2000, **122**, 166.
129. S. Acharya, A. Foldesi and J. Chattopadhyaya, *J. Org. Chem.*, 2003, **68**, 1906.

CHAPTER 3
Finding the Hammerhead Ribozyme Active Site

DOMINIC LAMBERT* AND JOHN M. BURKE

Department of Microbiology and Molecular Genetics, University of Vermont, 95 Carrigan Drive, 220B Stafford Hall, Burlington, Vermont 05405, USA

3.1 Introduction

The hammerhead ribozyme has long been the misunderstood problem child of the catalytic RNA family. Among the small ribozymes that catalyze RNA cleavage with formation of 2′,3′ cyclic phosphates, the hammerhead was one of the earliest-identified,[1] is certainly the most widely-studied, and yielded the first high-resolution crystal structures.[2,3] However, intensive biochemical investigations by many laboratories over a 20-year period yielded results that were consistently difficult to reconcile with the crystallographic structures.[4,5] We and others argued that the "classical" hammerhead crystal structures, while informative, were likely significantly different from the biologically-relevant active fold. In our opinion, the most conspicuous problem was that convincing and testable models of the active site and catalytic mechanism had not been identified.

Although the hammerhead remained poorly understood, significant strides had been made in elucidating the structures and catalytic activities of other ribozymes within the family, notably the hairpin and hepatitis delta virus ribozymes. Crystal structures of the delta and hairpin ribozymes[6,7] provided detailed structural information that was generally in close accordance with the extensive body of results obtained through biochemical, genetic, and biophysical investigations.[8,9] While delta and hairpin ribozymes will continue to provide important new avenues for exploration, we seem to have a basic grasp

---

*Current address: Departments of Chemistry and Biophysics, Johns Hopkins University, 3400 North Charles Street, Baltimore, Maryland 21218 USA.

of the fundamental features that govern catalytic function. Hence, they are regarded as the well-behaved children of the family.

In the analysis of the hairpin ribozyme, significant insights into the identity of nucleotides comprising the active site were obtained by this research group and others, together with insights into catalytic mechanism and secondary and tertiary folding, before the elucidation of the three-dimensional structure.[8] We reasoned that the application of analogous experimental strategies to the problem of the hammerhead ribozyme might be productive in working through the difficulties presented by the hammerhead.

Here, we summarize a series of studies that led to the identification of nucleotides at the hammerhead active site, together with a plausible mechanistic model. We show that these data lead to a folding model in which a coordinated large-scale conformational change transforms a ground-state structure (presumably similar to the initial or "classical" crystal structures) into an active fold. The recently-solved hammerhead X-ray structure of Martick and Scott (2006), described elsewhere in this volume (Chapter 4), represents a terrific advance in the field, provides substantial support for our models, and permits those in the field to focus further work concerning ribozyme folding and catalytic mechanism.

While the nature of our chosen problem, identification of an active site, is a very common one in enzymology, our experimental approach has been by necessity an atypical one. Here, we describe how the integration of results from biochemical, biophysical, phylogenetic and molecular modeling approaches have been used to identify the hammerhead active site and to map conformational changes implicated in hammerhead catalysis. Several important conclusions of this work have been borne out by the recent crystallographic work.[10]

3.2 Background

Hammerhead ribozymes were discovered as self-processing motifs within RNA replicons associated with plant RNA viruses and virusoids, and were first converted into trans-acting ribozymes during the late 1980s.[11,12] The hammerhead is widely distributed in nature, so that secondary structure models consisting of three conserved helices rapidly became available through phylogenetic analysis, and a set of strictly or relatively invariant nucleobases were recognized and proposed to form a "core" that was proposed to contain important determinants for catalytic activity of the ribozyme (Figure 3.1).

Biochemical and crystallographic analysis was greatly accelerated by the development of small, trans-acting hammerheads, yet this process deleted peripheral domains that proved, many years later, to be important for biological activity under physiological conditions,[13,14] and for the crystallographic analysis of structures that appear much closer to the active fold.[15]

Finding the Hammerhead Ribozyme Active Site 39

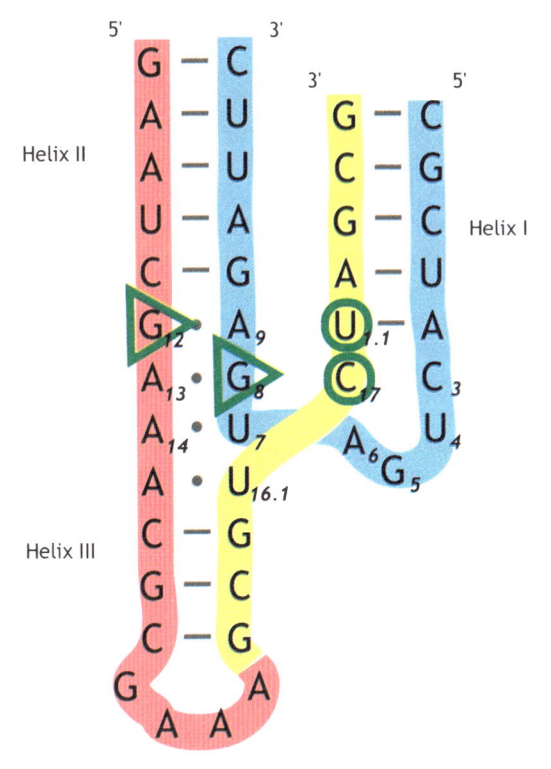

Figure 3.1 Ribozyme–substrate complex for minimal hammerhead construct alpha 1. Yellow indicates substrate sequences. Red and blue denote ribozyme sequences. Putative active site nucleotides are marked with green triangles. Circles denote cleavage site nucleotides. Residues within the conserved core are annotated and the substrate cleavage site is marked with an arrow. Helices are indicated Helix I, II and III.

While it is inarguable that the omission of peripheral domains impeded structural analysis and ribozyme biotechnology, the use of small trans-acting constructs was extremely valuable in that they enabled the use of solid-phase RNA synthetic chemistry to explore the functional consequences of a wide range of increasingly subtle nucleotide and functional group changes. Substitutions were generated and were analyzed by an expanding range of experimental methods over time. This work was greatly facilitated by (i) advances in the synthetic chemistry of oligoribonucleotides, (ii) a very large increase in the number of investigators interested in RNA chemistry, and (iii) concomitant increases in new and powerful methods, *e.g.*, combinatorial selection.

From the early days after the development of small ribozymes, biochemists and crystallographers combined their talents to attempt to probe the structure, active site and mechanism by applying experimental strategies that were well established in investigations of protein enzymes. In protein enzymology, the dominant paradigm involves (i) solving the 3D structure of the enzyme,

(ii) locating the active site *via* co-crystallization with substrates, inhibitors, and analogs, and (iii) probing mechanism through chemical and genetic analysis of the active site.

Publication of the initial hammerhead crystal structures[4,5] generated great excitement in the field, as they were the first high-resolution structures of catalytic RNA molecules. Unfortunately, they did not lead to a clear series of experiments identifying active site and mechanism, nor did they rapidly lead to crystal structures of other ribozymes. Discrepancies between crystallographic and biochemical studies have been well documented.[4,5] For example, many modifications that result in very large decreases in catalytic activity are distant from the reaction site in the crystal structures, while several modifications that were expected to be well-accommodated in the crystal structure strongly inhibit activity.[16] These results were thought by many biochemists to support arguments that the crystal structure represented a fold that was not representative of the catalytic fold. Opinions differed as to whether the crystallographic fold might represent a conformational ground state and, if so, whether a small- or large-scale conformational change might be required to achieve the transition state.

3.3 Experimental Data

3.3.1 Mechanistic Hypothesis Leads to Identification and Functional Test of Active Site Components

Models of hammerhead catalytic mechanism have evolved greatly over time. A significant number of early papers focused on the role of metal ions in the catalytic mechanism. This was not surprising, since RNA was well known to coordinate metal ions, and divalent metal ions were known to be essential components of the active site of several protein enzymes that use nucleic acids, mononucleotides, and related compounds as substrates. Catalytic models in which functional groups on the RNA itself were intimately involved in catalysis, *e.g.*, for general acid–base chemistry, were not seriously considered. Work on the hairpin ribozyme supported a general acid–base mechanism; a model supported in broad terms by modification–activity studies, theoretical analysis, and crystallography.[7,8,17,18]

Much of the hammerhead mechanistic work was predicated on the assumption that one or more magnesium ions were part of the catalytic mechanism, yet work in our laboratory and others showed that robust hammerhead reactions could proceed in the absence of divalent cations.[19–22] These results forced a reevaluation of prior assumptions concerning the mechanism of hammerhead catalysis.

We hypothesized that the characteristic log–linear increase in hammerhead reaction rate with pH might result from a general acid–base mechanism[23] in which both the general acid and the general base had very high pK_as (>9.0). Making the assumption that nucleobase N1 positions were most likely involved (because their pK_as were closest to neutrality in most oligoribonucleotides), we focused on conserved G and U's within the "catalytic core". This greatly simplified the problem, as there are only three such nucleobases – G5, G8, and G12. Nucleobase substitution experiments were carried out in which G5, G8 and G12 were substituted, individually and in combination, with inosine (I), 2-aminopurine (2AP), and 2,6-diaminopurine (diAP), guanosine analogues with altered base-pairing patterns and N1 pK_as. Subsequently, pH–activity relationships were monitored under conditions where cleavage was rate limiting. Results confirmed prior work showing that all three positions were important for optimal activity, but demonstrated that the N1 of nucleobases 8 and 12 influenced the pH–activity relationships in a manner that was sensitive to the nucleobase pK_a. The results are consistent with a model for general acid–base catalysis in which G8 and G12 function as active site components. In contrast, G5 variants significantly inhibited catalysis but did so in a pH-independent manner.[23]

3.3.2 Structural Hypothesis – Large-scale Conformational Changes are Required for Catalysis

Much controversy in the hammerhead field has centered around the degree to which conformational changes are involved in catalysis. Working with the classical crystal structures, our putative identification of G8 and G12 as active site components clearly required a large-scale conformational change for catalysis, since these nucleobases lie on the order of 15–20 Å from the attacking and leaving groups within the scissile phosphodiester linkage, using the classical crystal structures as a point of departure. Using hydroxyl-radical footprinting, we showed that minimal hammerheads undergo concerted cation-induced folding into a tertiary structure that appears to resemble the classical crystal structure, but that structure is not catalytically active.[22] Rather, substantially higher cation concentrations are required for the induction of catalysis; yet no further stable rearrangement of the tertiary fold was detected. Our inference was that the active fold of the minimal hammerhead formed only transiently in minimal constructs, even under optimal ionic conditions. Now that tertiary-stabilized native hammerheads have been developed and crystallized (see below), one might think that the issue of transient formation of the active fold might only be associated with minimal hammerheads. However, recent evidence suggests that this is not the case (Hampel, Heckman, Buskiewicz and Burke, unpublished results).

Despite their transient nature we were able to map these conformational changes in some detail using covalent zero-length photocrosslinking reactions triggered by UV-irradiation.[24,25] Depending on the hammerhead construct and the ionic conditions, UV-induced crosslinks can sometimes be obtained with unmodified RNA. More commonly, we used modified nucleobases, *e.g.*, 6-thio-guanosine and 2,6-diAP, pyrrolo-cytidine, or photosensitizing metal ion complexes, notably cobalt(III) hexammine, or both. In these studies, modified nucleobases were introduced into ribozymes or substrates by solid phase synthesis, and crosslinking reactions were initiated under conditions such that all of the radiolabeled strand was demonstrably incorporated into a ribozyme–substrate complex.

Crosslinking experiments need to be interpreted with caution, especially with RNA molecules where populations commonly consist of multiple folding isomers, and where RNA dynamics lead to interchanges between folding isoforms. To attempt to focus on crosslinks that reflected active folds of the hammerhead, we followed three strategies. First, we focused on those crosslinks where crosslinked ribozyme–substrate complex could be reconstituted and shown to retain catalytic activity. Second, we conducted crosslinking studies using multiple widely-studied high-activity hammerhead constructs, reasoning that crosslinks relevant to the active fold should be common to these constructs. Third, we conducted crosslinking under various ionic conditions, to determine if conditions that increased catalytic activity also increased crosslinking efficiency.

Key findings from the crosslinking analysis show that five conserved nucleobases within the hammerhead catalytic core were repeatedly and consistently found to be involved in crosslinks in multiple constructs under conditions where hammerheads have significant catalytic activity.[24,25] These are the nucleobases that make up the substrate cleavage site (C/U1.1 and C17), as well as the two nucleobases implicated in catalysis (G8 and G12), and a fifth nucleotide that is clearly important for folding and/or activity (G5). Crosslinks were observed between G8 and G12; these are consistent with the classical crystal structure, and likely reflect an inactive fold, possibly a solution ground state. Strikingly, crosslinks have repeatedly been found between the substrate cleavage site and nucleobases G8 and G12, results that appear to be consistent with our hypothesis that G8 and G12 may play a direct role in catalysis, as described above. Some of the most informative examples include crosslinks between (i) G8 and U1.1 in minimal hammerhead alpha 1, a UV crosslink that was obtained in the absence of modified nucleotides or photosensitizing metal ion complexes, (ii) 6sG8 and C1.1, (iii) and 6sG12 and C17 in the native *Schistosoma mansoni* hammerhead. Crosslinking studies show that G5 can form intimate interactions with several nucleotides within the catalytic core. These results appear to be consistent with models in which G5 plays an important role in tertiary folding, perhaps in steps leading to the formation of the catalytically active structure.

Taken together, these results are remarkably consistent with the two hypotheses described above. They provide direct evidence that the various hammerhead constructs can all adopt an active conformation that is significantly different from the classical crystal structures – one that brings G8 and G12 close to the scissile linkage, where they might function in general acid–base catalysis. The crosslinking results suggest that the active fold may be stabilized by stacking of the putative catalytic purine nucleobases on the pyrimidine nucleobases at the substrate cleavage site, reminiscent of the stacked arrangement of substrate and catalytic nucleobases in the hairpin ribozyme.

3.3.3 Molecular Modeling of a Hammerhead Active Fold that Satisfies Structural and Biochemical Constraints

Before the solution of the most recent crystal structures, which depict an active-like hammerhead fold, we undertook a molecular modeling exercise to ask if the crosslinking and mechanistic data described above, together with biochemical results from other laboratories, could be used to identify a plausible 3D structure that would satisfy available experimental data. Here, the goal was not to immediately generate a correct model of the active structure, but to ask if one or more biochemically-reasonable structures could be developed, and to understand what sort of conformational change would be required to move from a ground state (as described by the classical crystal structures) to the hypothetical active fold (Figure 3.2).

Modeling was conducted using constraint-satisfaction software MC-SYM. Strict A-form RNA was assumed for helices I, II and III while other residues in the core were allowed to adopt a greater range of C3' *endo anti* conformations. Nucleotides U16.1, C17, G5, A6 and U7, on the other hand, were allowed to search all sugar pucker conformations. Additional topological constraints derived primarily from experimental data where the constrained molecule retained significant catalytic activity were added to the MC-SYM script and applied all at once. These include (i) a single metal ion bridging of P9/G10.1 and a cleavage site thiophosphate[26] simulated by distance constraints; (ii) crosslinking of the cleavage site nucleobase C17/N1.1 to G824 introduced *via* a stacking constraint; (iii) in-line $S_N 2$ attack orientation of the scissile linkage was modeled using distance and angle constraints,[27] which positioned the attacking 2'-OH and leaving 5'-O groups in the proximity of N1 of G8 and G12 (Table 3.1).[23] Several features of the crystallographic structure were also included, such as coaxial stacking of helices II and III, and secondary structure, including the adjacent G8 · A13 and G12 · A9 sheared base pairs. This arrangement was confirmed by multiple reciprocal crosslinks between bases at positions 8 and 12 as well as between bases at positions 9 and 13.[25]

Other crosslinks that did not retain activity, but which appeared consistent with the interaction of the residues spanning the scissile bond and nucleobases G8 and G12, were introduced into the script *via* distance constraints or pairings instead of actual stacking relations. This allowed more flexibility to explore more dynamic regions for which no active species were isolated. For example, residue C17 was fixed in a base pair (Watson–Hoogsteen) with residue A14. This positioned it close to G12, to which it can also be crosslinked readily in different constructs although does not retain catalytic activity.[24,25] Furthermore, residue C3 was stacked on nucleobase A2.1, at the end of helix I. This oriented C3 for a putative interaction with the Hoogsteen face of G8. Residue U4 was stacked on C3, and located near residues U1.1, C17 and U16.1 in agreement with crosslinking data.[24,25] Similarly, the Watson–Crick face of G5 was oriented to interact with the 2'-OH group of U16.1 and O2 of C17. This also positions G5 in the neighborhood of residues U1.1, C17 and U16.1 to which it can be crosslinked.[24,25] Residue A6 was allowed to explore multiple

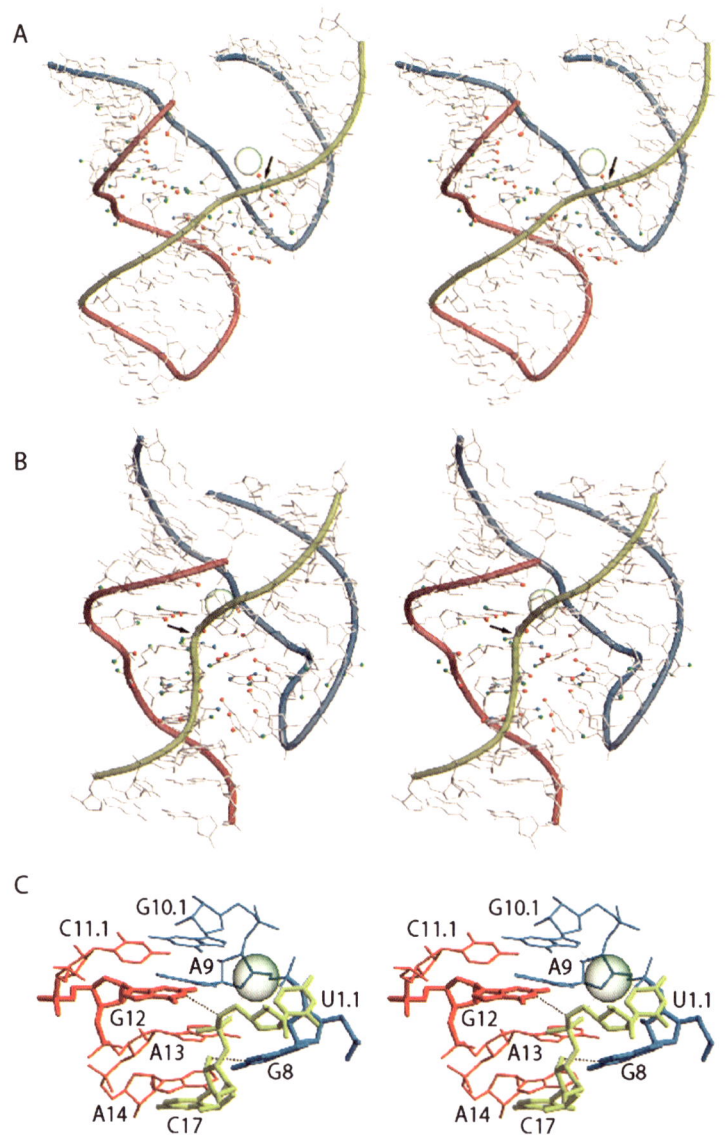

Figure 3.2 Three-dimensional structures of classical crystal structure and active fold model. Red and blue, ribozyme strands. Yellow, substrate strand. Gray sphere, magnesium ion. Arrow indicates substrate cleavage site. (*A*) Crystallographic (ground state) fold of the hammerhead.[3] Red dots indicate atoms and functional groups where substitution results in profound inhibition of catalytic activity; green dots indicate atoms and functional groups where substitutions result in little or no inhibition.[5] (*B*) Active fold model developed in this work. Results of atomic functional group substitutions are colored as in (*A*). (*C*) Proposed structure of the active site within the active fold model. MC-SYM scripts are available from the authors. PDB accession number 2GZC.

Table 3.1 Satisfaction of key distance and geometric constraints.

Interaction	Crystal structure	Model
Metal-ion bridge	18.55 Å	5.60 Å
C17–G8 crosslink	13.55 Å	5.60 Å
G8 N1 to C17 2'O	16.90 Å	2.93 Å
G12 N1 to U1.1 5'O	18.81 Å	3.16 Å
C17 2'O to U1.1 P	4.08 Å	3.05 Å
S_N2 angle	74.9°	164.5°

interactions with A14, whereas residue U7 was positioned to be bulged out. Solutions that were similar (≤ 2 Å rmsd) were combined by MC-SYM. Multiple models were visually compared and a representative model was selected. To optimize bond distances and angles, that model was then submitted to energy-minimization by the molecular simulation program Sander, from the Amber 7 suite of programs using the Amber 2002 force field for RNA. The S_N2 alignment was fixed using distance and angle restraints.

The active fold model retains several features of the classical crystallographic structure, including coaxial stacking of helices II and III, secondary structure (including the adjacent G8·A13 and G12·A9 sheared base pairs), and the A9/G10.1 metal-binding site. Key differences from the crystal structure at the active site include (i) cleavage site Mg^{2+} bridging to P9/G10.1; (ii) splaying of the scissile C17-U1.1 linkage stabilized by (iii) formation of a C17·A14 W/H base pair, (iv) stacking of U1.1 on G8, (v) close approximation of N1 of G8 to the attacking 2'-OH of C17, and (vi) close approximation of N1 of G12 to the leaving 5'-O of U1.1. Away from the active site, the model shows nucleobase C3 stacked on A2.1. U4 has its Watson–Crick face oriented towards the sugar of C17, while G5 forms a triplet with A15.1·U16. In the latter interaction, the exocyclic amino group of G5 hydrogen bonds with O4 of U16.1 while its keto group forms a hydrogen bond with the U16.1 ribose O2'. In addition, A6 forms a triplet with the proposed C17·A14 pair *via* a hydrogen bond between the exocyclic amino group of A14 and N1 of C17. U7 is bulged out of the ribozyme core. The structural elements proposed within the U-turn (part of domain 1) rationalize much of the biochemical data that are not explained by the crystal structures.[5] However, this region of the molecule is dynamic and relatively unconstrained by our crosslinking studies, so that the degree of structural uncertainty in the U-turn is somewhat greater than that of the rest of the model.

The modeling exercise provides insights into the conformational change necessary to convert the putative ground state structure into the active fold. The conformational change takes place on a large scale, with several residues (*e.g.*, G8, G12, and the reaction site) undergoing relative positional changes of 10–20 Å. Although this conformational change is quite complex from dynamic, kinetic and thermodynamic perspectives, it appears to be topologically straightforward, in that little remodeling of secondary structure takes place, and tertiary remodeling appears to be confined to a limited region of the ribozyme–substrate complex.

3.4 Current Status and Future Prospects

With the publication of the crystal structure of the native *S. mansoni* hammerhead, Martick and Scott[10] have obtained high-resolution evidence for a structure that appears much closer to the active fold of the hammerhead than did the classical crystal structures (Chapter 4). Like the structure of our model, the new crystal structure reconciles most, but not all, of the published biochemical and biophysical data. Obviously, the new crystal structure provides the best structural framework for further analysis of hammerhead folding, dynamics, and mechanism.

The new crystal structure provides further evidence for a role of G8 and G12 in catalysis, and for the stacking of these nucleobases on the cleavage site nucleobases, as hypothesized. In particular, the structure indicates that N1 protonation–deprotonation of G12 may indeed account for the pH–activity relationships of G12 variants that were noted by this laboratory. However, the unexpected discovery of a G8-C3 Watson–Crick base pair with an implied function of the 2′-OH of G8 in general acid–base catalysis is not obviously reconciled with our finding that the pH–activity relationships of G8 variants correlate with the N1 pK_a. Further analysis of possible pH-dependent conformational changes at the active site, together with analysis of the putative catalytic roles of functional groups at the active site, folding pathways and the effects of metal ions will provide opportunities for significant further work.

Acknowledgements

We are grateful to current and past members of the Burke research group for their contributions to this research, and to the U.S. National Institutes of Health for financial support.

References

1. A.C. Forster and R.H. Symons, *Cell*, 1987, **49**, 211.
2. H.W. Pley, K.M. Flaherty and D.B. McKay, *Nature*, 1994, **372**, 68.
3. W.G. Scott, J.T. Finch and A. Klug, *Cell*, 1995, **81**, 991.
4. J.E. Wedekind and D.B. McKay, *Annu. Rev. Biophys. Biomol. Struct.*, 1998, **27**, 475.
5. K.F. Blount and O.C. Uhlenbeck, *Annu. Rev. Biophys. Biomol. Struct.*, 2005, **34**, 415.
6. A.R. Ferré d' Amaré, K. Zhou and J.A. Doudna, *Nature*, 1998, **395**, 567.
7. P.B. Rupert and A.R. Ferré d' Amaré, *Nature*, 2001, **410**, 780.
8. J.M. Burke, *Nat. Struct. Biol.*, 2001, **8**, 382.
9. I.H. Shih and M.D. Been, *Annu. Rev. Biochem.*, 2002, **71**, 887.
10. M. Martick and W.G. Scott, *Cell*, 2006, **126**, 309.
11. J.R. Sampson, F.X. Sullivan, L.S. Behlen, A.B. DiRenzo and O.C. Uhlenbeck, *Cold Spring Harb. Symp. Quant. Biol.*, 1987, **52**, 267.

12. J. Haseloff and W.L. Gerlach, *Nature*, 1988, **334**, 585.
13. M. De la Pena, S. Gago and R. Flores, *EMBO J.*, 2003, **22**, 5561.
14. A. Khvorova, A. Lescoute, E. Westhof and S.D. Jayasena, *Nat. Struct. Biol.*, 2003, **10**, 708.
15. J.A. Nelson and O.C. Uhlenbeck, *Mol. Cell*, 2006, **23**, 447.
16. K.F. Blount, N.L. Grover, V. Mokler, L. Beigelman and O.C. Uhlenbeck, *Chem. Biol.*, 2002, **9**, 1009.
17. S.A. Strobel and S.P. Ryder, *Nature*, 2001, **410**, 761.
18. P.C. Bevilacqua, *Biochemistry*, 2003, **42**, 2259.
19. J.B. Murray, A.A. Seyhan, N.G. Walter, J.M. Burke and W.G. Scott, *Chem. Biol.*, 1998, **5**, 587.
20. J.L. O'Rear, S. Wang, A.L. Feig, L. Beigelman, O.C. Uhlenbeck and D. Herschlag, *RNA*, 2001, **7**, 537.
21. E.A. Curtis and D.P. Bartel, *RNA*, 2001, **7**, 546.
22. K.J. Hampel and J.M. Burke, *Biochemistry*, 2003, **42**, 4421.
23. J. Han and J.M. Burke, *Biochemistry*, 2005, **44**, 7864.
24. J.E. Heckman, D. Lambert and J.M. Burke, *Biochemistry*, 2005, **44**, 4148.
25. D. Lambert, J.E. Heckman and J.M. Burke, *Biochemistry*, 2006, **45**, 7140.
26. S. Wang, K. Karbstein, A. Peracchi, L. Beigelman and D. Herschlag, *Biochemistry*, 1999, **38**, 14363.
27. H. van Tol, J.M. Buzayan, P.A. Feldstein, F. Eckstein and G. Bruening, *Nucleic Acids Res.*, 1990, **18**, 1971.

CHAPTER 4
Hammerhead Ribozyme Crystal Structures and Catalysis

WILLIAM G. SCOTT

Center for the Molecular Biology of RNA, University of California, Santa Cruz, CA 95064, USA

4.1 Introduction

After the discovery of RNase P[1] (Chapter 9) and the Group I[2] (Chapter 10) intron ribozymes, both of which are comparatively large and complex catalytic RNAs, the identification of the hammerhead ribozyme offered hope that the phenomenon of RNA catalysis might be best understood within the framework of a smaller, more tractable RNA that catalyzed a simple phosphodiester isomerization reaction. Indeed, the first ribozyme crystal structures were in fact those of hammerhead ribozymes, but they seemed to create more questions than compelling explanations for RNA catalysis. The twelve years subsequent to the publication of these structures saw only increasing discord; the crystal structure analyses seemed hopelessly irreconcilable with a growing corpus of biochemical evidence. Meanwhile, crystal structures for many of the other ribozymes, including the Group I intron and RNase P, started to appear. Finally, a new crystal structure of the hammerhead ribozyme emerged, 20 years after the hammerhead's discovery. This structure included a set of distal tertiary contacts whose importance was largely unrecognized until 2003, but whose incorporation increased catalytic prowess by a factor of ~1000. The new crystal structure reveals that this remarkable rate enhancement is a direct consequence of localized yet dramatic active site conformational changes that are stabilized by a comparatively distant set of tertiary interactions. The new structure appears to reconcile twenty years of discord while offering some new insights into RNA structure and catalysis, as well as the foibles of experimental interpretation. The hammerhead ribozyme has indeed taught us much about RNA catalysis, despite (or, more likely, because of) major deviations from our seemingly best-devised lesson plans.

4.2 A Catalytic RNA Prototype

The discovery that RNA can be an enzyme[1,2] created the fundamental question of how RNA enzymes work. Before this discovery, it was generally assumed that proteins were the only biopolymers that had sufficient complexity and chemical heterogeneity to catalyze biochemical reactions. Clearly, RNA can adopt sufficiently complex tertiary structures that make catalysis possible. But how does the three-dimensional (3D) structure of an RNA endow it with catalytic activity? What structural and functional principles are unique to RNA enzymes (or ribozymes), and what principles are so fundamental that they are shared with protein enzymes?

By understanding how ribozymes work, we might also learn more about how life originated. RNA may have been the original self-replicating pre-biotic molecule, according to the "RNA World" hypothesis,[3] potentially catalyzing its own replication. Understanding the fundamental principles of ribozyme catalysis therefore may also give us new insights into the origin of life itself. The answer to the question of how ribozymes work also has practical consequences, as RNA enzymes are particularly well-suited for design as targeted therapeutics for various diseases (for a recent review, see ref. 4).

The hammerhead ribozyme has been thought of as a "prototype" ribozyme in the same sense that lysozyme and serine proteases have been thought of as "prototype" conventional protein enzymes, and for that reason the hammerhead ribozyme has attracted intense experimental scrutiny. The hammerhead ribozyme is a comparatively simple and well-studied ribozyme that in principle should be capable of revealing the secrets of its catalytic potential, if we are able to pose the right questions and carry out useful and informative experiments. Much attention has been focused upon this particular ribozyme with the hope that if its catalytic properties become well-understood our grasp of the phenomenon of RNA catalysis in general will become more comprehensive so that generalizations may appear that are applicable to the larger ribozymes, to RNA splicing and peptidyl transfer, and perhaps even beyond to a unified understanding of RNA and protein enzymology.

4.3 A Small Ribozyme

Hammerhead RNAs are small self-cleaving RNAs that have a conserved motif found in several of the viroids and satellite RNAs associated with plant RNA viruses[5-8] and other species[9-11] and that replicate *via* a rolling circle mechanism. The minimal hammerhead sequence that is catalytically active consists of three base-paired stems flanking a central core of 15 conserved (mostly invariant) nucleotides, as depicted in Figure 4.1.[8,12,13] The conserved central bases, with few exceptions, are essential for ribozyme's catalytic activity.

The hammerhead ribozyme is arguably the best-characterized ribozyme. Its small size, thoroughly-investigated cleavage chemistry, known crystal structure, and its biological relevance make the hammerhead ribozyme particularly

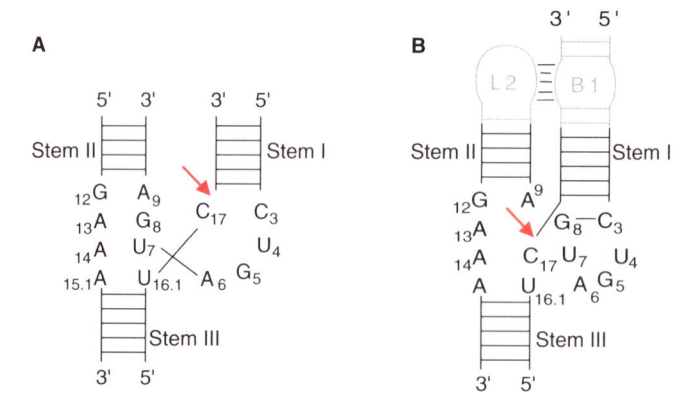

Figure 4.1 Schematic diagrams of the secondary structures of two hammerhead ribozyme constructs. The minimal hammerhead (*A*) consists of a conserved, mostly invariant core region that contains the cleavage site, as well as three flanking A-form RNA helices whose sequence is only restricted by the need to maintain base-pairing. The full-length hammerhead (*B*) contains additional nucleotides as depicted in grey. Although the sequences of the stem II loop (L2) and stem I bulge (B1) are not restricted in any obvious way that is apparent from the RNA sequence, the tertiary interaction that forms between these two regions is critically important and enhances catalysis by approximately three orders of magnitude.

well-suited for biochemical and biophysical investigations into the fundamental nature of RNA catalysis. Despite the extensive structural and biochemical characterization of the hammerhead ribozyme (reviewed in refs. 14–16) many important questions have remained concerning how this RNA molecule's structure enables it to have catalytic activity. Our understanding of the relationship between the structure of the hammerhead RNA and its catalytic activity has enjoyed a particularly tumultuous history.

4.4 Chemistry of Phosphodiester Bond Isomerization

The cleavage reaction is a phosphodiester isomerization reaction that is initiated by abstraction of the 2′-hydroxyl proton from the 2′-oxygen, which then becomes the attacking nucleophile in an "in-line" or $S_N2(P)$-like reaction (Figure 4.2),[17–19] although it is not known whether this proton is removed before or during the chemical step of the hammerhead cleavage reaction. (The cleavage reaction is technically not bimolecular, but behaves in the same way a genuine $S_N2(P)$ reaction does; it undergoes inversion of configuration subsequent to forming an associative transition-state consisting of a pentacoordinated oxyphosphorane.) The attacking and leaving group oxygens will both occupy the two axial positions in the trigonal bipyramidal transition-state structure as is required for an S_N2-like reaction mechanism.

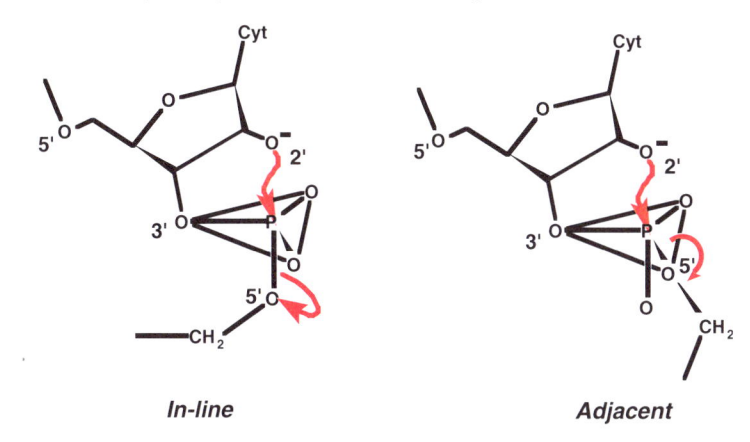

Figure 4.2 Different phosphate backbone conformations required for an "in-line" vs. an "adjacent" cleavage mechanism.

The 5′-product, as a result of this cleavage reaction mechanism, possesses a 2′,3′-cyclic phosphate terminus, and the 3′-product possesses a 5′-OH terminus,[20–21] as with non-enzymatic alkaline cleavage of RNA. The reaction is therefore, in principle, reversible, as the scissile phosphate remains a phosphodiester, and may thus act as a substrate for hammerhead RNA-mediated ligation without a requirement for ATP or a similar exogenous energy source. The hammerhead ribozyme-catalyzed reaction, unlike the formally identical non-enzymatic alkaline cleavage of RNA, is a highly sequence-specific cleavage reaction with a typical turnover rate of approximately one molecule of substrate per molecule of enzyme per minute at pH 7.5 in 10 mM Mg^{2+} (so-called "standard reaction conditions" for the minimal hammerhead RNA sequence), depending upon the sequence of the particular hammerhead ribozyme construct measured. This represents an approximately 10 000-fold rate enhancement over the non-enzymatic cleavage of RNA.

4.5 Hammerhead Ribozyme Structure

The first hammerhead ribozyme 3D structure was published in 1994 by Pley, Flaherty and McKay.[22] This structure consisted of a 34 nucleotide hammerhead enzyme strand consisting of unmodified RNA, and a 13 nucleotide hammerhead substrate-analogue composed of DNA that functions as a competitive inhibitor of the ribozyme, preventing cleavage due to the lack of a nucleophile at the active site (Figure 4.3A and B). It was the first RNA whose structural complexity approached that of tRNA published in the 20 years subsequent to the elucidation of the yeast phenylalanine tRNA structures and, more importantly, it was the first structure of a catalytic RNA. This breakthrough was celebrated in an accompanying *News and Views* highlight in *Nature*, authored by Thomas Cech and Olke Uhlenbeck, entitled

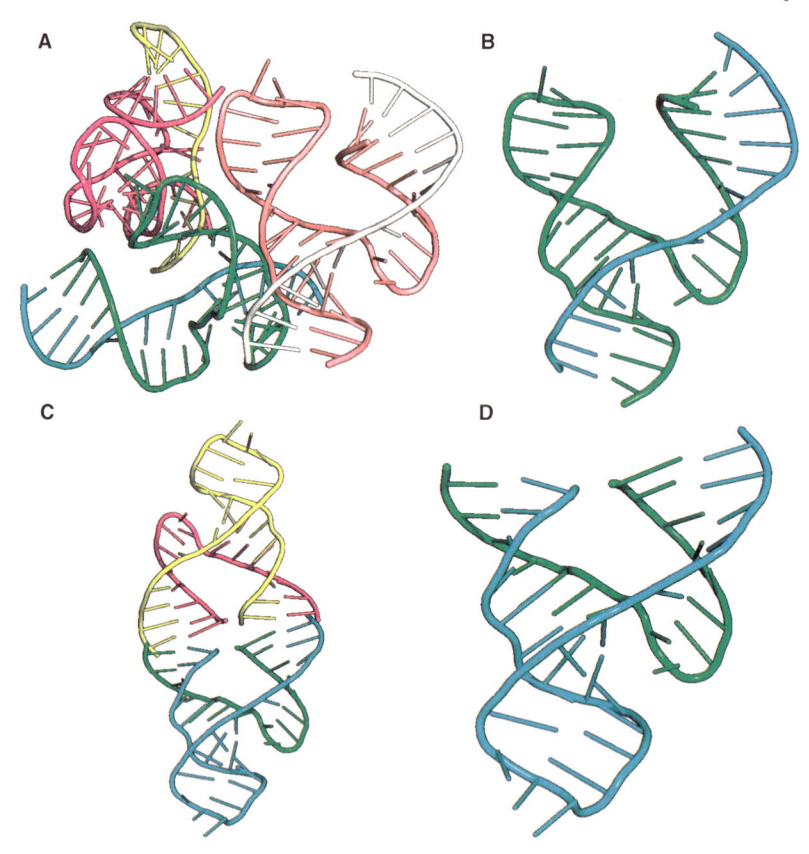

Figure 4.3 Three crystallographically independent hammerhead ribozymes (*A*) occupied the asymmetric unit in the first hammerhead crystal structure; one of these is displayed in what has become the conventional orientation (*B*) such that the RNA enzyme strand (green) and DNA substrate-analogue (cyan) are clearly visible. Two crystallographically independent all-RNA hammerheads occupied the asymmetric unit (*C*) of the second hammerhead ribozyme crystal structure. One of these (*D*) is chosen and displayed to facilitate comparison with (B). Here, the shorter strand is (nominally) the enzyme strand and the longer strand (cyan) is the substrate, where a 2'-OMe nucleotide occupies the cleavage site. A subsequent crystal form having but one molecule per asymmetric unit (not shown) is otherwise identical, despite the absence of the 2'-OMe modification.

"Hammerhead nailed down".[23] Subsequent publication of two all-RNA hammerhead ribozyme crystal structures, the first with a 2'-OMe replacing the nucleophile at the active site,[24] and the second with an unmodified nucleotide at the active site,[25] revealed the same 3D structure of the invariant core nucleotides in the ribozyme well within experimental error (Figure 4.3C and D), despite significant differences in the sequence within the nonessential regions, DNA *vs.* RNA substrates, presence *vs.* absence of divalent metal ions and active nucleophile, and crystal packing schemes. In essence, the spatial positioning of the invariant region of the hammerhead ribozyme appeared to be immune to a series of potential perturbations in at least six crystallographically independent molecules, further corroborating, it seemed, the initial assessment that the structure had indeed been "nailed down" in 1994.

4.6 Catalysis in the Crystal

The hammerhead ribozyme sequence that contained all unmodified RNA, including the active-site nucleophile, was catalytically active both in solution and in the crystal.[25,26] The crystals of the active hammerhead ribozyme thus enabled us to test for cleavage activity in the crystalline state. We initiated the cleavage reaction by flooding the crystal with divalent cations while raising the pH above the apparent kinetic pK_a using a soaking solution buffered at pH 8.5.[25-27] In conditions with a Mg^{2+} concentration of 50 mM and pH 8.5, the self-cleavage rate in the crystal is approximately 0.4 min^{-1}. Under similar conditions in solution, this same hammerhead ribozyme construct, a sequence that was optimized for purposes of growing crystals rather than for catalytic prowess, cleaves at a rate of approximately 0.08 min^{-1}, permitting us to suggest that the crystal lattice did more to aid in the proper folding of the ribozyme than it did to inhibit the hammerhead ribozyme's cleavage activity. Moreover, the extent of cleavage of the substrate in the crystal was almost complete.[26]

Although this collection of minimal hammerhead ribozyme crystal structures provided rationalizations for several of the previously reported experimental observations, many important problems remained unresolved (see, for example, ref. 14). One of the most important and immediately recognized[22] problems was that the scissile phosphate in all of the structures just described was observed to be in a conformation that is completely incompatible[28] with an "in-line" mechanism. Hence the need to bring the scissile phosphate into a conformation amenable to an in-line attack can be (and in fact was) taken as *prima facie* evidence for a required conformational change prior to bond cleavage.

4.7 Making Movies from Crystallographic Snapshots

The fact that the minimal hammerhead RNA sequence can cleave after being crystallized[25] presented the opportunity to capture various states, including pre-catalytic conformational changes, along the reaction pathway, using crystallographic freeze-trapping techniques.

Although conventional X-ray crystallography is essentially static and by necessity involves both a spatial and a temporal averaging of a macroscopic number of molecules and a time period of hours (the duration of data collection), time-resolved crystallographic experiments may be possible under the right conditions. Two options for time-resolved experiments exist that permit one to obtain crystallographic snapshots during catalysis.

The Laue method uses polychromatic X-rays to enable collection of a fairly complete X-ray dataset in a matter of milliseconds or less, and has been used with success in the case of many protein enzymes that catalyze rapid reactions.[29,30] Laue experiments require ideal crystals that possess very low mosaicity, as well as a way of very rapidly initiating a reaction simultaneously throughout the crystal lattice. Once the reaction has been initiated, each enzyme–substrate complex evolves stochastically with respect to time. Intermediates can be observed only when the majority of molecules in a crystal occupy the same intermediate state at the same time. Lower-occupancy intermediates remain essentially unobservable.

When combined with flash-freezing (*e.g.*, immersion in liquid nitrogen) as a physical trapping method, conventional monochromatic X-ray data collection also offers the opportunity to conduct time-resolved experiments using a second approach. Instead of recording live snapshots, as with the Laue method, the reaction is simply initiated in the crystal, allowed to evolve, and the molecular contents of the crystal are immobilized by freeze-trapping prior to standard data collection.[31] Since most conventional X-ray data collection is already performed using crystals frozen at 100 K, this procedure in practice involves very little additional experimental modification, and possesses the added advantage of being much more tolerant of mosaicity and the other small crystal imperfections that often accompany reactivity in the crystal. The utility of this approach is confined to comparatively slowly reactive enzymes. Fortunately, the cleavage rate of the minimal hammerhead ribozyme not only is on the order of 1 min^{-1}, it can be further modulated as a function of pH and presence of divalent cations, making it an ideal candidate for the freeze-trapping approach.

To better understand the structural basis of the proposed conformational change required to activate the hammerhead ribozyme for catalysis, we performed monochromatic time-resolved crystallographic freeze-trapping studies with the aim of observing conformational intermediates preceding catalysis.[25,27,32,33] To trap the structure of a precatalytic structural intermediate, a hammerhead ribozyme having a "kinetic bottleneck" at the final or bond-breaking point on the reaction pathway was synthesized using a modified leaving-group, (Figure 4.4A). This modified hammerhead RNA was used to capture a conformational intermediate that began to approach an in-line conformation[27] that evolved subsequent to triggering the reaction (Figure 4.4B). Prior to triggering the reaction, the modified RNA possessed a structure indistinguishable from that of the initial-state structure of the unmodified RNA, an important control that was conducted and reported,[27] contrary to some claims that have been made.[34] Two other pre-catalytic conformational changes were also captured using complementary approaches that did not involve modification of the leaving group, and appeared consistent in that they made up a trajectory toward an in-line structure.[25,33] (In retrospect, this structure still would require significant additional but unobserved changes to reach the transition-state.)

The structure of the enzyme–product complex (Figure 4.4C) was also obtained.[32] In this structure, it is apparent that C-17 has rotated fully away

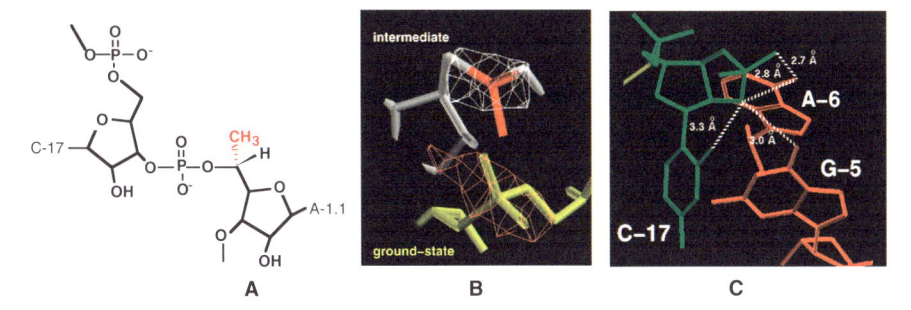

Figure 4.4 (A) Depiction of the talo-5′-methyl-ribose "kinetic bottleneck" modification of the hammerhead ribozyme leaving group. (B) Initial (yellow) and intermediate (gray) structures, with the scissile phosphate of the intermediate modeled into positive difference density (white), and the initial-state phosphate occupying negative difference density (red). (C) Structure of the cleavage product.

from the CUGA uridine turn and occupies a site in which the newly generated 2′,3′-cyclic phosphate can potentially interact with the Watson–Crick pairing face of the invariant core residues G5 and A6. The crystals that permitted us to observe the product structures were actually mixtures of (at least) two partially occupied states, so we were reluctant to ascribe too much significance to any but the most well-defined structural changes.

When placed together in succession, the initial-state structure, three pre-catalytic conformational intermediate structures, and then the cleavage product, enabled construction of a five-frame "movie" that revealed a plausible reaction trajectory that we suggested might represent "at least a subset"[26,33] of the conformational movements required for catalysis.

4.8 An Ever-growing List of Concerns

Despite the observations of hammerhead ribozyme catalysis in a crystal in which the crystal lattice packing contacts by necessity confined the global positions of the distal termini of all three flanking helical stems,[25–27,32,33,35] many biochemical experiments designed to probe transition-state interactions and the chemistry of catalysis appeared to be irreconcilable with the crystal structures.

For example, the invariant core residues G5, G8, G12 and C3 in the minimal hammerhead ribozyme were each observed to be so fragile that changing even a single exocyclic functional group on any one of these nucleotides results in a dramatic reduction or abolition of catalytic activity,[14] yet few of these appeared to form hydrogen bonds involving the Watson–Crick faces of these nucleotide bases in any of the minimal hammerhead structures, apart from a G5 interaction in the product structure.

A particularly striking and only recently observed example consisted of G8 and G12, which were identified[36] as possible participants in acid–base catalysis (Chapter 3). Once it was demonstrated that the hammerhead RNA does not require divalent metal ions for catalysis,[37,38] it gradually became apparent that that the RNA itself, rather than passively bound divalent metal ions, must play a direct chemical role in any acid–base chemistry in the hammerhead ribozyme active site. It was, however, completely unclear how G12 and G8 could accomplish this, given the original structures of the minimal hammerhead ribozyme.

Other concerns included an NOE between U4 and U7 of the cleaved hammerhead ribozyme[39] that had also been observed during NMR characterization, which suggested that these nucleotide bases must approach one another closer than about 6 Å, although close approach of U7 to U4 did not appear to be possible from the crystal structure. Finally, as previously discussed, the attacking nucleophile in the original structures, the 2′-OH of C17, was not in a position amenable to in-line attack upon the adjacent scissile phosphate.[22]

Perhaps most worrisome were experiments that suggested the A9 and scissile phosphates must come within about 4 Å of one anther in the transition-state, based upon double phosphorothioate substitution and soft metal ion rescue experiments;[40] the distance between these phosphates in the crystal structure was about 18 Å, with no clear mechanism for close approach if the stem II and stem I A-form helices were treated as rigid bodies. Taken together, these results appeared to suggest that a fairly large-scale conformational change must take place to reach the transition-state within the minimal hammerhead ribozyme structure.

For these reasons, the two sets of experiments (biochemical *vs.* crystallographic) appeared not only to be at odds, but to be completely and hopelessly irreconcilable, generating a substantial amount of discord in the field.[34] No compelling evidence for dismissing either set of experimental results was ever made successfully, although many claims to the contrary[34,40,41] were made in favor of each.

4.9 Occam's Razor Can Slit Your Throat

The principle of parsimony, attributed to William of Occam, states that one should favor simple scientific explanations and hypotheses that make a minimum number of assumptions over more complicated alternatives. The "Occam's Razor" principle, in which one shaves away extraneous and irrelevant decorations to obtain the most parsimonious hypothesis, has also been extended to the physical investigation of nucleic acids and proteins: study the simplest macromolecular assemblies that possess the biological function of interest, and cut everything else off. This "reductionist paradigm" has been employed with such success in molecular biology that one can easily lose sight of the larger, physiological context.

When the hammerhead RNA was first discovered, it was observed in a ~370 nucleotide single-stranded genomic satellite RNA, most of which could be deleted while preserving the RNA's catalytic properties.[6–8] Eventually, it was found that about 13 core nucleotides and a minimal number of flanking helical nucleotides were all that was required for a respectable catalytic turnover rate of 1 to 10 min^{-1}, and this "minimal" hammerhead construct became the focus of almost all of the biochemical, biophysical and crystallographic investigations, as described earlier.

It thus came as a great surprise to most in the field when, in 2003, it was finally pointed out that for optimal activity, the hammerhead ribozyme requires the presence of sequences in stems I and II that interact to form tertiary contacts[42,43] (Figure 4.1B) that were removed in the process of shaving seemingly superfluous structures from the hammerhead ribozyme. Once the full ramifications of this revelation became apparent, *i.e.*, that the entire field had been studying the residual catalytic activity of an over-zealously truncated version of the full-length ribozyme, attention shifted away from the minimal constructs.[44] It also quickly became clear to us that a crystal structure of the full-length hammerhead ribozyme, in which these distal tertiary contacts were present, might be of considerable interest.

4.10 Structure of a Full-length Hammerhead Ribozyme

After several years of struggle, in 2006 we finally obtained a 2.2 Å resolution crystal structure of the full-length hammerhead ribozyme.[45] This new structure (Figure 4.5) appears to resolve the most worrisome of the previous discrepancies. In particular, C17 is now positioned for in-line attack, and the invariant residues C3, G5, G8 and G12 all appear involved in vital interactions relevant to catalysis. Moreover, the A9 and scissile phosphates are observed to be 4.3 Å apart, which is consistent with the idea that, when modified, these phosphates could bind a single thiophilic metal ion. The structure also reveals how two invariant residues, G12 and G8, are positioned within the active site – consistent with their previously proposed[36] role in acid–base catalysis. G12 is within hydrogen bonding distance to the 2′-O of C17, the nucleophile in the cleavage reaction, and the ribose of G8 hydrogen bonds to the leaving group 5′-O (Figure 4.6), while the nucleotide base of G8 forms a Watson–Crick pair with the invariant C3. This arrangement permits us to suggest that G12 is the general base in the cleavage reaction, and that G8 may function as the general acid, which is consistent with previous biochemical observations.[36] G5 hydrogen bonds to the furanose oxygen of C17, helping to position it for in-line attack. U4 and U7, as a consequence of the base-pair formation between G8 and C3, are now positioned such that an NOE between their bases is easily explained.

The crystal structure of the full-length hammerhead ribozyme thus clearly addresses all of the major concerns that appeared irreconcilable with the previous crystal structure.[46]

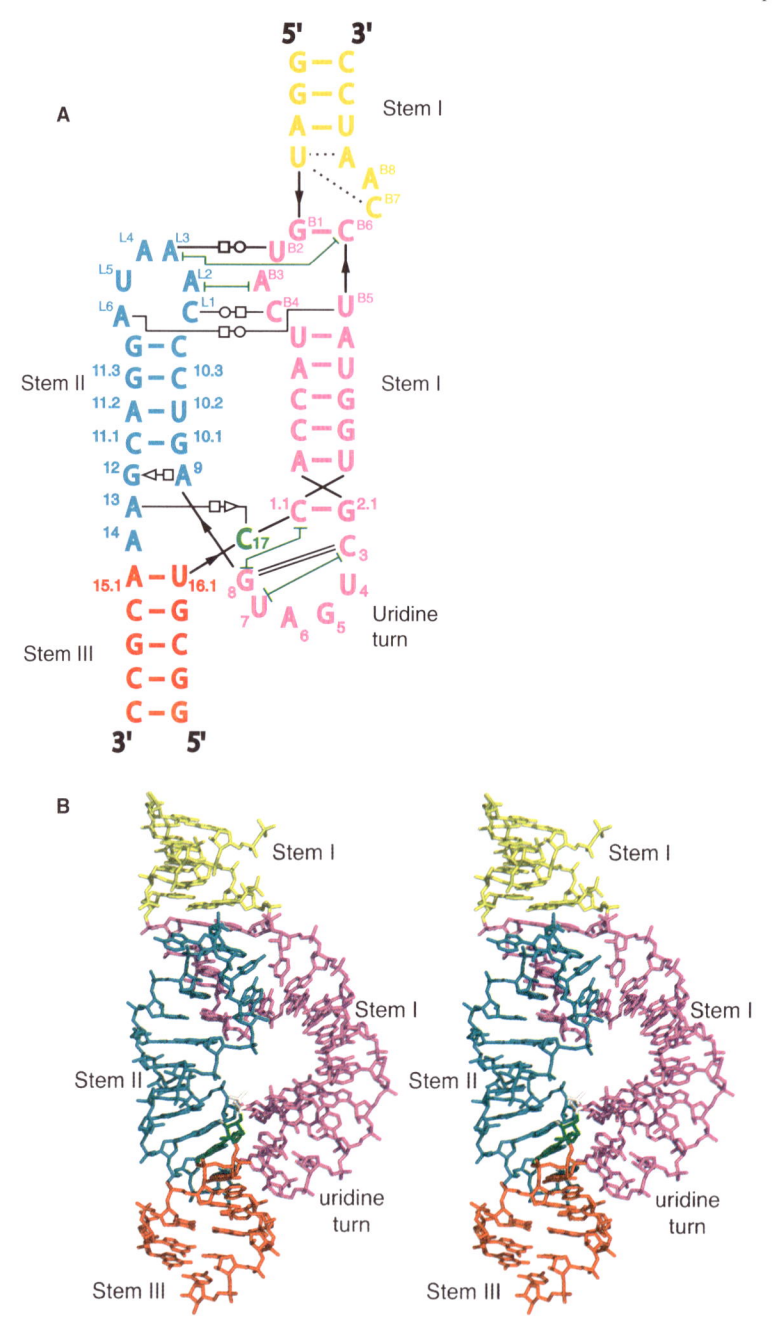

Figure 4.5 Secondary (*A*) and tertiary (*B*) structure of the full-length hammerhead ribozyme. The color-coding of the secondary structure corresponds to that of the tertiary structure. The cleavage-site nucleotide, C-17, is depicted in green in both figures.

Hammerhead Ribozyme Crystal Structures and Catalysis 59

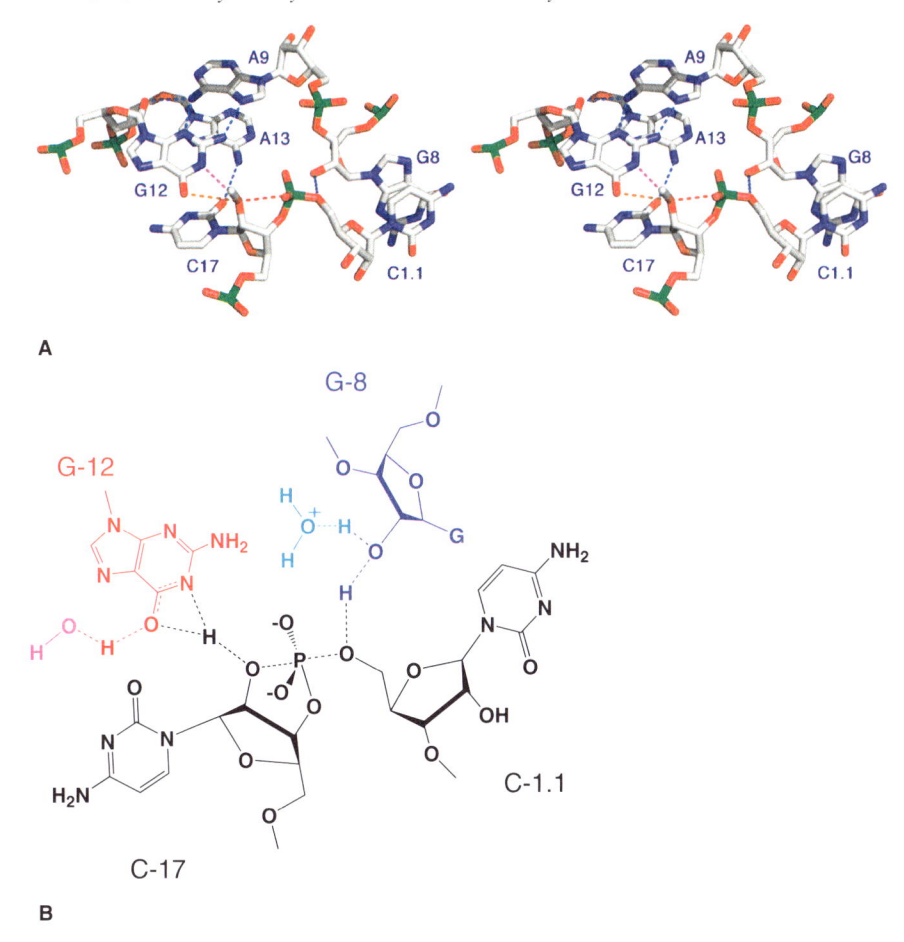

Figure 4.6 Wall-eyed stereo close-up view of the active site (*A*) of the full-length hammerhead ribozyme, and a corresponding mechanistic hypothesis (*B*) based upon the observed arrangement of nucleotides in (*A*).

4.11 Do the Minimal and Full-length Hammerhead Crystal Structures have Anything in Common?

Although the structure of the full-length hammerhead ribozyme, at first glance, appears to be radically different from the minimal hammerhead, both share some similarities in structure in both the global fold and in detail. The similarities are best seen by comparing the set of nucleotides both share in common. Specifically, if one compares the core residues and the first five basepairs of stem I, as well as the shared residues of stems II and III, while omitting the capping loops, the similarities become most apparent. A side-by-side comparison of the folds of the minimal and full-length hammerheads is shown in Figure 4.7, each with a yellow substrate strand that includes the cleavage-site

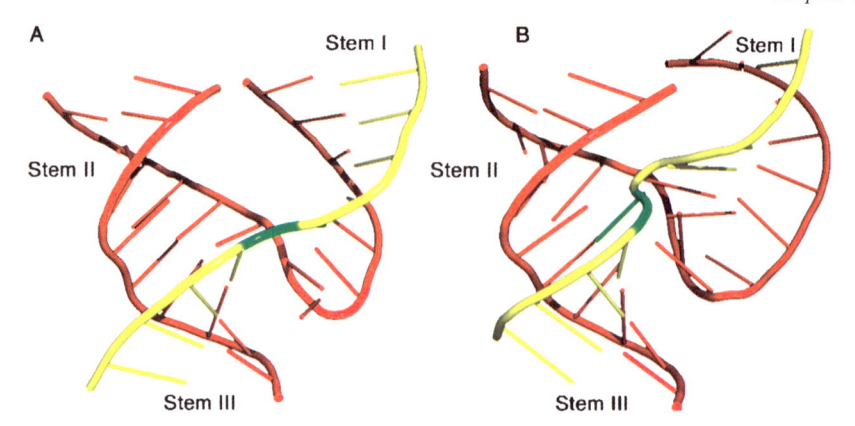

Figure 4.7 Backbone diagrams of the minimal (A) and full-length (B) hammerhead ribozyme core regions, showing the similarity of the local fold.

nucleotide highlighted in green. What is apparent from the comparison made in this manner, in which only the shared nucleotides are considered, is that the folds are strikingly similar, the largest difference being the kink in the substrate strand at the cleavage site that accompanies rearrangement of the active site nucleotides.

The significance of this observation is that it explains why the hammerhead ribozyme cleavage reaction could take place in crystals of the minimal construct.[25] These crystals, being 78% solvent by volume, permit molecular motions to take place subject to constraints imposed by the crystal lattice contacts. The lattice contacts of the minimal hammerhead restrict the distal termini of stems I, II and III.[25]

4.12 How Does the Minimal Hammerhead Work?

In solution, the simplest explanation (Occam's Razor is more safely applied to scientific hypotheses than to macromolecules) for all of the observed minimal hammerhead biochemistry (including the invariance of G5, G8, G12, C3 and the proximity of the A9 and scissile phosphates, as well as the 1000-fold slower cleavage rate of the minimal hammerhead) is that the active conformational state, which resembles the structure of the full-length hammerhead, occurs only transiently, such that only about 0.1% of the uncleaved molecules occupy this state at any given time. Thus for cleavage to occur, a transient conformational change must occur that deforms the structure observed in the minimal hammerhead crystals into that resembling the full-length hammerhead, in which the nucleotides critical for catalysis are correctly positioned.

This rearrangement can in fact also take place within the confines of the crystal lattice, because only the distal ends of the three helical stems are restricted in movement by the crystal contacts. Alternative hypotheses, including the suggestion that the minimal hammerhead cleaves *via* a different pathway than that of the full-length hammerhead, and that the minimal hammerhead structure in solution is identical to the full-length hammerhead conformation observed in the crystal structure, have considerably less explanatory power. The first hypothesis cannot explain the requirement for the invariant residues, and the second hypothesis cannot account for the observed 1000-fold rate enhancement. Hence it seems most likely that in solution, the minimal hammerhead has nearly the same structure as it does in the crystal, and that in both cases the minimal ribozyme only occasionally visits the conformation that is stabilized and therefore dominates in the full-length hammerhead construct.

4.13 A Movie Sequel with a Happy Ending

In silico adiabatic morphing[47] of the minimal into the full-length hammerhead ribozyme structure is possible when the nucleotides shared in common by both hammerhead constructs are interpolated.[48] The structure observed in the minimal hammerhead ribozyme can be continuously deformed *via* low energy-barrier torsion angle conformational changes into the structure observed in the full-length hammerhead. This process is best represented as a series of consecutive structures viewed as a movie (link to Quicktime movie: http://tinyurl.com/2vly4d). It is likely that the first (1994 and 1995), initial-state hammerhead ribozyme crystal structures, represent more or less accurately the dominant structure of the minimal hammerhead ribozyme in the crystal – consistent with the minimal hammerhead being 1000-fold less active in solution than the full-length hammerhead. The product, or cleaved state, of the minimal hammerhead in some ways resembles the full-length hammerhead to a greater extent than does the uncleaved minimal hammerhead structure. Specifically, in the cleaved structure,[32] the cleavage-site base, C-17, is observed to make contacts with G5 and A6 that are similar to those observed in the full-length structure, and the interactions with C3 are completely absent in both cases.

The more developed cleavage intermediates, in retrospect, appear to resemble a torsion angle conformational change of only about 1/3 of what is required to morph the structure from the minimal hammerhead to the full-length hammerhead active-site conformation. The crystallographic observations of various states along the cleavage reaction pathway thus appear in retrospect to be more incomplete than erroneous, in that they appear consistent with the first ~1/3 of the conformational change ultimately required to reach a structure similar to that of the full-length hammerhead. Missing by necessity from the set of snapshots was the low-occupancy transient conformation that is stabilized by the distal tertiary contacts in the full-length hammerhead ribozyme. In crystallographic experiments, one can only hope to resolve the dominant, high-occupancy, species in the crystal, so it is likely that the true pre-catalytic intermediate would never be observed crystallographically in the context of the minimal hammerhead construct.

4.14 Concluding Remarks

In summary, it appears that the actual experimental data obtained from the crystallographic analyses and the biochemical characterizations, which were performed on high-occupancy, near-ground-state and transient near-transition-state structures, respectively, were sound within the confines imposed by the minimal hammerhead structure. The mutually held interpretation that acceptance of one set of experimental results precluded acceptance of the other, however, was based on the flawed assumption that the two sets of observations were incommensurate and irreconcilable. In our case, the flawed assumption manifested itself most explicitly as the claim that unwinding and unpairing of helical elements was unlikely to take place.[41] In the other case, the flawed assumption manifested itself with the claim that any cleavage observed in the crystal must be due to an off-pathway artifact or experimental incompetence.[34,40] In retrospect, neither dismissal was justified, nor compelling. The resolution of the apparent paradox came with the structure of the full-length hammerhead, which reconciles and permits explanations of both sets of experimental results.

Acknowledgements

Sung-Hou Kim, Aaron Klug, Olke Uhlenbeck, David McKay, Barry Stoddard, David Lilley, Fritz Eckstein, Eric Westhof, Darrin York, John Burke, Harry Noller, George Bruening, Monika Martick, Christine Dunham, James Murray, Michael Gait, Jane Grasby, Dan Herschlag, Jay Pandit, Ann Caviani Pease, Rosalind Kim, Elizabeth Holbrook, Young-In Chi and many others have been enormously helpful as collaborators, advisors, friends, and scientific sparring partners on many aspects of hammerhead ribozyme structure and catalysis from 1987 to the present. The American Cancer Society, the MRC Laboratory for Molecular Biology, The RNA Center at the University of California at Santa Cruz, The William Keck Foundation, the National Science Foundation and the National Institutes of Health have all provided crucial funding for hammerhead ribozyme structural studies in one way or another at some point during that 20 year period.

References

1. C. Guerrier-Takada, K. Gardiner, T. Marsh, N. Pace and S. Altman, The RNA moiety of ribonuclease P is the catalytic subunit of the enzyme, *Cell*, 1983, **35**, 849–857.
2. A.J. Zaug and T.R. Cech, The intervening sequence RNA of Tetrahymena is an enzyme, *Science*, 1986, **231**, 470–475.

3. R.F. Gesteland and J.F. Atkins (eds.), *The RNA World*, Cold Spring Harbor Laboratory Press, Plainview, NY, 1993.
4. N.K. Vaish, A.R. Kore and F. Eckstein, Recent developments in the hammerhead ribozyme field, *Nucleic Acids Res.*, 1998, **26**, 5237–5242.
5. G.A. Prody, J.T. Bakos, J.M. Buzayan, I.R. Schneider and G. Bruening, Autolytic processing of dimeric plant virus satellite RNA, *Science*, 1986, **231**, 1577–1580.
6. A.C. Forster and R.H. Symons, Self-cleavage of virusoid RNA is performed by the proposed 55-nucleotide active site, *Cell*, 1987, **50**, 9–16.
7. J. Haseloff and W.L. Gerlach, Sequences required for self-catalysed cleavage of the satellite RNA of tobacco ringspot virus, *Gene*, 1989, **82**, 43–52.
8. O.C. Uhlenbeck, A small catalytic oligoribonucleotide, *Nature*, 1987, **328**, 596–600.
9. G. Ferbeyre, J.M. Smith and R. Cedergren, Schistosome satellite DNA encodes active hammerhead ribozymes, *Mol. Cell Biol.*, 1998, **18**, 3880–3888.
10. V. Bourdeau, G. Ferbeyre, M. Pageau, B. Paquin and R. Cedergren, The distribution of RNA motifs in natural sequences, *Nucleic Acids Res.*, 1999, **27**, 4457–4467.
11. A.C. Forster, C. Davies, C.C. Sheldon, A.C. Jeffries and R.H. Symons, Self-cleaving viroid and newt RNAs may only be active as dimers, *Nature*, 1988, **334**, 265–267.
12. D.E. Ruffner, G.D. Stormo and O.C. Uhlenbeck, Sequence requirements of the hammerhead RNA self-cleavage reaction, *Biochemistry*, 1990, **29**, 10695–10702.
13. C.C. Sheldon and R.H. Symons, Mutagenesis analysis of a self-cleaving RNA, *Nucleic Acids Res.*, 1989, **17**, 5679–5685.
14. D.B. McKay, Structure and function of the hammerhead ribozyme: An unfinished story, *RNA*, 1996, **2**, 395–403.
15. J.E. Wedekind and D.B. McKay, Crystallographic structures of the hammerhead ribozyme: Relationship to ribozyme folding and catalysis, *Annu. Rev. Biophys. Biomol. Struct.*, 1998, **27**, 475–502.
16. W.G. Scott, Biophysical and biochemical investigations of RNA catalysis in the hammerhead ribozyme, *Q. Rev. Biophys.*, 1999, **32**, 241–284.
17. H. van Tol, J.M. Buzayan, P.A. Feldstein, F. Eckstein and G. Bruening, Two autolytic processing reactions of a satellite RNA proceed with inversion of configuration, *Nucleic Acids Res.*, 1990, **18**, 1971–1975.
18. G. Slim and M.J. Gait, Configurationally defined phosphorothioate-containing oligoribonucleotides in the study of the mechanism of cleavage of hammerhead ribozymes, *Nucleic Acids Res.*, 1991, **19**, 1183–1188.
19. M. Koizumi and E. Ohtsuka, Effects of phosphorothioate and 2-amino groups in hammerhead ribozymes on cleavage rates and Mg^{2+} binding, *Biochemistry*, 1991, **30**, 5145–5150.
20. J.M. Buzayan, A. Hampel and G. Bruening, Nucleotide sequence and newly formed phosphodiester bond of spontaneously ligated satellite tobacco ringspot virus RNA, *Nucleic Acids Res.*, 1986, **14**, 9729–9743.

21. C.J. Hutchins, P.D. Rathjen, A.C. Forster and R.H. Symons, Self-cleavage of plus and minus RNA transcripts of avocado sunblotch viroid, *Nucleic Acids Res.*, 1986, **14**, 3627–3640.
22. H.W. Pley, K.M. Flaherty and D.B. McKay, Three-dimensional structure of a hammerhead ribozyme, *Nature*, 1994, **372**, 68–74.
23. T.R. Cech and O.C. Uhlenbeck, Ribozymes. Hammerhead nailed down, *Nature*, 1994, **372**, 39–40.
24. W.G. Scott, J.T. Finch and A. Klug, The crystal structure of an all-RNA hammerhead ribozyme: A proposed mechanism for RNA catalytic cleavage, *Cell*, 1995, **81**, 991–1002.
25. W.G. Scott, J.B. Murray, J.R. Arnold, B.L. Stoddard and A. Klug, Capturing the structure of a catalytic RNA intermediate: The hammerhead ribozyme, *Science*, 1996, **274**, 2065–2069.
26. J.B. Murray, C.M. Dunham and W.G. Scott, A pH-dependent conformational change, rather than the chemical step, appears to be rate-limiting in the hammerhead ribozyme cleavage reaction, *J. Mol. Biol.*, 2002, **315**, 121–130.
27. J.B. Murray, D.P. Terwey, L. Maloney, A. Karpeisky, N. Usman, L. Beigelman and W.G. Scott, The structural basis of hammerhead ribozyme self-cleavage, *Cell*, 1998, **92**, 665–673.
28. W.G. Scott, Ribozyme catalysis via orbital steering, *J. Mol. Biol.*, 2001, **311**, 989–999.
29. K. Moffat, D. Bilderback, W. Schildkamp, D. Szebenyi and T.Y. Teng, Laue photography from protein crystals, *Basic Life Sci.*, 1989, **51**, 325–330.
30. L.N. Johnson, Time-resolved protein crystallography, *Protein Sci.*, 1992, **1**, 1237–1243.
31. B.L. Stoddard, Intermediate trapping and laue X-ray diffraction: Potential for enzyme mechanism, dynamics, and inhibitor screening, *Pharmacol. Ther.*, 1996, **70**, 215–256.
32. J.B. Murray, H. Szoke, A. Szoke and W.G. Scott, Capture and visualization of a catalytic RNA enzyme-product complex using crystal lattice trapping and X-ray holographic reconstruction, *Mol. Cell*, 2000, **5**, 279–287.
33. C.M. Dunham, J.B. Murray and W.G. Scott, A helical twist-induced conformational switch activates cleavage in the hammerhead ribozyme, *J. Mol. Biol.*, 2003, **332**, 327–336.
34. K.F. Blount and O.C. Uhlenbeck, The structure-function dilemma of the hammerhead ribozyme, *Annu. Rev. Biophys. Biomol. Struct.*, 2005, **34**, 415–440.
35. W.G. Scott, Visualizing the structure and mechanism of a small nucleolytic ribozyme, *Methods*, 2002, **28**, 302–306.
36. J. Han and J.M. Burke, Model for general acid-base catalysis by the hammerhead ribozyme: pH–activity relationships of G8 and G12 variants at the putative active site, *Biochemistry*, 2005, **44**, 7864–7870.

37. J.B. Murray, A.A. Seyhan, N.G. Walter, J.M. Burke and W.G. Scott, The hammerhead, hairpin and VS ribozymes are catalytically proficient in monovalent cations alone,, *Chem. Biol.*, 1998, **5**, 587–595.
38. W.G. Scott, RNA structure metal ions, and catalysis, *Curr. Opin. Chem. Biol.*, 1999, **3**, 705–709.
39. J.P. Simorre, P. Legault, A.B. Hangar, P. Michiels and A. Pardi, A conformational change in the catalytic core of the hammerhead ribozyme upon cleavage of an RNA substrate, *Biochemistry*, 1997, **36**, 518–525.
40. S. Wang, K. Karbstein, A. Peracchi, L. Beigelman and D. Herschlag, Identification of the hammerhead ribozyme metal ion binding site responsible for rescue of the deleterious effect of a cleavage site phosphorothioate, *Biochemistry*, 1999, **38**, 14363–14378.
41. J.B. Murray and W.G. Scott, Does a single metal ion bridge the A-9 and scissile phosphate groups in the catalytically active hammerhead ribozyme structure?, *J. Mol. Biol.*, 2000, **296**, 33–41.
42. M. De la Pena, S. Gago and R. Flores, Peripheral regions of natural hammerhead ribozymes greatly increase their self-cleavage activity, *EMBO J.*, 2003, **22**, 5561–5570.
43. A. Khvorova, A. Lescoute, E. Westhof and S.D. Jayasena, Sequence elements outside the hammerhead ribozyme catalytic core enable intracellular activity, *Nat. Struct. Biol.*, 2003, **10**, 708–712.
44. D.M. Lilley, Ribozymes–a snip too far?, *Nat. Struct. Biol.*, 2003, **10**, 672–673.
45. M. Martick and W.G. Scott, Tertiary contacts distant from the active site prime a ribozyme for catalysis, *Cell.*, 2006, **126**, 309–320.
46. J.A. Nelson and O.C. Uhlenbeck, When to believe what you see, *Mol. Cell*, 2006, **23**, 447–450.
47. W.G. Krebs and M. Gerstein, The morph server: A standardized system for analyzing and visualizing macromolecular motions in a database framework, *Nucleic Acids Res.*, 2000, **28**, 1665–1675.
48. W.G. Scott, Morphing the minimal and full-length hammerhead ribozymes: implications for the cleavage mechanism, *Biol. Chem.*, 2007, **388**, 1–7.

CHAPTER 5
The Hairpin and Varkud Satellite Ribozymes

DAVID M.J. LILLEY

Cancer Research UK Nucleic Acid Structure Research Group, MSI/WTB Complex, The University of Dundee, Dow Street, Dundee DD1 5EH, UK

5.1 Nucleolytic Ribozymes

Nucleolytic ribozymes carry out cleavage at a specific site by transesterification reactions in which the 2′-oxygen attacks the 3′-phosphorus, with departure of the 5′-oxygen to leave a cyclic 2′-3′-phosphate. The reaction follows an S_N2 mechanism, with inversion of configuration at the phosphorus. These ribozymes will also perform the reverse ligation reaction provided that the required substrates are both held in place. The reactions are accelerated by $\geq 10^5$-fold in the context of the active ribozyme, although significantly greater rate enhancements have been demonstrated recently for specific forms of the hammerhead and VS ribozymes.[1,2]

5.2 Hairpin Ribozyme

The tobacco ringspot virus has a circular single-stranded satellite RNA of 359 nt. In its replication cycle, the negative strand of the satellite RNA is produced as a concatemeric transcript that is processed into monomeric circular molecules by sequential cleavage and ligation reactions catalysed by the hairpin ribozyme that is contained within the RNA.

The hairpin ribozyme consists of two internal loops that are present on adjacent arms of a four-way helical junction (Figure 5.1). Most of the nucleotides critical for catalytic activity are contained within the two loops, and it was long suspected that these would therefore interact to generate the local environment in which catalysis could occur. This was confirmed in solution using fluorescence resonance energy transfer (FRET) between fluorophores attached to the ends of the two arms.[3,4] A minimal version of the

The Hairpin and Varkud Satellite Ribozymes

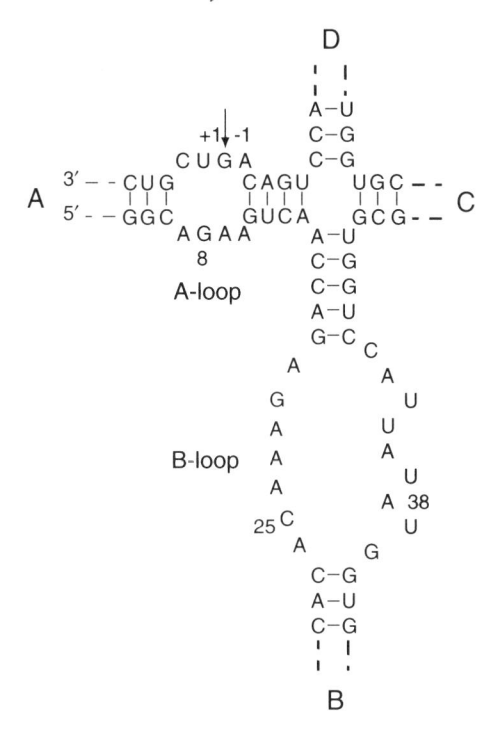

```
                          D
                          ' |
                          ' '
                          A-U
              +1↓-1       C-G
              CUGA        C-G
    3'- -CUG       CAGU       UGC- -
  A  |||       ||||       |||       C
    5'- -GGC       GUCA       GCG- -
              AGAA        A-U
               8          C-G
           A-loop         C-G
                          A-U
                          G-C  C
                  A            A
                  G            U
                  A            U
          B-loop  A            A
                  A            U
                  A          A  38
                25 C            U
                   A        G
                        C-G
                        A-U
                        C-G
                        ' |
                        ' '
                          B
```

Figure 5.1 Nucleotide sequence of the hairpin ribozyme. The ribozyme is shown in its junction form, with arms sequentially labelled A through D around the four-way junction. Internal loops A and B contain all the critical nucleotides for catalytic activity, and key positions discussed in the text are indicated. The position of cleavage or ligation is indicated by the arrow. Helices C and D are not present in the hinged form of the ribozyme, which is therefore held together by the covalent continuity of the strand running between the A and B arms. In the folded conformation of the ribozyme, helices A and D are coaxially stacked, as are helices B and C.

ribozyme (termed here the hinged form) in which the four-way junction is replaced by a simple phosphodiester linkage is active,[5] but requires a 1000-fold higher Mg^{2+} ion concentration to induce folding.[6,7] Moreover, the internal equilibrium between cleavage and ligation is shifted relative to the natural form.[8]

5.2.1 Structure of the Hairpin Ribozyme

The crystal structure of the hairpin ribozyme in its four-way junction form was solved by Rupert and Ferré d'Amaré.[9] In agreement with the earlier solution studies,[3] the junction exhibited coaxial stacking of helices A on D and B on C, and was rotated in an antiparallel manner to allow intimate association between the two loops (Figure 5.2A). Unsurprisingly, the interaction changed

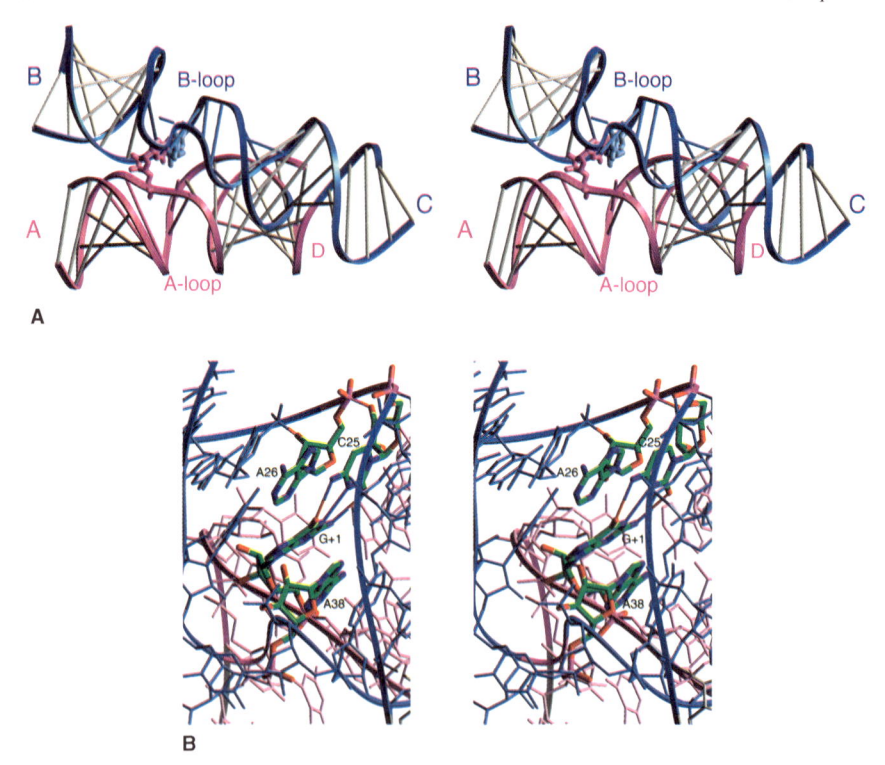

Figure 5.2 Crystal structure of the hairpin ribozyme. (*A*) Parallel-eye stereo image of the overall structure of the ribozyme. Helices A and D are shown in magenta, while B and C are in cyan. The four-way junction is on the right-hand side of this view; the intimate association between the A and B loops occurs towards the left. Extrusion of G+1 from the A loop and formation of a basepair with C25 within the B loop is highlighted. (*B*) Parallel-eye stereo image of the pocket into which G+1 is inserted, where it is sandwiched between nucleobases of A26 and A38, and Watson–Crick paired with C25.

the local conformation of both loops from those of the isolated structures determined by NMR.[10,11] The interaction involved a large (1579 Å^2) and complex interface, including a predicted[12,13] ribose zipper between adenine N3 atoms and 2′-hydroxyl groups of A10, G11, A24, and C25, and multiple hydrogen bonds formed with the nucleobase of U42. An especially important interaction is formed between G+1 (the guanine adjacent to the scissile phosphate in the A loop) and C25 (in the B loop). A G+1A variant is catalytically impaired, but it had been shown that activity could be restored by a compensatory C25U substitution.[14] In the crystal structure it was observed that the nucleobase of G+1 was extruded from loop A, and inserted into a pocket within loop B where it is stacked between two adenine bases, and forms a Watson–Crick basepair with C25[9] (Figure 5.2B). This interaction is very important, because the extrusion of G+1 distorts the conformation of the backbone such that the 2′-OH of A-1 becomes aligned for nucleophilic attack (see below).

5.2.2 Metal Ion-dependent Folding of the Hairpin Ribozyme

In view of the polyelectrolyte character of RNA it is not surprising that folding of the hairpin ribozyme into its active conformation requires metal ions. In the absence of added metal ions, both the hinged and junction forms of the ribozyme adopt extended structures, preventing loop–loop interaction.[3,4,15] However, the hinged form requires a 1000-fold higher Mg^{2+} ion concentration to induce folding,[6,7] so that it would not be active under physiological conditions. Metal ion-induced folding of the natural form of the ribozyme can be regarded as a scissor-like rotation of the junction, juxtaposing the loops to allow interaction to generate the active ribozyme. Folding was found to be strongly affected by substitutions that prevent contacts important in loop–loop interaction,[16] *e.g.* folding of a G+1A variant was seriously impaired, and required 20-fold higher Mg^{2+} concentrations.

The simple junction generated by replacing the loops with Watson–Crick base pairing also exhibits coaxial stacking of arms with the same stacking preference.[17] However, the helical axes of the simple junction were found to be apparently perpendicular at sub-millimolar Mg^{2+} concentrations. Single-molecule studies showed that the junction was structurally very polymorphic, and the structure observed in bulk reflected rapid interconversion between several different conformations.[18] The chirality has been analysed by an electrophoretic approach.[19]

Bulk studies of the ion-induced folding of the natural ribozyme using FRET between fluorophores attached to the ends of the loop-carrying arms revealed the process to be complex,[7] apparently occurring in two stages. The first occurred in the micromolar range of Mg^{2+}, and involved the cooperative binding of several ions. However monovalent ions could bring about the same change, so it was probable that diffuse binding was responsible. A second non-cooperative binding occurred in the millimolar range of Mg^{2+}, and required divalent metal ions.

Single-molecule studies have been very informative in the analysis of the folding of the hairpin ribozyme.[20,21] We have explored the role of the four-way junction in the folding of the hairpin ribozyme by means of single-molecule FRET studies,[21] using surface-attached ribozymes with donor and acceptor fluorophores attached to the 5′-termini of the A and B arms respectively. We have also studied the corresponding simple four-way junction. Records of FRET efficiency (E_{app}) with time for single molecules of the hairpin ribozyme (Figure 5.3) and its simple junction show that both species exhibit fluctuations between states of high and low efficiency. The high FRET state for the complete ribozyme has $E_{app} \sim 0.9$, while that for the junction has $E_{app} \sim 0.5$. The ribozyme becomes increasingly stabilized in the high FRET state as the concentration of Mg^{2+} is increased. This state is assigned to the species in which the loops are docked together as found in the crystal.[9] The simple

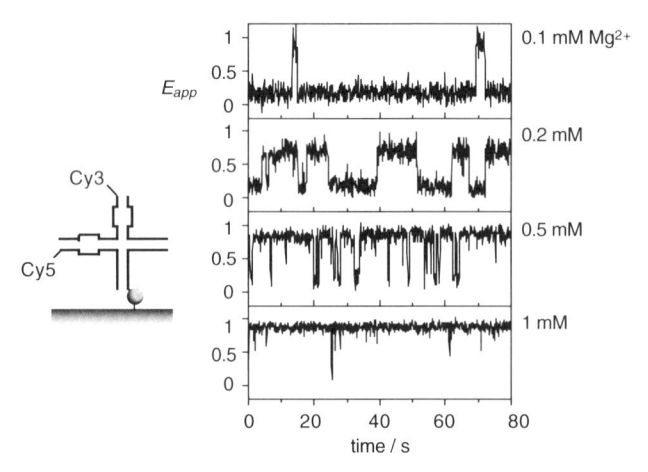

Figure 5.3 Time records of conformational transitions in the hairpin ribozyme as a function of Mg^{2+} concentration. The schematic shows a hairpin ribozyme molecule terminally-labelled with Cy3 and Cy5 fluorophores on the loop-bearing arms, and tethered to a quartz surface *via* a biotin–streptavidin linkage. Loop–loop interaction brings the fluorophores into closer proximity, leading to high efficiency of energy transfer. Time records are shown of single ribozymes in the presence of the indicated Mg^{2+} concentrations.

junction exhibited a rather different dependence on Mg^{2+} ion concentration. Below 20 mM the relative fractions of the two FRET states remained constant, but their rate of interconversion became slower as Mg^{2+} concentration increased. At concentrations below 1 mM Mg^{2+}, the fluctuations of the simple junction were too fast to be resolved, but rates could be measured using cross-correlation analysis. Donor and acceptor intensities remained anti correlated at sub-millimolar Mg^{2+} concentrations, with rates in the low-millisecond range.

This behaviour clearly raises an interesting question about the dynamics of the complete ribozyme. Cross-correlation analysis applied to the low-FRET state of the ribozyme[21] revealed a similar rapid fluctuation between two states, with interconversion rates that were similar of those for the simple junction. Data for the low FRET state of a single ribozyme molecule binned at 3 ms could be deconvoluted into two Gaussian distributions, centred at $E_{app} = 0.15$ and 0.4, corresponding closely to the two states of the simple junction. Thus the undocked ribozyme fluctuates between two states (termed U_D and U_P where U_D is the lowest FRET state) very similar to the simple junction. This suggests that the complete ribozyme exists in three possible states (U_D and U_P and F, where F is the fully-docked state) (Figure 5.4A). Direct evidence for a three-state system was obtained for a C25U variant of the ribozyme (Figure 5.4B). Time traces show that the molecule exists in three states, and that the intermediate-FRET state U_P is an obligate intermediate state for passage between the high- and low-FRET states.

The rapid exchange within the undocked states ($U_D \leftrightarrow U_P$) repeatedly juxtaposes the two loops, increasing the probability of interaction between them and thus folding. The complex loop–loop interface may require multiple

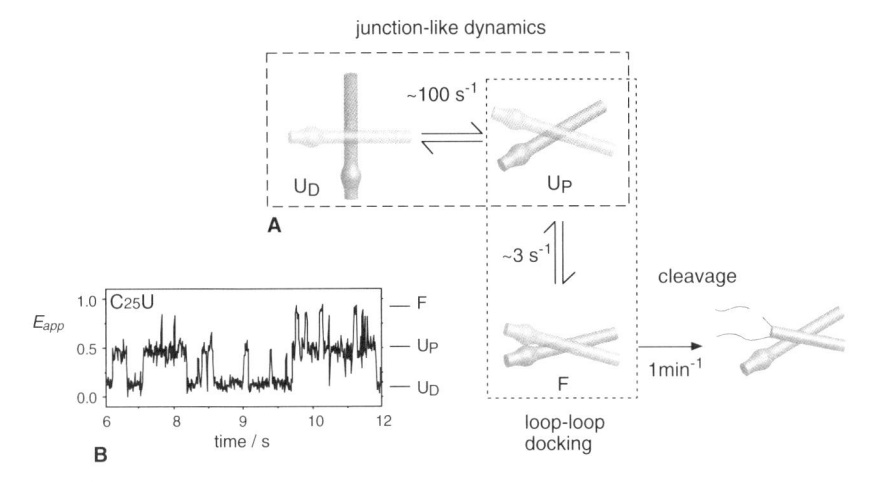

Figure 5.4 Three-state folding of the hairpin ribozyme. (*A*) Schematic of the three-state folding process and cleavage reaction. The junction-like folding process between the undocked U_D and U_P is boxed (short-dash line), as is the loop docking process between U_P and F (broken line). (*B*) Time record for the C25U hairpin ribozyme variant. Three states are clearly seen, from which it was determined that the intermediate-FRET state is an obligate intermediate between the high- and low-FRET states.

conformational adjustments to achieve the active state. Docking rates exhibit significant heterogeneity between molecules, suggesting that the loops exist in persistent, multiple conformations.[21,22] The $U_D \leftrightarrow U_P$ transition occurs at a rate of $\sim 100\,\text{s}^{-1}$ in 0.5 mM Mg^{2+}, whereas the rate of formation of the F state is two orders of magnitude slower at $\leq 3\,\text{s}^{-1}$. The folding under these conditions is limited by the conformational changes involved in loop–loop interaction rather than by the junction dynamics.

The rate of loop docking in the complete ribozyme is around 500-fold faster than that measured for the minimal hinged form that lacks the four-way junction.[20] Thus the junction accelerates the rate of folding of the ribozyme into the active conformation, with the result that folding is no longer rate limiting as it is for the minimal form. As previously noted, the junction also allows folding to occur in physiological ionic conditions.[6] The helical junction is not essential for activity, but it is important for efficient activity under cellular conditions. A functionally similar situation is found in the hammerhead ribozyme, where loops in two helical stems promote activity[23] and facilitate folding[24] at physiological ionic concentrations.

5.2.3 Observing the Cleavage and Ligation Activities of the Hairpin Ribozyme

Single-molecule FRET has also been used to follow the cleavage and ligation reactions in the hairpin ribozyme, by exploiting the distinct dynamics of the

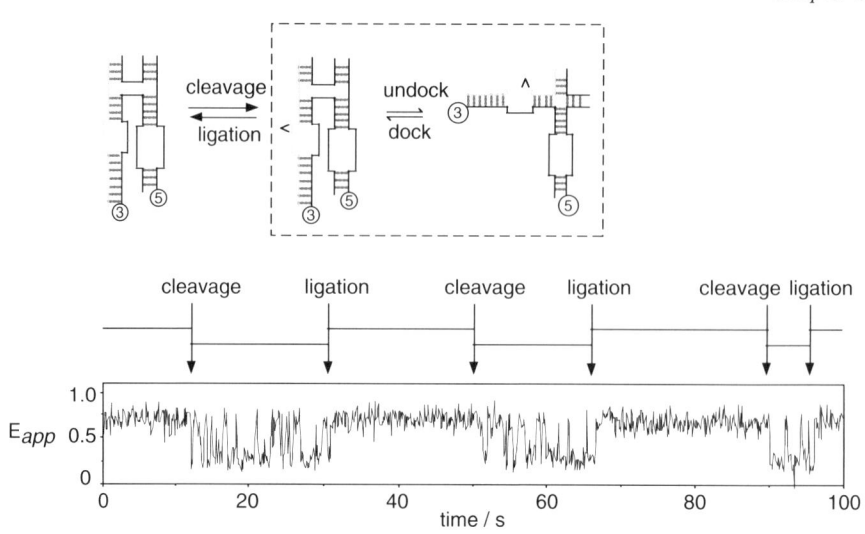

Figure 5.5 Multiple cycles of cleavage and ligation in a single hairpin ribozyme in the presence of 1 mM Mg^{2+} at 20 °C. The ribozyme has a terminal helix of arm A of 7 bp, ensuring retention of the product of cleavage, which can therefore serve as the substrate for ligation. The ligated form of the ribozyme is stably docked under the conditions of the experiment. Upon cleavage the product ribozyme can undergo multiple docking/undocking transitions at $\sim 2.5\,s^{-1}$. Since the molecule was generated by ribozyme cleavage it has a cyclic 2′,3′ phosphate (triangle). It is, therefore, competent in ligation, and ligation events may be identified by the change in dynamics back to a stable docked conformation.

intact ribozyme and its cleaved product.[25] Figure 5.5 shows an experiment in which a single ribozyme molecule undergoes repeated cycles of cleavage and ligation. The ribozyme switches between two distinct modes. The molecule remains stably docked for a period before changing into a form that undergoes rapid docking and undocking. We assigned the two states as the ligated and cleaved forms of the ribozyme, respectively. The docking and undocking rates within the cleaved state are $k^C_{dock} = 2.5 \pm 0.1\,s^{-1}$ and $k^C_{undock} = 2.3 \pm 0.1\,s^{-1}$, significantly slower than the junction dynamics under the same conditions.

Analysis of data from many molecules provides the internal conversion rates for the hairpin ribozyme. The cleavage rate (k_C) is $\sim 0.6\,min^{-1}$ in the presence of 1 mM Mg^{2+}, while the ligation rate (k_L = the rate of ligation in the docked ribozyme) is $\sim 0.3\,s^{-1}$. The reaction is significantly biased towards ligation, with an internal equilibrium constant of $K_{int} = k_L/k_C = 34$. This property maintains the integrity of the circular (−) strand, allowing it to act as a template for (+) strand synthesis. However, cleavage of the concatenated product of replication is made possible by of the rapid undocking that follows cleavage.

5.2.4 Mechanism of the Hairpin Ribozyme

The alignment of the 2′-O nucleophile observed in the crystal structure[9] suggests that facilitation of the trajectory into the transition state could be a significant part of the catalytic mechanism.[26] However, this could only contribute a factor of ≤100-fold to the catalytic rate enhancement – a relatively small contribution to the observed acceleration of ≥10^5. Clearly other mechanisms must play their part.

At one time it was believed that bound metal ions would play the major role in accelerating the transesterification reactions of the nucleolytic ribozymes, but this is clearly not the case for the hairpin ribozyme. It was found that significant levels of activity could be attained in the hairpin and VS ribozymes in the presence of high concentrations of monovalent ions such as Li$^+$ and NH$_4^+$,[27] apparently excluding a requirement for site-specific binding of the ion. Moreover, the hairpin ribozyme remained active when Mg^{2+} ions were replaced by substitutionally-inert hexammine cobalt(III) ions.[28–30] This excludes a role for metal ion-bound water molecules in general acid–base catalysis, as well as acting as a site-bound Lewis acid. The cleavage rates measured in single-molecule experiments were also found to depend very weakly on Mg^{2+} ion concentration.[25] It remains possible, however, that a degree of electrostatic stabilization of the dianionic transition state could be achieved by diffusely-bound metal ions.

This leaves nucleobase participation to provide the remaining rate acceleration. Early evidence pointed to an important role for G8 in loop A, located on the opposite strand from the scissile phosphate. Substitution by other nucleotides led to rates of cleavage being reduced by several orders of magnitude in both the natural[16] and hinged form[31] of the ribozyme, without affecting ion-induced folding. Complete removal of the nucleobase (creating an abasic site) also led to a major loss of ribozyme activity.[32] Strong evidence for nucleobase participation by G8 came from the pH dependence of variants with alternative nucleotides substituted at position 8. The cleavage rate was found to decrease with pH when 2,6-diaminopurine was substituted at this position for example, showing that the catalysis was now dependent on proton transfer involving a group with a pK_a of 6.9.[31] Thomas and Perrin[33] have recently demonstrated alkylation of G8 when the 2′-OH nucleophile was substituted by bromoacetamide, further supporting the direct participation of this nucleobase.

Further insight was provided into the nature of the active site by a crystal structure of a modified hairpin ribozyme in which the scissile phosphate was replaced by vanadate to provide a model of a pentacoordinate phosphorane transition state[34] (Figure 5.6). It was found that G8 is hydrogen bonded to the 2′-O and the *pro*-S non-bridging O of the scissile phosphate, well positioned to participate in the catalytic chemistry. As well as confirming the importance of G8, the structure also revealed the presence of a second nucleotide juxtaposed with the scissile phosphate. The nucleobase of A38 (contributed by loop B) was found to form hydrogen bonds to the 5′-O and the *pro*-R O. Removal of the

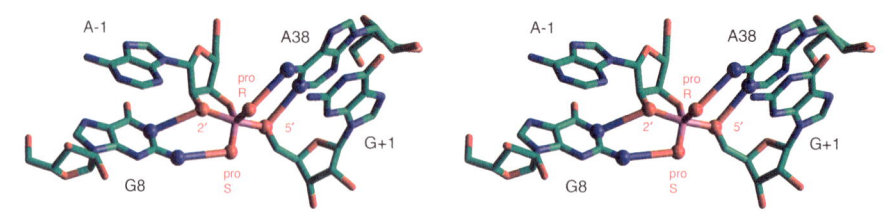

Figure 5.6 Parallel-eye stereo view of the structure of the active site of the hairpin ribozyme site observed in the crystal structure of a vanadium-containing transition state analogue.[34] G8 and A38 are hydrogen bonded to the 2′ and 5′ oxygen atoms respectively, as well as to the non-bridging O atoms. G8 and A38 are therefore well positioned to serve as general base and acid, respectively, in the cleavage reaction.

nucleobase at position 38 led to a 10 000-fold loss of activity,[35] while ligation activity was shown to be sensitive to substitution by inclusion of modified nucleotides in nucleotide analog interference mapping (NAIM) experiments.[36] The nucleobases of G8 and A38 should provide some stabilization of the transition state, and consequently a potential rate enhancement. However, they are also well placed to participate in coordinated general acid–base catalysis (Chapter 2). G8 is positioned to deprotonate the 2′-O, while A38 could protonate the 5′-oxyanion leaving group in the cleavage reaction. By the principle of microscopic reversibility, G8 and A38 would act as general acid and base, respectively, in the ligation reaction. The pH dependence of hairpin ribozyme rates are consistent with the participation of two bases with pK_as of ∼6.2 and >9.5.[25,31,32,37] The lower value is within the range of an adenine base with a pK_a perturbed by an electronegative environment.

Perhaps one of the most persuasive demonstrations of nucleobase participation in ribozyme catalysis was the restoration of activity in a C76U variant of the HDV ribozyme (Chapter 6) by free imidazole base in the solvent by Been and co-workers.[38] However, this experiment requires that the base has access to the active site, and this is not always the case. Fedor and co-workers[39] have investigated a large range of bases for their ability to restore activity to a G8 abasic hairpin ribozyme, but found that imidazole is ineffective. We therefore adopted a new chemo-genetic strategy in which imidazole base was introduced as a pseudo-nucleoside by chemical synthesis,[40] making a substituted ribozyme in which the imidazole nucleoside was introduced at position 8 in place of guanine.[41] The substitution did not perturb folding, and the modified ribozyme was active in both cleavage and ligation. The cleavage product was indistinguishable from that of the natural ribozyme. The rate of cleavage measured at pH 7 in the presence of 10 mM MgCl$_2$ at 25 °C was 0.003 min^{-1}. This is slower than the natural ribozyme under these conditions (k_{obs} = 0.7 min^{-1}), but is ten-fold faster than that for the G8U variant.[16] One of the major reasons for substitution with imidazole is to look for the expected pH dependence if it is participating in the chemistry. If G8 acts as the general base in the cleavage reaction, it can only do so in the deprotonated form. Since its pK_a is expected to

The Hairpin and Varkud Satellite Ribozymes

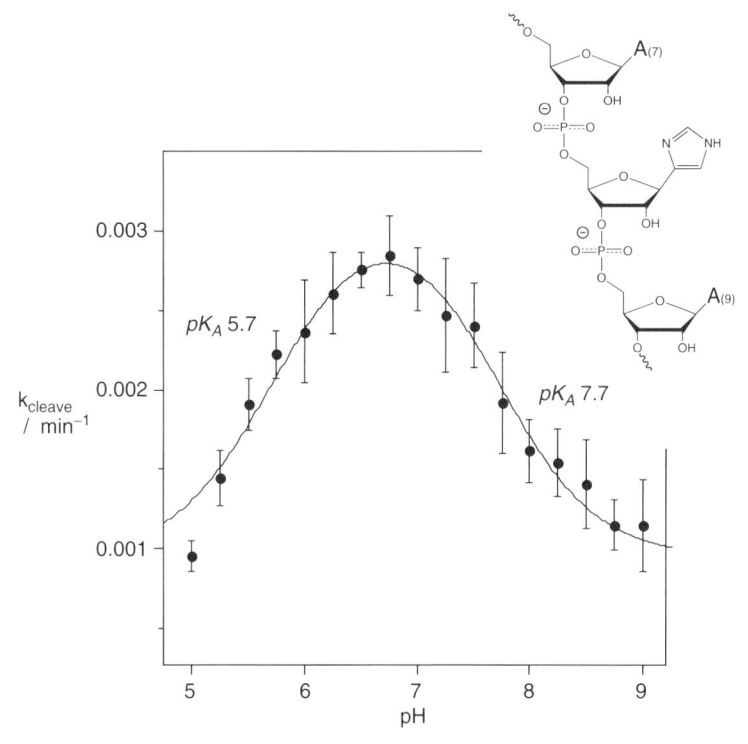

Figure 5.7 pH dependence of the cleavage reaction of a hairpin ribozyme with guanine at position 8 replaced by imidazole. The bell-shaped curve has been fitted to a model involving two proton exchanges with pK_as of 5.7 and 7.7. The local structure of the imidazole nucleotide incorporated into the a strand at position 8 is also shown.

be >9, the concentration of the deprotonated form increases linearly with pH over the experimentally observable range. However, deprotonation of A38 occurs over the same range, and this compensates so that the concentration of active ribozyme remains constant. This is why a plateau of rate with pH is observed. But if the function of this guanine is taken by imidazole with a lower pK_a, the threshold for reduction in rate will occur at lower pH due to complete deprotonation of the imidazole in the region of pH 8, leading to a bell-shaped pH profile. This was observed experimentally[41] (Figure 5.7), and the data were fitted to a model based on two proton transfers, giving estimated pK_as of 5.7 ± 0.1 and 7.7 ± 0.1. The higher pK_a should be that of the imidazole. A bell-shaped pH dependence was also observed for the ligation reaction, with pK_as of 6.1 ± 0.3 and 6.9 ± 0.3.

In summary, the framework of the hairpin ribozyme facilitates the docking of the A- and B-loops to generate the active form of the ribozyme. This has two main effects that are directly related to catalysis. The extrusion of G+1 into its binding pocket in the B-loop changes the backbone conformation so that the 2′-O nucleophile is well aligned for S_N2 attack. Second, G8 and A38 are juxtaposed with the scissile phosphate so that they can both stabilize the transition state and participate in general acid–base catalysis.

5.3 VS Ribozyme

The Varkud satellite (VS) RNA is an abundant transcript that occurs in the mitochondria of several natural isolates of *Neurospora*. Collins and co-workers found that the VS RNA contains an element capable of self-cleavage,[42] with a role in the processing of replication intermediates.[43] At ~150 nt, the VS ribozyme is the largest of the nucleolytic ribozymes, and is the only member of this group for which there is no crystal structure.

The sequence of the VS ribozyme is unrelated to those of the other nucleolytic ribozymes. The secondary structure (Figure 5.8A) was deduced

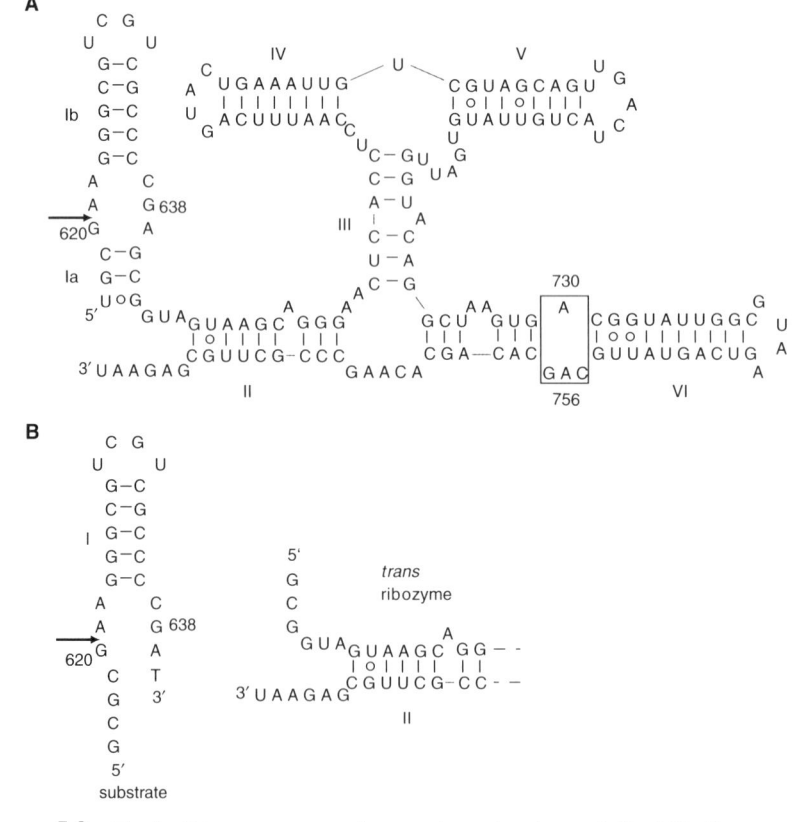

Figure 5.8 Nucleotide sequence and secondary structure of the VS ribozyme. (*A*) Natural cis form of the ribozyme. The position of cleavage is indicated by the arrow, and the A730 loop (the putative active site) is boxed. (*B*) Division into a trans-acting ribozyme and its substrate. In this form helix I (substrate) is physically separate from the ribozyme (consisting of helices II–VI).

by Collins and co-workers by a combination of probing and mutagenesis.[44] A 5′ stem-loop structure (helix I) contains an internal loop in which the ribozyme cleavage occurs. This is linked to an H-shaped structure consisting of the five helices II–VI, where helix III forms the bar of the H. Stem-loop I can be disconnected from the rest of the RNA, leaving a ribozyme (helices II–VI) that acts in trans on the substrate.[45] The interaction between the ribozyme and its substrate consists of tertiary contacts alone. An important component is the formation of loop–loop pairing between the GAC triplet in the terminal loop of stem V, and a GUC triplet in the substrate.[46] The latter is largely contained within the terminal loop, but exposure of the 3′ cytosine would require the opening of the terminal basepair, and it is proposed that this results in a change in the conformation of the substrate that is important for the cleavage reaction.[47]

In its trans-acting form the ribozyme is very well behaved, especially if basepairing in helix Ia is absent (Figure 5.8B). Cleavage reaction progress is well fitted by a single exponential, and the kinetics conform to an enzyme-type scheme involving formation of a non-covalent complex followed by the chemical step, with values of $K_d = 1\,\mu M$ and $k_2 = 2\,min^{-1}$.[48] The trans-acting ribozyme can also carry out an efficient ligation reaction, provided the lower helix (Ia) of the substrate stem-loop is extended to ∼10 bp to hold the substrate strand with its terminal cyclic 2′3′-phosphate.[49] Rates of $\geq 7\,min^{-1}$ were measured for this reaction.

5.3.1 Structure of the VS Ribozyme

Despite several valiant attempts, no-one has yet obtained an atomic resolution crystallographic structure of the VS ribozyme. However, the global fold of the ribozyme has been deduced using lower resolution approaches in solution. The H-shaped secondary structure (Figure 5.9) is essentially organized by two three-way helical junctions; the 2-3-6 junction (between helices II, III and VI) and the 3-4-5 junction (between helices III, IV and V). The hammerhead and hairpin ribozymes also contain prominent helical junctions, and such junctions are common features of small, autonomously-folding RNA species. The overall secondary structure is required for the function of the ribozyme, but most substitutions are tolerated so long as the secondary structure is preserved. Most of helix IV and the distal end of helix VI can be substantially deleted without significant loss of function.[50] The lengths (but not sequence) of helices III and V are very important, and the sequence of the terminal loop of stem V cannot be changed without a major loss of activity. The base bulges in helices II and VI must be present for full activity, although nucleotide substitutions are tolerated so that their role is probably structural. However, ribozyme activity is strongly dependent on the local sequence of the two three-way junctions.[50–52] It therefore seemed probable that the structures of the two junctions essentially determine the overall fold of the VS ribozyme, and we therefore adopted a "divide-and-conquer" approach to the global structure.

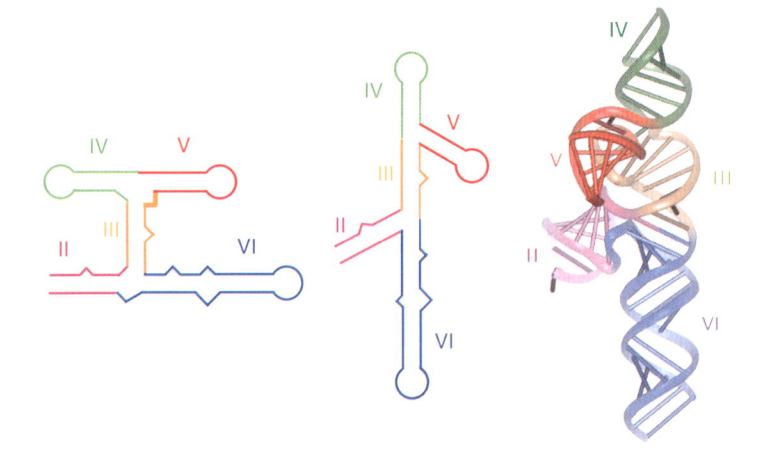

Figure 5.9 Structure of the VS ribozyme assembled from those of its component junctions. Schematic of the secondary structure in the conventional form (like that of Figure 5.8) and in a form indicating the deduced coaxial helical stacking. This generates a single coaxial stack of helices IV, II and VI, with helices V and II projecting away from this axis. A three-dimensional model of the structure is shown on the right-hand side.

The global structures of the 2-3-6 and 3-4-5 junctions were determined by a combination of comparative gel electrophoresis and FRET measurements of end-to-end distances.[50,51] Both junctions undergo coaxial stacking of two arms, induced by the non-cooperative binding of magnesium ions, with $[Mg^{2+}]_{1/2}$ concentrations in the micromolar region. Helices III and VI are coaxially stacked in the 2-3-6 junction, with an acute angle subtended between helices VI and II.[51] A three-way junction of very similar sequence and global structure is present in 23S rRNA. When transplanted into the VS ribozyme, the modified ribozyme was active. The nucleotides A656 and G768 interact in a wedge-like manner that sets up the trajectory of helix II. These are highly conserved in 23S rRNA,[53] and nucleotide substitutions at corresponding positions in the ribozyme were found to be very detrimental to cleavage or ligation activity.[51] Helices IV and III are coaxially stacked in the 3-4-5 junction, and the smallest angle is subtended between helices III and V. Once again a junction of closely similar sequence and global structure occurs in 23S rRNA. It has been suggested that the formally single-stranded sequence between helices V and III adopts a UNR-type turn, and it has been found that activity is preserved if this is replaced by a stem-loop, *i.e.* turning the 3-4-5 junction into a four-way helical junction.[52]

The two junctions are linked through the common helix III, generating a long tube consisting of the coaxially stacked helices IV, III and VI, from which helices V and II radiate. The dihedral angle between helices V and II was estimated at $\sim 75°$ from electrophoretic experiments involving variation of the length of helix III.[50] Putting all the data together with homology modelling of

The Hairpin and Varkud Satellite Ribozymes

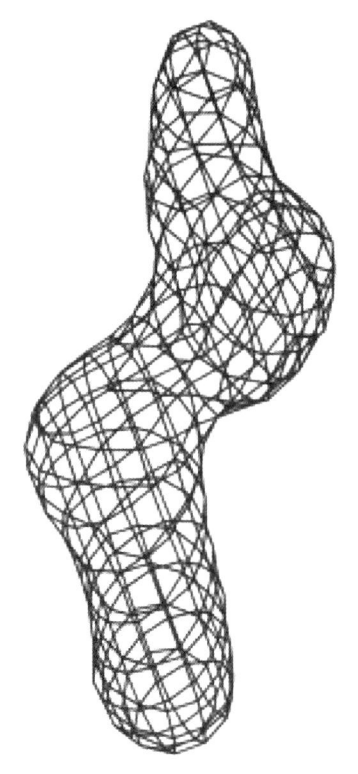

Figure 5.10 Electron density map of the VS ribozyme derived from small-angle X-ray scattering. Scattering profiles obtained from the trans-acting form of the ribozyme folded in the presence of Mg^{2+} ions were analysed by bead reconstructions [using the program DAMMIN[80]]. Structure factors calculated from averaged reconstructions were then transformed into electron density map as shown.

the 2-3-6 junction provided a model for the global structure of the complete ribozyme (Figure 5.9).

In view of the continued lack of crystallographic data, we have taken an alternative approach to analysing the structure of the VS ribozyme, which provides information on the complete ribozyme without resorting to assembly from its component parts. We have used small-angle X-ray scattering (SAXS) in solution to provide size and shape information on the five-helix (helices II through VI) ribozyme (J. Lipfert, J. Ouellet, D.G. Norman, S. Doniach and D.M.J. Lilley, unpublished results). Analysis of scattering profiles recorded in the presence of Mg^{2+} ions reveal that the structure is a long tube of ~ 130 Å, with two lateral bulges (Figure 5.10). The location of the laterally-projecting arms II and V are harder to identify with confidence at present. However, the overall shape fits the FRET-based model quite well, with its IV-III-VI end-to-end distance of 130 Å.

5.3.2 Structure of the Substrate

The substrate consists of a stem-loop, with an internal loop that contains the scissile phosphate. This has been the subject of several studies by NMR. In the first, the upper stem-loop (helix Ib) was truncated and terminated with a tetraloop,[54] while a complete substrate strand was analysed in the second study.[55] In both structures the internal loop consisted of two sheared G·A basepairs similar to domain 2 of the hammerhead ribozyme,[56,57] together with a protonated $A^+ \cdot C$ pair.

Upon interaction with the ribozyme it is highly improbable that the substrate structure will remain the same as the free state, and experience with the hairpin ribozyme (see above) provides an excellent precedent. From covariation analysis it has been proposed that loop–loop interaction between the substrate and stem V (see below) results in a rearrangement of the basepairing in stem Ib of the substrate.[47] This involves the opening of the terminal basepair, the flipping out of C634, and rearrangement of basepairing in stem Ib such that CCC (635–637) pairs with GGG (623–625). On interacting with stem-loop V, C634 exhibits enhanced reactivity to dimethyl sulfate.[58] In an NMR structure of a stabilized rearranged structure of the internal loop[59] the sheared G620·A639 pair is preserved, but G638 interacts with both A621 and A622 in a non-coplanar manner, while A621 is cross-strand stacked onto A639 on the minor groove side.

While these structures are a useful guide to the ground state structure, further rearrangement of the internal loop is highly likely as the substrate interacts fully with the ribozyme, as we now discuss.

5.3.3 Location of the Substrate

Clearly the substrate must dock into the ribozyme for cleavage or ligation to be catalysed. The location of the substrate is constrained by two known points of attachment. First, the 3′ end of the substrate helix Ia is connected by three nucleotides to the 5′ end of helix II in the natural cis-acting form of the ribozyme, so these helical ends must be physically close. Second, at the other end of the substrate the terminal loop interacts with that of helix V. These twin requirements suggest that the substrate will dock into the cleft formed between helices II and VI, as indicated in Figure 5.11.[50] This places the cleavage site close to the A730 loop of stem VI, which we believe to be the active site of the ribozyme (see below). Helix Ia of the substrate protrudes unhindered from the complex; we have found that this helix can be extended without loss of catalytic activity.[49] Helix I can be attached to the ribozyme in several different ways with preservation of activity. It can be attached to the 3′ end of helix II (instead of the normal 5′ attachment),[60] and to the end of helix VI *via* a linker (D.A. Lafontaine and D.M.J. Lilley, unpublished data). These observations are readily accommodated within the model.

Recent data from SAXS are also in agreement with this structure. X-Ray scattering from the simple H-shaped trans-acting ribozyme (*i.e.* no substrate)

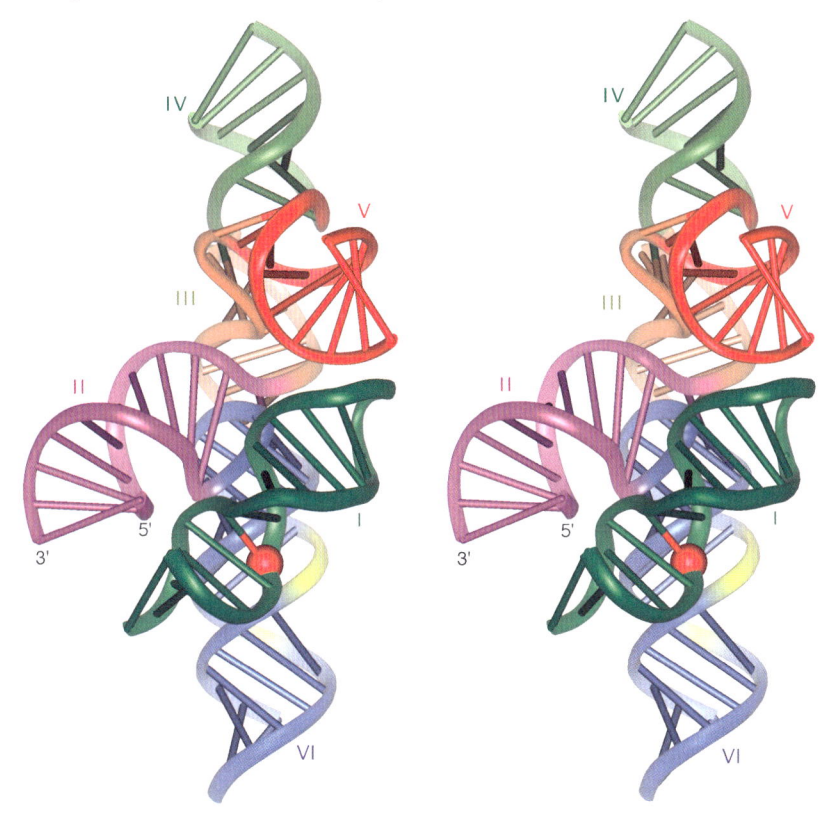

Figure 5.11 Parallel-eye stereoscopic view of a model of the VS ribozyme with the substrate bound.

and the natural form with its substrate stem-loop covalently linked to the 5′-end of the ribozyme were compared in the presence of sufficient Mg^{2+} to achieve docking (J. Lipfert, J. Ouellet, D.G. Norman, A.C. McLeod, S. Doniach and D.M.J. Lilley, unpublished results). Under these conditions the two species had similar radii of gyration, indicating that the substrate must be located quite close to the centre of mass. Analysis of the shape of the cis form from scattering profiles was also consistent with the deduced location of the substrate, although this could not be regarded as definitive proof.

A prominent region of protection against hydroxyl radical attack was observed on both strands of helix II adjacent to the 2-3-6 junction upon folding.[61] This is consistent with a close association between the substrate and helix II in this region, as could be expected if the substrate is located in the cleft between helices II and VI. The interaction might be mediated by 2′-hydroxyl groups on the substrate and helix II;[62] NAIM experiments revealed that the removal of 2′-hydroxyl groups at the junction-proximal end of helix II was deleterious.[63] Jones and Strobel have suggested the formation of an A-minor interaction.[64]

Most of the sequence requirements for the ribozyme function can be understood in terms of the requirements for substrate binding. Many of the sequence changes that affect activity are located in the two junctions, which organize the global architecture of the ribozyme. There is a strong correlation between sequence changes in the 2-3-6 junction that affect its ion-induced folding and those that affect the activity of the complete ribozyme.[51] Both cleavage and ligation activity are highly sensitive to the lengths of helices III and V. The length of helix III is important because it determines the dihedral angle between helices II and V, both of which interact with the substrate. The dependence of cleavage rate on the length of helix III can be fitted on the assumption that the increase in activation energy results from the requirement to distort that dihedral angle back to its optimal value, which should be proportional to the square of the displacement angle.[50] The length of helix V determines the spatial and rotational setting of its terminal loop; only a 9 bp helix allows it to interact productively with the substrate loop.

5.3.4 Active Site of the VS Ribozyme

Most of the sequence changes in the ribozyme that affect catalytic activity do so because they alter the structure. These susceptible positions are located in the helical junctions or the bulges; changes in the lengths of critical helices III and V are also deleterious. However, there is one major exception to this, *i.e.* the internal loop of helix VI containing A730, termed the A730 loop (Figure 5.12). Most single-base changes introduced into this loop result in significant loss of cleavage activity (observed cleavage rate generally lowered by 50-fold or more) in trans[48] and in cis,[65] yet these effects do not appear to result from perturbation of folding.[48] The A730 loop was also found to be sensitive to ethylation interference, and contained sites of interference by phosphorothioate incorporation, and suppression by thiophilic manganese ions. Sequence changes in the loop also affected ligation activity,[49] which was also very sensitive to the introduction of nucleotide analogs.[64]

In our coarse grained model of the ribozyme, the scissile phosphate (shown as a red sphere in Figure 5.11) is readily juxtaposed with the A730 loop (coloured yellow in that figure). Taken together with the major sensitivity of ribozyme activity to changes in this region suggests that the A730 loop is likely to be a significant part of the active site of the ribozyme. Thus we expect that the ribozyme functions by docking the substrate into the cleft between helices II and VI, to facilitate an intimate interaction between the cleavage site and the A730 loop.

5.3.5 Candidate Catalytic Nucleobases

Nucleobase participation is emerging as a common theme in the nucleolytic ribozymes, as the hairpin ribozyme demonstrates. What, then, are the candidates for functional nucleobases in the VS ribozyme? Within the A730 loop, one particular nucleotide stands out. Substitution of A756 by G, C or U leads to ≥ 300-fold loss of cleavage[48] and ligation activity.[49] These changes have only

The Hairpin and Varkud Satellite Ribozymes 83

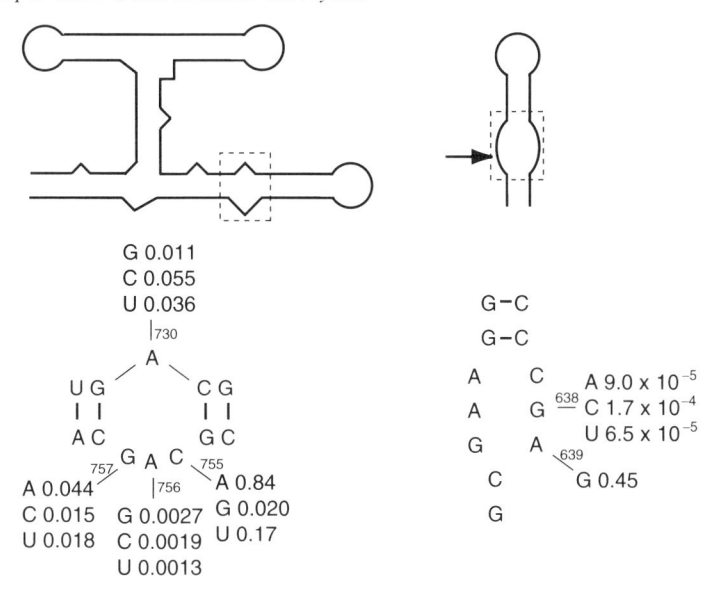

```
            G 0.011
            C 0.055
            U 0.036
             |730
              A
            /   \                                    G–C
      U G       C G                                  G–C
      | |       | |                         A    C          A 9.0 x 10⁻⁵
      A C       G C                         A    G ⁶³⁸ ═  C 1.7 x 10⁻⁴
          G  A  C                           G    A          U 6.5 x 10⁻⁵
        757 / | \ 755                                    ⁶³⁹
      A 0.044 |756  A 0.84                  C           G 0.45
      C 0.015 G 0.0027 G 0.020               G
      U 0.018 C 0.0019 U 0.17
              U 0.0013
```

Figure 5.12 Probable active components of the VS ribozyme. The A730 loop is highlighted within helix VI of the ribozyme (left-hand side). The internal loop of the substrate stem-loop is indicated on the right. Both have been subjected to extensive sequence variation, and cleavage rates measured in trans under single turnover conditions. These are shown for each variant, measured in min^{-1}. Nucleotide substitution of A756 or G638 is particularly deleterious.

a small effect on substrate binding affinity ($\Delta K_D \leq 5$ fold), and most of the effect arises from a reduced rate of central conversion of the substrate into product (k_2).[48] There are also crosslinking data that place A756 physically close to the cleavage site in the substrate; a UV crosslink was obtained between 4-thiouridine at position 621, adjacent to the cleavage position, and A756.[66]

The role of A756 has been analysed by functional group modification.[67] Removal of the 2′-hydroxyl group resulted in a small reduction in observed cleavage rate, whereas removal of the nucleobase (creating an abasic site at position 756) lowered the activity $\geq$1000-fold. Removal of the exocyclic amine from the 6 position, translocation to the 2 position or addition of a 2-amino group all led to 1000-fold slower cleavage. By contrast, replacement of N7 by CH (7-deaza adenosine) had a negligible effect on activity. It appears that the Watson–Crick edge of the nucleobase of A756 is important for catalytic activity. In NAIM experiments using a wider range of analogs,[64] the 756 position was the most sensitive nucleotide in the entire ribozyme in terms for the ligation reaction. Perhaps disappointingly, no restoration of activity of an A756Abasic VS ribozyme could be observed with high concentrations of imidazole base present in the medium.[67] However, none was observed with free adenine either, so it seemed likely that the catalytic core is inaccessible to exogenous bases. Using our imidazole nucleoside phosphoramidite, however, we synthesized a variant ribozyme with imidazole at position 756, and observed significant rates of cleavage and ligation reactions.[68]

Thus there is persuasive evidence pointing to a direct role for A756 in the catalytic chemistry. Studies on other ribozymes, notably of course the hairpin ribozyme, show that two nucleobases can act in concert. If A756 is one member of such a pair, what then might be the other? We can exclude most of the ribozyme nucleotides, as substitutions can be tolerated so long as secondary structure is preserved, together with the lengths of helices III and V.[50,51] Furthermore, our current view of the global structure of the ribozyme and the location of the substrate would severely restrict positions that would have physical access. The basepairs flanking the A730 loop can be inverted without large loss of activity. Within the A730 loop itself, while substitutions of either G757 or A730 lead to significant loss of cleavage and ligation activity, these are nevertheless more active than A756 variants, suggesting that neither of these nucleotides play a critical role in the catalytic chemistry.

This leaves one region for consideration, the internal loop of the substrate stem-loop. Within this loop we have identified a guanine base at position 638 that is a strong candidate for the second nucleobase acting in the catalysis.[69] Replacement by any other nucleotide results in four orders of magnitude impairment of cleavage, yet substrate folding is not perturbed. Moreover, the variant substrates bind the ribozyme with similar affinity, acting as competitive inhibitors of the natural sequence substrate. Thus G638A substrate binds to the ribozyme in an apparently normal manner, but is catalytically incompetent. This suggests that G638 plays a direct role in the function of the ribozyme. Since its 2′-hydroxyl group can be substituted by a proton at this position[70] it is probable that the guanine nucleobase is the key participant.

5.3.6 Mechanism of the VS Ribozyme

In the absence of a crystal structure for the VS ribozyme we are unlikely to arrive at a full understanding of the complete catalytic mechanism. Like other ribozymes, it probably derives its catalytic power from the combination of several processes. Since the VS ribozyme is active in monovalent ions,[27] the direct participation of a site-bound metal ion as a Lewis acid, or in general acid–base catalysis, is unlikely, although some electrostatic stabilization of the transition state by diffusely-bound ions remains possible. Orientational effects are quite likely, but it is not possible to evaluate their significance at present.

In the light of the data presented above it seems quite probable that A756 and G638 play a direct role in the chemistry of the VS ribozyme. Nucleobases can participate by stabilizing the transition state, *e.g.* by hydrogen bonding the phosphorane. A positively-charged protonated adenine base might provide electrostatic stabilization of the dianionic transition state.

A756 and G638 might also function in general acid–base catalysis. Recent analysis of kinetic isotope effects in cis-acting VS ribozymes show that proton transfer occurs in the transition state of the cleavage reaction.[71] General acid–base catalysis seems to be a common theme emerging in the nucleolytic ribozymes. As discussed above, considerable crystallographic and chemical evidence supports the role of G8 and A38 as acting as the general base and acid, respectively, in the cleavage reaction of the hairpin ribozyme, and recent crystallographic[72] and chemical[73] data on the extended form of the hammerhead ribozyme suggest a similar role for a guanine base (G12) in that system (Chapters 3 and 4). General acid–base catalysis also seems to be very important in the hepatitis delta virus ribozyme, but the participants are different (Chapter 6). A cytosine base (C75) is juxtaposed with the scissile phosphate;[74,75] substitution of this base leads to a marked loss of activity that can be partially restored by exogenous imidazole.[38] The second participant appears to be

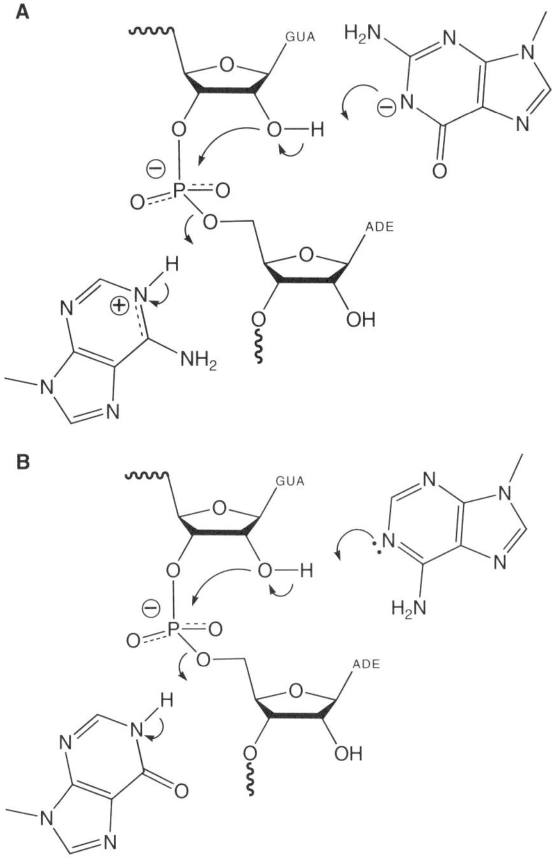

Figure 5.13 Possible chemical mechanisms of the VS ribozyme. Catalytic mechanisms based on the participation of G638 and A756 in general acid–base catalysis. In (A) the guanine acts as the general base, deprotonating the 2′-OH, while the adenine protonates the 5′-oxyanion leaving group. In (B) the roles of the two nucleobases are reversed. Either mechanism would account for the available data on the pH dependence, and we cannot distinguish between them at this time.

hydrated metal ion in this ribozyme.[76] Labilization of the leaving group by phosphorothiolate substitution indicates that the cytosine nucleobase is the general acid in the cleavage reaction.[77]

Figure 5.13 shows chemical mechanisms consistent with the combined roles of G638 and A756.[69] Present data do not allow us to assign the roles of acid and base between the two nucleobases. Nevertheless, these mechanisms require the ribozyme to be in the correct state of protonation to be active, and this will be reflected by the pH dependence of the observed rate of cleavage or ligation. An early study of the cleavage reaction showed very little pH dependence between pH 6 and 9.[45] However, a later study of the pH dependence of the ligation reaction in trans found that the rate increased at low pH in a log–linear manner until reaching a plateau around neutrality. These data fitted a pK_a of 5.6, and would be consistent with the participation of an adenine (with an elevated pK_a) and a guanine (with a pK_a > 9.5). We reinvestigated the pH dependence of the cleavage reaction in the presence of a high concentration of Mg^{2+} ions, and obtained a bell-shaped pH dependence (Figure 5.14), which was fitted to a double-ionization model with pK_a values of 5.2 and 8.4.[69] The upper value would be consistent with a guanine base if the pK_a were reduced by proximity to metal ions. NMR chemical shift mapping has suggested that Mg^{2+} ion binding occurs in the isolated substrate loop.[59]

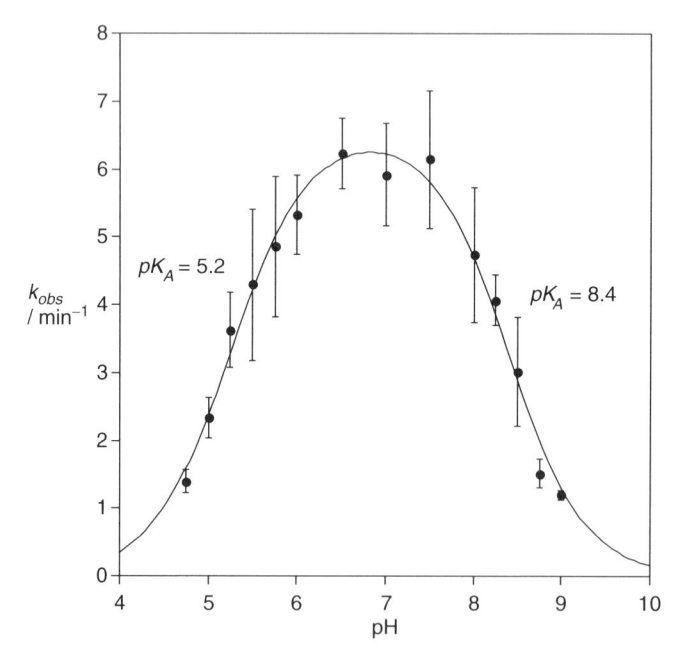

Figure 5.14 pH dependence of the VS ribozyme in the presence of a high concentration of Mg^{2+} ions. The cleavage rates in trans exhibit a bell-shaped pH dependence. The data are well fitted by the general acid–base catalysis model (R = 0.991), giving pK_a values of 5.2 and 8.4, and an intrinsic rate of $6.6\,min^{-1}$.

The Hairpin and Varkud Satellite Ribozymes 87

Jones and Strobel[64] found that impairment of ligation by introduction of either purine or 8-aza-adenosine at position 756 was largely reversed by lowering the pH to 5.4, but this position has not yet been systematically investigated. A more detailed analysis of pH effects has been carried out for position 638. The expected pK_a of inosine is ~0.5 unit lower than that of guanine; the G638I substrate gave a bell-shaped dependence of cleavage rate on pH, with an upper pK_a reduced by 0.3 units. The normal pK_a of 2,6-diaminopurine (DAP) is 5.1. The rate of cleavage of a G638DAP substrate was observed to be lowered over 1000-fold at pH 8, but significantly restored at

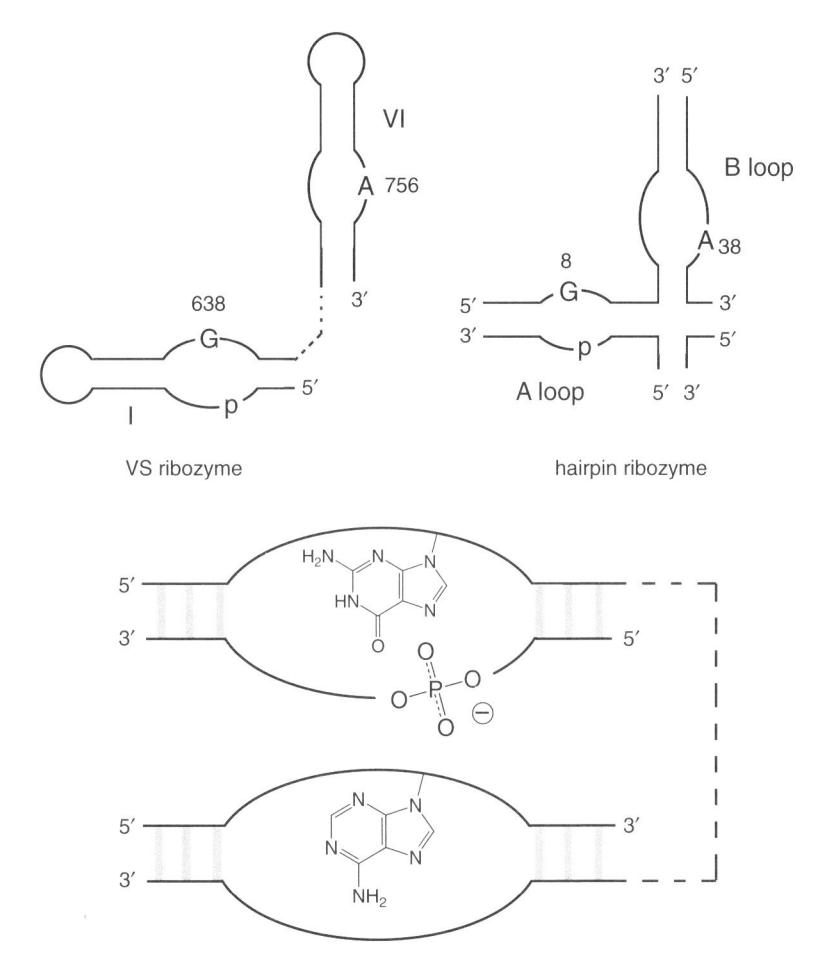

Figure 5.15 Comparison of hairpin and VS ribozymes. In the upper section, the ribozymes have been drawn oriented so as to underline the similar topology of the loops and their active components. This is further illustrated below, where both ribozymes can be summarized using the same scheme. The active geometry of both is generated by loop–loop interaction, and the topological organization of scissile phosphate and the putative catalytic nucleobases is identical in both ribozymes.

lower pH, corresponding to a pK_a of 5.6. The rate for this species was found to be log–linear with pH over the range 6–8, with a unit gradient. This indicates that general acid–base catalysis contributes more than two orders of magnitude to the catalytic power of the VS ribozyme.

5.4 Some Striking Similarities between the Hairpin and VS Ribozymes

Putting the conclusions for the two ribozymes together brings out some striking similarities. While there is nothing structurally in common discernible, each generates the active form by an intimate association between two internal loops. In both ribozymes the scissile phosphate is located on the opposing strand within the same loop as the active guanine (the A loop of the hairpin and the substrate loop of the VS ribozymes), while the adenine is a component of the other loop (the B loop and A730 loop of the hairpin and VS ribozymes, respectively). If the polarity of the strands is included, the topology of the two ribozymes is seen to be identical (Figure 5.15). Furthermore, the relative positioning of the scissile phosphate and guanine is similar in both ribozymes.

The likely topological identity of the hairpin and VS ribozymes is probably the result of convergent evolution, and has perhaps arisen because there are relatively few ways to build an active site for nucleolysis in RNA. While these ribozymes seem to have adopted very similar solutions to the problem, the HDV ribozyme uses alternative functionalities to carry out otherwise similar general acid–base catalysis. The *glmS* ribozyme (Chapter 8) appears to have adopted a yet different solution, using a glucosamine-6-phosphate cofactor in its catalytic chemistry.[78,79] Nevertheless, the probable similarities between the different ribozymes suggest common mechanisms to achieve their function.

Acknowledgements

The author thanks his co-workers and collaborators Timothy Wilson, Taekjip Ha (UIUC), Michelle Nahas (UIUC), Shinya Harusawa (Osaka), Daniel Lafontaine, David Norman, Aileen McLeod, Jonathan Ouellet, Jan Lipfert (Stanford) and Sebastian Doniach (Stanford) for their major experimental and intellectual contributions to the study of the hairpin and VS ribozymes, and Cancer Research UK and the BBSRC for financial support.

References

1. M.D. Canny, F.M. Jucker, E. Kellogg, A. Khvorova, S.D. Jayasena and A. Pardi, *J. Am. Chem. Soc.*, 2004, **126**, 10848–10849.
2. R. Zamel, A. Poon, D. Jaikaran, A. Andersen, J. Olive, D. De Abreu and R.A. Collins, *Proc. Natl. Acad. Sci. U.S.A.*, 2004, **101**, 1467–1472.

3. A.I.H. Murchie, J.B. Thomson, F. Walter and D.M.J. Lilley, *Mol. Cell*, 1998, **1**, 873–881.
4. N.G. Walter, J.M. Burke and D.P. Millar, *Nat. Struct. Biol.*, 1999, **6**, 544–549.
5. A. Berzal-Herranz, J. Simpson, B.M. Chowrira, S.E. Butcher and J.M. Burke, *EMBO J.*, 1993, **12**, 2567–2574.
6. Z-Y. Zhao, T.J. Wilson, K. Maxwell and D.M.J. Lilley, *RNA*, 2000, **6**, 1833–1846.
7. T.J. Wilson and D.M.J. Lilley, *RNA*, 2002, **8**, 587–600.
8. M.J. Fedor, *Biochemistry*, 1999, **38**, 11040–11050.
9. P.B. Rupert and A.R. Ferré-D'Amaré, *Nature*, 2001, **410**, 780–786.
10. Z.P. Cai and I. Tinoco, *Biochemistry*, 1996, **35**, 6026–6036.
11. S.E. Butcher, F.H. Allain and J. Feigon, *Nat. Struct. Biol.*, 1999, **6**, 212–216.
12. D.J. Earnshaw, B. Masquida, S. Müller, S.T. Sigurdsson, F. Eckstein, E. Westhof and M.J. Gait, *J. Mol. Biol.*, 1997, **274**, 197–212.
13. S.P. Ryder and S.A. Strobel, *J. Mol. Biol.*, 1999, **291**, 295–311.
14. R. Pinard, D. Lambert, N.G. Walter, J.E. Heckman, F. Major and J.M. Burke, *Biochemistry*, 1999, **38**, 16035–16039.
15. F. Walter, A.I.H. Murchie, J.B. Thomson and D.M.J. Lilley, *Biochemistry*, 1998, **37**, 14195–14203.
16. T.J. Wilson, Z-Y. Zhao, K. Maxwell, L. Kontogiannis and D.M.J. Lilley, *Biochemistry*, 2001, **40**, 2291–2302.
17. F. Walter, A.I.H. Murchie and D.M.J. Lilley, *Biochemistry*, 1998, **37**, 17629–17636.
18. S. Hohng, T.J. Wilson, E. Tan, R.M. Clegg, D.M.J. Lilley and T. Ha, *J. Mol. Biol.*, 2004, **336**, 69–79.
19. T.A. Goody, D.M.J. Lilley and D.G. Norman, *J. Am. Chem. Soc.*, 2004, **126**, 4126–4127.
20. X.W. Zhuang, H.D. Kim, M.J.B. Pereira, H.P. Babcock, N.G. Walter and S. Chu, *Science*, 2002, **296**, 1473–1476.
21. E. Tan, T.J. Wilson, M.K. Nahas, R.M. Clegg, D.M.J. Lilley and T. Ha, *Proc. Natl. Acad. Sci. U.S.A.*, 2003, **100**, 9308–9313.
22. B. Okumus, T.J. Wilson, D.M.J. Lilley and T. Ha, *Biophys. J.*, 2004, **87**, 2798–2806.
23. A. Khvorova, A. Lescoute, E. Westhof and S.D. Jayasena, *Nat. Struct. Biol.*, 2003, **10**, 1–5.
24. J.C. Penedo, T.J. Wilson, S.D. Jayasena, A. Khvorova and D.M.J. Lilley, *RNA*, 2004, **10**, 880–888.
25. M.K. Nahas, T.J. Wilson, S. Hohng, K. Jarvie, D.M.J. Lilley and T. Ha, *Nat. Struct. Molec. Biol.*, 2004, **11**, 1107–1113.
26. G.A. Soukup and R.R. Breaker, *RNA*, 1999, **5**, 1308–1325.
27. J.B. Murray, A.A. Seyhan, N.G. Walter, J.M. Burke and W.G. Scott, *Chem. Biol.*, 1998, **5**, 587–595.
28. A. Hampel and J.A. Cowan, *Chem. Biol.*, 1997, **4**, 513–517.
29. S. Nesbitt, L.A. Hegg and M.J. Fedor, *Chem. Biol.*, 1997, **4**, 619–630.

30. K.J. Young, F. Gill and J.A. Grasby, *Nucleic Acids Res.*, 1997, **25**, 3760–3766.
31. R. Pinard, K.J. Hampel, J.E. Heckman, D. Lambert, P.A. Chan, F. Major and J.M. Burke, *EMBO J.*, 2001, **20**, 6434–6442.
32. Y.I. Kuzmin, C.P. Da Costa and M.J. Fedor, *J. Mol. Biol.*, 2004, **340**, 233–251.
33. J.M. Thomas and D.M. Perrin, *J. Am. Chem. Soc.*, 2006, **128**, 16540–16545.
34. P.B. Rupert, A.P. Massey, S.T. Sigurdsson and A.R. Ferré-D'Amaré, *Science*, 2002, **298**, 1421–1424.
35. Y.I. Kuzmin, C.P. Da Costa, J.W. Cottrell and M.J. Fedor, *J. Mol. Biol.*, 2005, **349**, 989–1010.
36. S.P. Ryder, A.K. Oyelere, J.L. Padilla, D. Klostermeier, D.P. Millar and S.A. Strobel, *RNA*, 2001, **7**, 1454–1463.
37. P.C. Bevilacqua, *Biochemistry*, 2003, **42**, 2259–2265.
38. A.T. Perrotta, I. Shih and M.D. Been, *Science*, 1999, **286**, 123–126.
39. L.L. Lebruska, I. Kuzmine and M.J. Fedor, *Chem. Biol.*, 2002, **9**, 465–473.
40. L. Araki, S. Harusawa, M. Yamaguchi, S. Yonezawa, N. Taniguchi, D.M.J. Lilley, Z. Zhao and T. Kurihara, *Tetrahedron Lett.*, 2004, **45**, 2657–2661.
41. T.J. Wilson, J. Ouellet, Z.Y. Zhao, S. Harusawa, L. Araki, T. Kurihara and D.M. Lilley, *RNA*, 2006, **12**, 980–987.
42. B.J. Saville and R.A. Collins, *Cell*, 1990, **61**, 685–696.
43. J.C. Kennell, B.J. Saville, S. Mohr, M.T. Kuiper, J.R. Sabourin, R.A. Collins and A.M. Lambowitz, *Genes Dev.*, 1995, **9**, 294–303.
44. T.L. Beattie, J.E. Olive and R.A. Collins, *Proc. Natl. Acad. Sci. U.S.A.*, 1995, **92**, 4686–4690.
45. H.C.T. Guo and R.A. Collins, *EMBO J.*, 1995, **14**, 368–376.
46. T. Rastogi, T.L. Beattie, J.E. Olive and R.A. Collins, *EMBO J.*, 1996, **15**, 2820–2825.
47. A.A. Andersen and R.A. Collins, *Mol. Cell*, 2000, **5**, 469–478.
48. D.A. Lafontaine, T.J. Wilson, D.G. Norman and D.M.J. Lilley, *J. Mol. Biol.*, 2001, **312**, 663–674.
49. A.C. McLeod and D.M.J. Lilley, *Biochemistry*, 2004, **43**, 1118–1125.
50. D.A. Lafontaine, D.G. Norman and D.M.J. Lilley, *EMBO J.*, 2002, **21**, 2461–2471.
51. D.A. Lafontaine, D.G. Norman and D.M.J. Lilley, *EMBO J.*, 2001, **20**, 1415–1424.
52. V.D. Sood and R.A. Collins, *J. Mol. Biol.*, 2001, **313**, 1013–1019.
53. R.R. Gutell, S. Subashchandran, M. Schnare, Y. Du, N. Lin, L. Madabusi, K. Muller, N. Pande, N. Yu, Z. Shang, S. Date, D. Konings, V. Schweiker, B. Weiser and J.J. Cannone, 2000, http://www.rna.icmb.utexas.edu/.
54. P.J.A. Michiels, C.H.J. Schouten, C.W. Hilbers and H.A. Heus, *RNA*, 2000, **6**, 1821–1832.
55. J. Flinders and T. Dieckmann, *J. Mol. Biol.*, 2001, **308**, 665–679.
56. H.W. Pley, K.M. Flaherty and D.B. McKay, *Nature*, 1994, **372**, 68–74.

57. W.G. Scott, J.T. Finch and A. Klug, *Cell*, 1995, **81**, 991–1002.
58. A.A. Andersen and R.A. Collins, *Proc. Natl. Acad. Sci. U.S.A.*, 2001, **98**, 7730–7735.
59. B. Hoffmann, G.T. Mitchell, P. Gendron, F. Major, A.A. Andersen, R.A. Collins and P. Legault, *Proc. Natl. Acad. Sci. U.S.A.*, 2003, **100**, 7003–7008.
60. T. Rastogi and R.A. Collins, *J. Mol. Biol.*, 1998, **277**, 215–224.
61. S.L. Hiley and R.A. Collins, *EMBO J.*, 2001, **20**, 5461–5469.
62. V.D. Sood, S. Yekta and R.A. Collins, *Nucleic Acids Res.*, 2002, **30**, 1132–1138.
63. S.P. Ryder and S.A. Strobel, *Methods*, 1999, **18**, 38–50.
64. F.D. Jones and S.A. Strobel, *Biochemistry*, 2003, **42**, 4265–4276.
65. V.D. Sood and R.A. Collins, *J. Mol. Biol.*, 2002, **320**, 443–454.
66. S.L. Hiley, V.D. Sood, J. Fan and R.A. Collins, *EMBO J.*, 2002, **21**, 4691–4698.
67. D.A. Lafontaine, T.J. Wilson, Z-Y. Zhao and D.M.J. Lilley, *J. Mol. Biol.*, 2002, **323**, 23–34.
68. Z. Zhao, A. McLeod, S. Harusawa, L. Araki, M. Yamaguchi, T. Kurihara and D.M.J. Lilley, *J. Am. Chem. Soc.*, 2005, **127**, 5026–5027.
69. T.J. Wilson, A.C. McLeod and D.M.J. Lilley, *EMBO J.*, 2007, **26**, 2489–2500.
70. A.B. Tzokov, I.A. Murray and J.A. Grasby, *J. Mol. Biol.*, 2002, **324**, 215–226.
71. M.D. Smith and R.A. Collins, *Proc. Natl. Acad. Sci. U.S.A.*, 2007, **104**, 5818–5823.
72. M. Martick and W.G. Scott, *Cell*, 2006, **126**, 309–320.
73. J. Han and J.M. Burke, *Biochemistry*, 2005, **44**, 7864–7870.
74. A.R. Ferré-d'Amaré, K. Zhou and J.A. Doudna, *Nature*, 1998, **395**, 567–574.
75. A. Ke, K. Zhou, F. Ding, J.H. Cate and J.A. Doudna, *Nature*, 2004, **429**, 201–205.
76. S. Nakano, D.M. Chadalavada and P.C. Bevilacqua, *Science*, 2000, **287**, 1493–1497.
77. S.R. Das and J.A. Piccirilli, *Nat. Chem. Biol.*, 2005, **1**, 45–52.
78. T.J. McCarthy, M.A. Plog, S.A. Floy, J.A. Jansen, J.K. Soukup and G.A. Soukup, *Chem. Biol.*, 2005, **12**, 1221–1226.
79. D.J. Klein and A.R. Ferré-D'Amaré, *Science*, 2006, **313**, 1752–1756.
80. D.I. Svergun, *Biophys. J.*, 1999, **76**, 2879–2886.

CHAPTER 6
Catalytic Mechanism of the HDV Ribozyme

SELENE KOO,[a,c] THADDEUS NOVAK[a,c] AND
JOSEPH A. PICCIRILLI[a,b]

[a] Department of Biochemistry and Molecular Biology, The University of Chicago, 929 East 57th Street, Chicago, Illinois 60637, USA; [b] Howard Hughes Medical Institute, Department of Chemistry, The University of Chicago, 929 East 57th Street, Chicago, Illinois 60637, USA; [c] These authors contributed equally to this work

6.1 Introduction

In the 25 years since the discovery of RNA-based enzymes (ribozymes), the naturally occurring ribozymes have been found to fall primarily into two classes: large ribozymes that function to splice together flanking sequences, and small ribozymes that cleave a conserved site within their own sequence. Obvious exceptions exist, including such trans-acting ribozymes as RNase P and the ribosome. The prototype of the large ribozymes is the *Tetrahymena* group I intron, the first ribozyme discovered, while the small ribozymes include the hammerhead, hairpin, Varkud satellite, and hepatitis delta virus (HDV) ribozymes. In this chapter, we discuss the HDV ribozyme, and the extensive body of structural and biochemical experiments that have attempted to elucidate its catalytic mechanism.

6.1.1 Hepatitis Delta Virus Biology

The hepatitis delta virus (HDV) is a satellite virus of the human pathogen hepatitis B; coinfection with HDV exacerbates the symptoms of hepatitis B.[1] HDV contains a circular single-stranded RNA genome; because much of one half of the genome is complementary to the other half, this RNA can base-pair intramolecularly to form an unbranched rod-like structure. This genome encodes a protein known as hepatitis delta antigen and contains the sequence for a ribozyme essential for viral replication. Genome replication occurs by a rolling-circle mechanism, believed to be largely or entirely performed by host cell RNA polymerase II.[1,2] Transcription off the genomic RNA released from the invading viral particles results in the synthesis of linear multimeric antigenomic RNA. The encoded antigenomic HDV ribozyme site-specifically cleaves the multimeric RNA into unit-length linear products, some of which are circularized by ligation. The circles of antigenomic RNA are themselves the templates for synthesis of linear multimeric genomic RNA, which is cleaved by the genomic ribozyme into unit-length products that are ligated and packaged into new viral particles. The genomic and antigenomic ribozymes necessary for viral genome replication are closely related because of the high degree of self-complementarity in the viral genome and share similar secondary sequences[3] (Figure 6.1). Ribozyme activity requires about 85 nucleotides 3′ of the cleavage site and only one nucleotide 5′ of the cleavage site; sequences further downstream or upstream of the minimal ribozyme sequence may be involved in modulating ribozyme activity.[3–7]

Recently a sequence that shares many structural and biochemical features with the HDV ribozyme was isolated *via in vitro* selection of self-cleaving RNA from the human genome[8] (Figure 6.1C). This sequence appears to be conserved in mammals and absent from non-mammal vertebrates. Currently, only the HDV ribozyme itself and this newly-isolated sequence, found within an intron in CPEB3 (cytoplasmic polyadenylation element-binding protein 3) mRNA, are known to adopt the HDV-like secondary structure (Chapter 7), in contrast to the much more sequence-diverse families of hammerhead and hairpin ribozymes. Because of the apparent structural similarity of the two ribozymes and numerous suggestive similarities in their biochemistry, this result provides for the interesting possibility that the HDV ribozyme and the virus itself originally arose from the human genome.[8]

6.1.2 Cleavage Reactions of Small Ribozymes

Like other small ribozymes, both the genomic and antigenomic forms of the HDV ribozyme generate products containing 2′,3′-cyclic phosphate and 5′-OH termini.[9,10] These products imply a cleavage mechanism with attack of a 2′-hydroxyl group on the adjacent phosphate, breaking the P-5′–O bond (Figure 6.2). Studies on the hammerhead[11] and hairpin[12] ribozymes have demonstrated inversion of configuration at the scissile phosphate in the cleavage product, suggesting an in-line, S_N2-type reaction; these results have been assumed to apply to the HDV ribozyme as well. One notable difference between HDV and other small ribozymes is that the HDV ribozyme has never been demonstrated to catalyze ligation (reverse cleavage). Although the reason for this apparent inability remains unclear, several explanations have been proposed. The absence of base pairing 5′ to the cleavage site may prevent the 5′ cleavage product from binding productively to the enzyme, or a proposed conformational

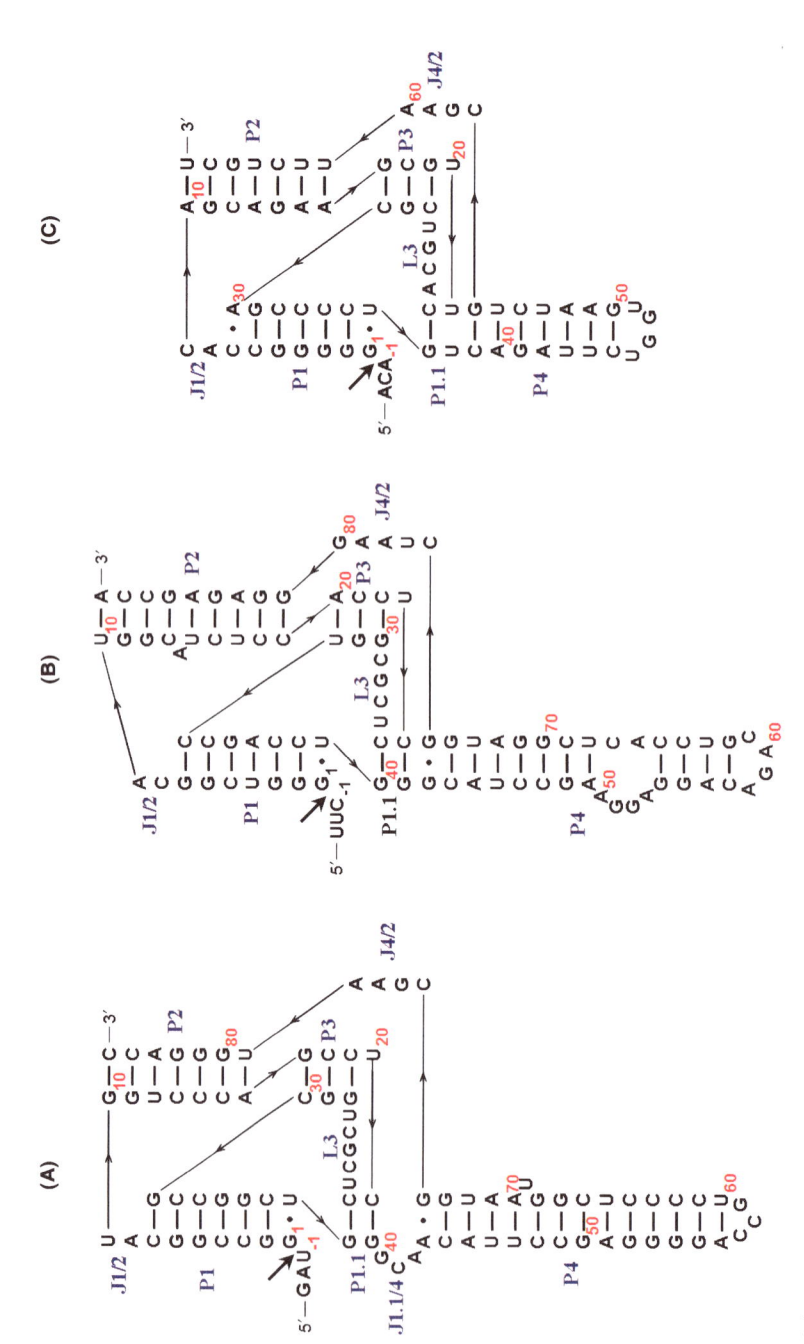

Figure 6.1 Secondary structure of HDV and HDV-like ribozymes. Numbering is 5′ to 3′, with nucleotide 1 immediately downstream of the cleavage site. Secondary structure elements are labeled in blue; the cleavage site is indicated by a bold arrow. (*A*) Genomic HDV ribozyme; (*B*) antigenomic HDV ribozyme; (*C*) HDV-like ribozyme isolated from the CPEB3 intron.[8]

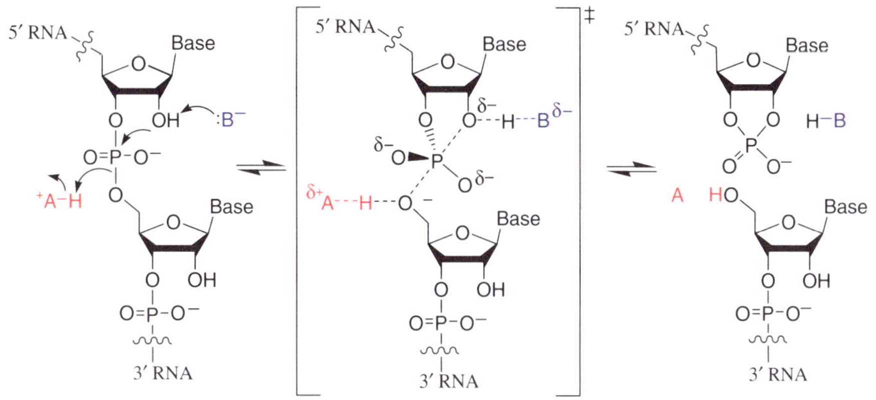

Figure 6.2 Cleavage reaction catalyzed by the HDV ribozyme. The reaction may proceed in a stepwise fashion *via* a phosphorane intermediate, or in a concerted fashion *via* a single transition state as shown. General acids (A, red) and bases (B, blue) or specific acids and bases catalyze the reaction *via* proton transfer to activate the leaving group and nucleophile, respectively. The non-bridging oxygens also may engage in interactions (not shown) that stabilize the phosphorane transition state.

change during catalysis may disrupt the active site[13,14] (see Section 6.2.3 for further discussion).

6.2 HDV Ribozyme Structure

6.2.1 Determination of Crystal Structures

A pivotal contribution to characterization of the HDV ribozyme came from the crystallization and structural determination of the genomic ribozyme.[13,15] This breakthrough was facilitated by a novel crys"fig2"tallization strategy featuring installation of the U1A RNA stem-loop into the functionally irrelevant P4 stem and co-crystallization with the U1A RNA-binding protein.[16] Initially, a structure of the ribozyme after it had undergone self-cleavage, terminating in a 5′-hydroxyl group and termed the "product" structure, was obtained[15] (Figure 6.3B). This structure supported several predictions from previous modeling and biochemical studies, but also revealed the existence of a two-base-pair stem, P1.1, whose importance was subsequently confirmed biochemically.[17,18] It also revealed that the functionally important C75 nucleobase was positioned within hydrogen-bonding distance of the 5′-oxygen at the cleavage site, suggesting that this nucleobase might participate directly in catalysis. This possibility initiated a paradigm shift in the ribozyme field, as nucleobase-mediated catalysis had previously been dismissed as implausible due to both the minimal variety of nucleobase functional groups and their apparently unfavorable pK_a values.

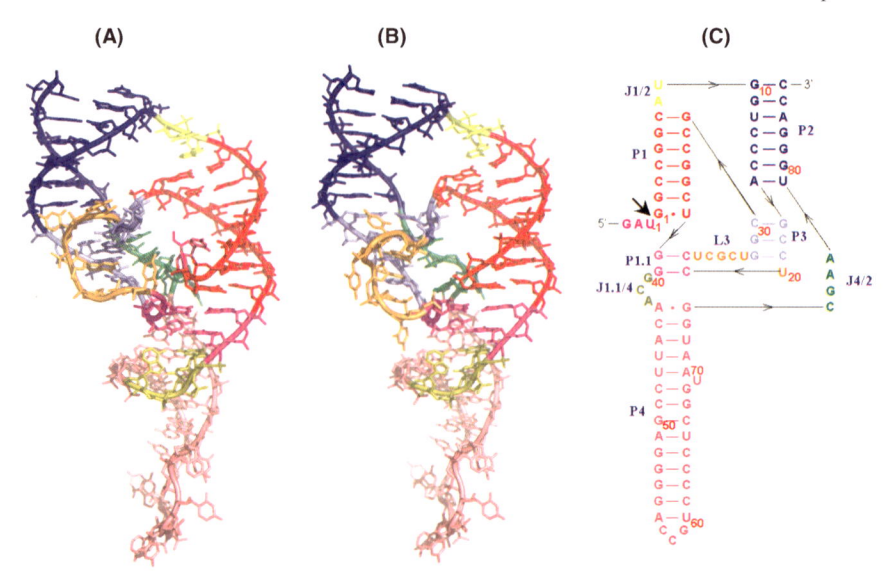

Figure 6.3 Global view of crystal structures obtained for the (A) precursor[13] (PDB code 1SJ3) and (B) product[15] (PDB code 1CX0) forms of the genomic HDV ribozyme. Secondary structure elements are colored according to the secondary structure diagram in (C); the cleavage site is indicated by a bold arrow. This figure (and all subsequent figures derived from crystal structures) was made using Pymol molecular graphics software (http://pymol.sourceforge.net/).

Subsequently, several crystal structures of the genomic ribozyme prior to cleavage, termed "precursor" structures, were determined[13] (Figure 6.3A). Ke et al. employed various strategies to inactivate the ribozyme, including sequestration of divalent metal ions with EDTA and mutation of the catalytically essential C75 to a functionally inactive U. The product and precursor crystal structures adopt similar global folds but exhibit several key differences near the active site (see Section 6.2.3).

Extrapolating from either the precursor or product structures to the structure of the transition state, when the catalytic strategies of the ribozyme truly become evident, is not straightforward, as both forms are ground state structures. Because the product structure does not contain the scissile phosphate, accommodation of this group and of the remainder of the substrate 5' of the leaving group is an important consideration when attempting to obtain a view of the transition state based on the product structure. To obtain the precursor structure, various strategies were undertaken to inactivate the ribozyme, most notably C75U mutations in the highest-resolution precursor structures.[13] This raises the possibility that the precursor structure may represent a conformation of the ribozyme that is incapable of proceeding through the transition state to the product because key functionalities are positioned incorrectly. In that case, the precursor structures may not provide much verification for catalytic strategies the ribozyme employs.

6.2.2 Structure Overview

The structure of the HDV ribozyme consists primarily of five helical segments arranged into two parallel stacks[15] (Figure 6.3). Helices P1, P1.1, and P4 form one stack, while the other consists of P2 and P3. Topologically, these segments form a double-nested pseudoknot, with the two base pairs of P1.1 forming the second pseudoknot and stitching together the two helical stacks.

This complex fold sequesters the scissile phosphate in a buried cleft resembling the active-site pockets of protein enzymes. This fold is sufficiently compact to protect residues in the core from hydroxy-radical cleavage,[19] suggesting that it serves to create a defined environment in the active site of the ribozyme. Further highlighting the importance of the global fold to activity, mutations that disrupt either of the two pseudoknots reduce cleavage activity by one to five orders of magnitude.[20–22]

6.2.3 Active Site

The active site of the ribozyme (Figure 6.4) is surrounded by the following structural elements (for this discussion, the active site cleft faces the viewer, and

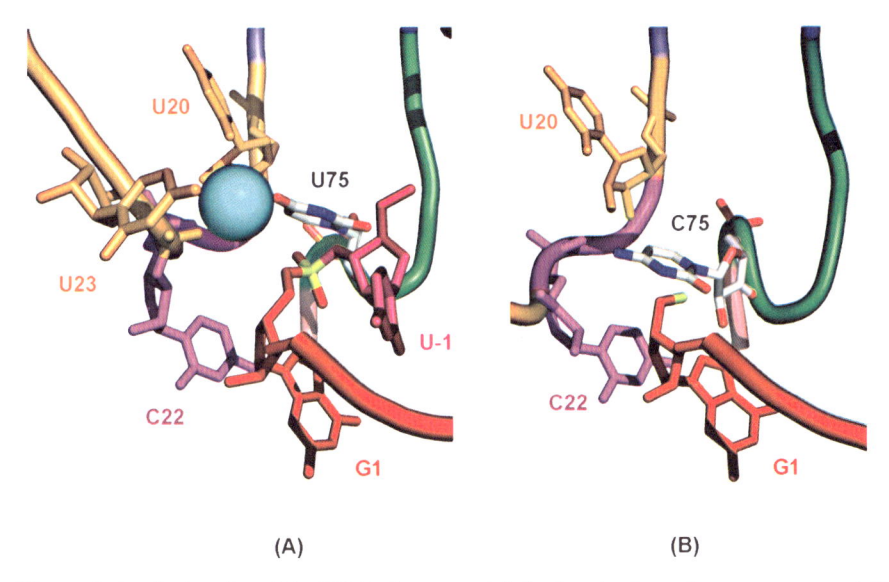

Figure 6.4 Active site of the HDV ribozyme. (*A*) Active site from the precursor crystal structure. U75, which replaces C75 in this structure, is shown as sticks colored by element (white = C, blue = N, red = O, orange = P), while other bases in the active-site region are shown as sticks colored as in Figure 6.3. The cyan sphere shows the location of the active site metal ion. The scissile phosphate is highlighted in yellow. (*B*) Active site from the product crystal structure. C75 is shown as sticks colored by element as in (*A*). The 5′-OH, highlighted in yellow, marks the site where the scissile phosphate had been.

P4 points downward): above, by the strand crossover from P3 to P1; below, by the two-base-pair helix P1.1, which stacks on P1; behind, by J4/2, which contains the catalytically-important C75; and on the sides, by P3 and the substrate-bearing helix P1. A sharp, nearly 180° turn in the backbone 5′ of the cleavage site is observed in the precursor structures[13] (Figure 6.5). J4/2 contains a trefoil turn that positions the base C75 in the active site (Figure 6.6). C75 stacks with A77, and, in the product structure, N4 of C75 forms potential hydrogen bonding interactions with the *pro*-R_p oxygen of C22 and the 2′-hydroxyl group of U20; neither of those groups resides within hydrogen bonding distance of the O4 of U75 in the C75U mutant precursor structure. Although divalent metal ions are required for maximal ribozyme activity,[10,23–26] no evidence for a divalent metal ion at or near the active site was found in the product structure. However, several of the inactive precursor structures indicated the presence of a divalent metal ion outer-sphere coordinated to various functional groups in the ribozyme, including the 5′-phosphate of U-1 and the 5′-oxygen leaving group of G1. Comparison of the product and precursor crystal structures led to the hypothesis that the active site collapses upon cleavage, resulting in dissociation of the active site metal ion.[13] This active site collapse was proposed to explain the irreversibility of the cleavage reaction and the absence of an observed ligation reaction.

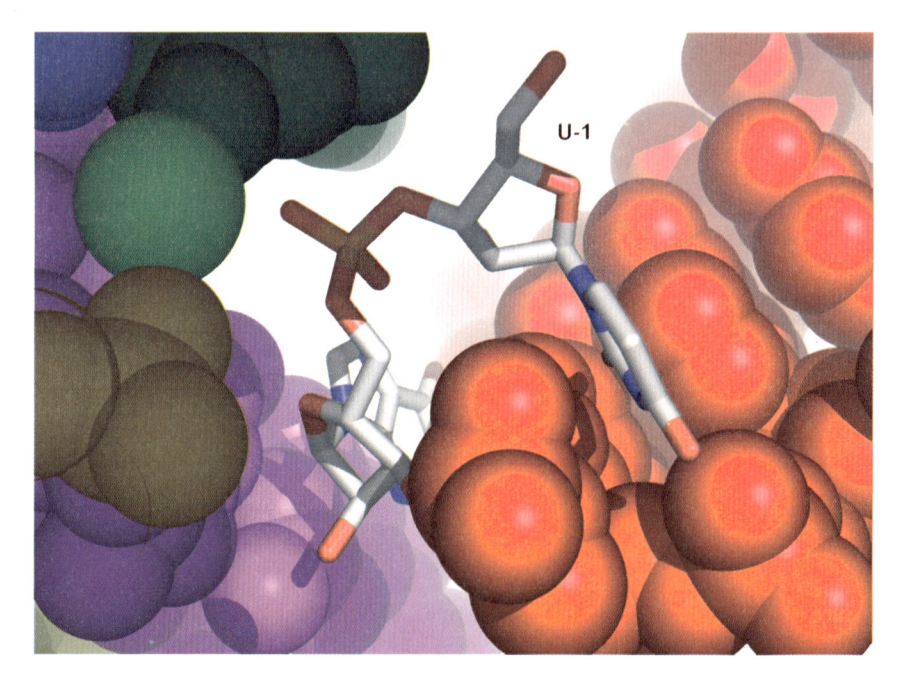

Figure 6.5 The RNA backbone around the scissile phosphate adopts an unusual structure in the precursor. U-1, the scissile phosphate and G1 are shown in stick form and colored by element as in Figure 6.4. The surrounding elements of the ribozyme are shown in space-filling form and colored as in Figure 6.3.

Catalytic Mechanism of the HDV Ribozyme 99

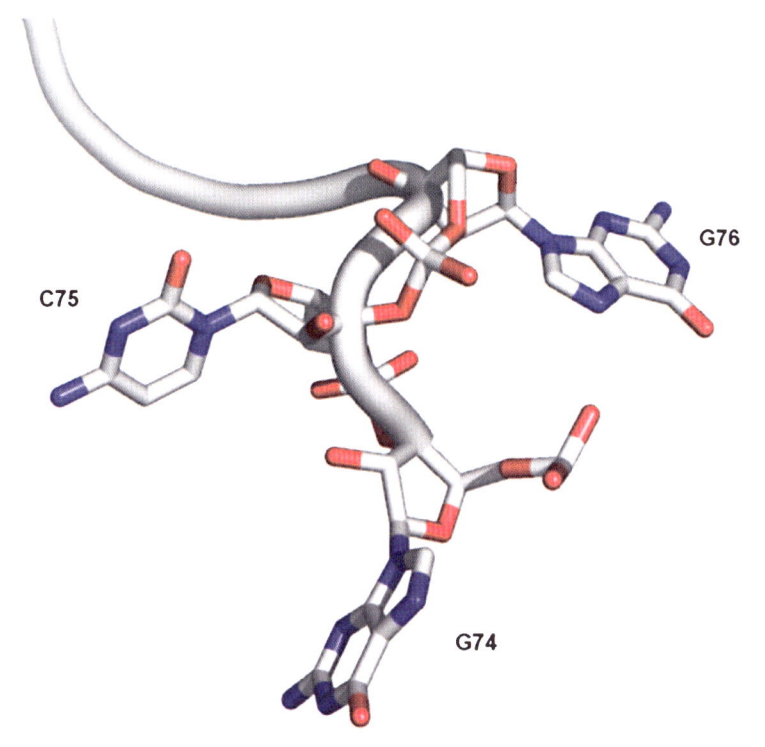

Figure 6.6 Trefoil turn in J4/2, as seen in the product crystal structure. The unusual backbone conformation in this region pushes C75 into the active site while exposing G76 to solvent. The bases are shown in stick form, colored by element as in Figure 6.4.

6.3 Catalytic Strategies for RNA Cleavage

Non-enzymatic reactions of RNA have been studied extensively; these analyses illustrate the considerable complexity and ambiguity associated with the mechanism of RNA cleavage.[27,28] Nevertheless, they allow us to write the book on "How to Catalyze RNA Cleavage." Mechanistic investigations of small ribozymes generally attempt to define one or more of the following catalytic strategies (see Figure 6.2 for transition state).

Strategy 1: The reactants (the 2'-OH and the adjacent phosphate) must adopt the proper alignment for reaction. Of course, intrinsic geometric restrictions depend upon whether the reaction proceeds through a concerted (single-step) mechanism or a two-step mechanism that involves a phosphorane intermediate. Assuming that an enzyme active site might constrain a phosphorane intermediate from pseudorotation, the nucleophile would have to attack in-line with the P-5'-O bond regardless of whether the reaction occurs by a stepwise or concerted pathway.[29]

Strategy 2: Because a 2'-oxyanion has dramatically greater nucleophilic power than the neutral 2'-hydroxyl group,[30,31] proton transfer from the 2'-hydroxyl group either before (specific base) or concomitant with (general base) nucleophilic attack provides one obvious strategy to catalyze the reaction.

Strategy 3: In the ground state a formal negative charge resonates over the phosphorus atom and its non-bridging oxygen atoms *via* pπ-dπ orbital overlap. Bond formation between the nucleophile and the phosphorus in the transition state disrupts this orbital overlap, causing electron density to build up on the non-bridging oxygens. Reflecting this increase in electron density, the phosphoryl group ionizes with pK_a ~1 in the ground state, whereas a dianionic phosphorane ionizes with an estimated pK_a >11.[28] It follows that hydrogen bond donation, proton donation, metal ion coordination, or electrostatic interactions with a positively charged group (either before or during nucleophilic attack) could provide catalysis by stabilizing the charge build-up on the non-bridging oxygens in the transition state. The 10^5-fold faster reaction of ribonucleoside 3'-alkyl phosphotriesters[32,33] illustrates the extent to which neutralization of the non-bridging phosphoryl oxygens can enhance the reaction rate.

Strategy 4: Polar substituent effects reveal that significant negative charge also builds up on the leaving group during the reaction. Under alkaline conditions RNA cleavage exhibits a dramatic effect of leaving group pK_a [β_{lg} = –1.28 (ref. 27)], indicating a rate-limiting transition state in which the leaving group bears considerable negative charge and experiences considerable bond scission. Analogous to catalytic interactions with the non-bridging oxygen atoms, strategies that stabilize the build-up of negative charge on the leaving group could provide catalysis. Attenuation of β_{lg} by acid catalysis and divalent metal ions supports this assertion.

Consistent with these expectations, general acids and bases, metal ions, and appropriately positioned proton donors catalyze the non-enzymatic cleavage of ribonucleoside 3'-phosphodiesters.[27,28] These strategies need not operate exclusively of one another, especially in an enzyme active site, which may engage in multiple catalytic strategies by appropriate positioning of catalytic agents. Moreover, these strategies need not operate simultaneously, especially if the reaction involves a phosphorane intermediate. For example, an acid and base might catalyze the formation of the phosphorane by protonating a non-bridging oxygen and deprotonating the 2'-hydroxyl group, respectively. Analogously, a base and acid might facilitate breakdown of the intermediate by removing the proton from the phosphorane and protonating the leaving group, respectively.

6.4 The Active Site Nucleobase: C75

Early biochemical data suggested that C75 of the genomic ribozyme (γC75, used interchangeably with αC76 of the antigenomic ribozyme) provided a significant contribution to catalysis; in a ribozyme based on the genomic sequence, mutation of C75 to A, G, or U abolished activity,[22] and in an antigenomic trans ribozyme, deoxythiouridine at the -2 position crosslinked to C76.[19] When the X-ray analysis of the genomic product form of the ribozyme revealed that the C75 base protrudes into the active site[15] (Figure 6.4), within hydrogen bonding distance of the 5′-oxygen leaving group, the catalytic relevance of C75 took center stage as the focus of several biochemical studies. More recently, ribozymes selected *in vitro* from a pool of antigenomic ribozymes containing random mutations in L3, J1/4, and J4/2 all had cytosine either at the wild-type position 76 or, in about one-third of the mutants, at the adjacent position 77.[34]

6.4.1 Exogenous Base Rescue Reactions

Several lines of evidence strongly implicate a catalytic role for the C75 nucleobase in proton transfer during the chemical step of the cleavage reaction. A series of studies by Been and co-workers demonstrated that addition of exogenous cytosine and related heterocycles partially rescues the deleterious effect of C76U or C76Δ (Δ = deletion) mutations[35–37] (Figure 6.7A,B). Presumably the added nucleobase supplants the role of the missing C76, raising the intriguing possibility that in the wild-type ribozyme C76 itself participates in the reaction. Significantly, imidazole, the quintessential general acid/base used in model studies and in the active sites of protein enzymes (as histidine), mimics the rescuing function of the exogenous cytosine. The availability of numerous imidazole analogues enabled Been and co-workers to investigate the relationship between the reaction rate and the pK_a of the rescuing imidazole.[37] The reaction rate reflects the free energy change on going from the ground state of free RNA (R) and free imidazole (R + Im) to the rate-limiting transition state [R · Im]‡ so that the Brønsted correlation reflects the sum of pK_a effects on imidazole binding and catalysis. The Brønsted analysis showed that as the pK_a of the imidazole analogue increases, the second-order rate constant for imidazole rescue increases linearly with slope $\beta = 0.5$.[37] In addition, the rescue reaction was observed to have a kinetic solvent isotope effect of 2.7 and a proton inventory suggesting one proton is transferred in the rate-limiting step. Taken together, these observations provide definitive evidence that (1) the rate-limiting step of the imidazole rescue reaction involves proton transfer, (2) the imidazole analogue participates directly in this proton transfer *via* its imino nitrogen, and (3) the chemical step limits the rate of the imidazole rescue reaction.

Extrapolation of these conclusions to the wild-type reaction requires that imidazole mimics the chemical role of the C76 nucleobase. As the imino nitrogen of imidazole is implicated in catalysis, the simplest model presumes that the corresponding imino nitrogen on C76 performs an equivalent catalytic function. The imidazole rescue experiments therefore implicate the imino nitrogen of C76 in proton transfer. How does imidazole-mediated proton

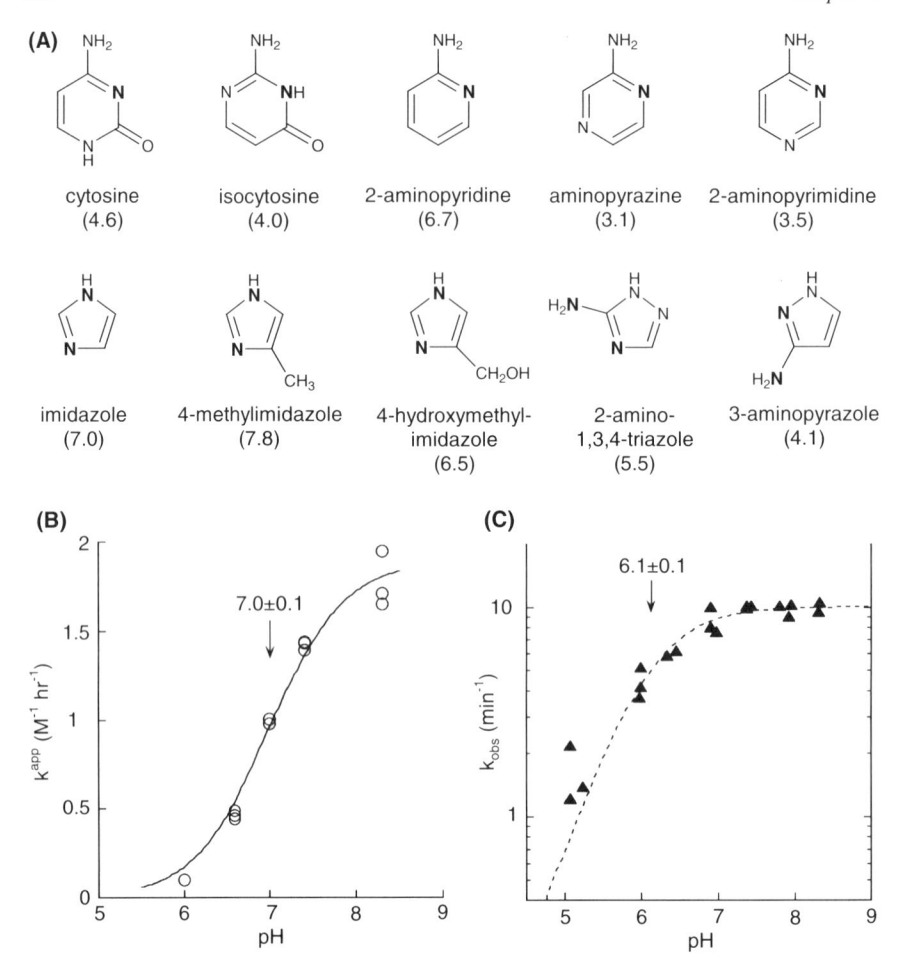

Figure 6.7 pH–rate profiles of the antigenomic HDV ribozyme. (*A*) Cytosine and imidazole analogues used by Been and co-workers to rescue the reaction of the C76Δ ribozyme. (*B*) pH–rate profile for imidazole rescue of the C76U ribozyme. (*C*) pH–rate profile for the wild-type (wt) antigenomic ribozyme. (Panel *A* is adapted with permission from ref. 36; panels *B* and *C* are adapted with permission from ref. 35.)

transfer facilitate cleavage at the scissile phosphodiester bond? Lessons from non-enzymatic reactions elicit multiple possibilities: general acid catalysis, in which the imidazolium ion donates a proton to either the leaving group or a non-bridging oxygen (the points where negative charge builds up) in the transition state, or general base catalysis, in which neutral imidazole accepts a proton from the nucleophilic 2′-hydroxyl group in the transition state. Establishing which mode of catalysis actually governs the imidazole rescue reaction presents significant experimental challenges.

Each of these models for imidazole catalysis can account for the observed pH–rate profile for imidazole rescue of the C76Δ antigenomic ribozyme.

At lower pH values, the second-order rate constant for imidazole catalysis increases log–linearly with pH and then becomes pH-independent at higher pH values[35] (Figure 6.7B). The transition from a pH-dependent rate to a pH-independent rate could reflect a change in rate-limiting step or the actual titration of a functional group important for the reaction. With imidazole analogues, the pH–rate profile of HDV rescue shifts in a manner that matches the pK_a of the imidazole analogue, strongly suggesting that the apparent pK_a reflects titration of the free imidazole and ruling out the possibility that the rate-limiting step changes. Additionally, the kinetic solvent isotope effect remains constant throughout the pH–rate profile (after accounting for the equilibrium isotope effect on the imidazole pK_a),[37] indicating that the same step remains rate limiting throughout the pH dependence. By the same token, had a non-chemical step limited the reaction rate at some point, the proton inventory experiments would likely suggest a non-integral number of protons in flight,[38,39] instead of the single proton actually suggested by these experiments.

In the simplest model that accounts for the data, neutral imidazole acts as a general base catalyst. At pH values below the imidazole pK_a, most of the imidazole populates the ionized imidazolium state. As the pH increases, the concentration of neutral imidazole increases in a log–linear fashion. Consistent with the prediction of the general-base model, the reaction rate tracks with the concentration of free imidazole over the pH range. However, a model involving specific base/general acid catalysis accounts equally well for the observed pH dependence.[36,37] In this model, the imidazolium cation acts as a general acid, and the log–linear increase in reaction rate with pH reflects pre-equilibrium ionization of the 2′-hydroxyl group (specific base catalysis). At pH values below the imidazole pK_a, the concentration of imidazolium ion changes minimally. As the pH increases above the imidazolium pK_a, specific base catalysis continues to contribute to the reaction but the concentration of imidazolium ion starts to decrease in a log–linear manner. The two titrations (the 2′-hydroxyl group and the imidazolium ion) offset each other, and the reaction becomes pH-independent over the basic limb of the pH–rate profile. Thus, the observed pH dependence could arise equally well from titration of a single functional group or two functional groups.

6.4.2 Role of C75 in HDV Catalysis

The precise role of C75 in the wild-type reaction remains a matter of debate. Clearly the crystal structures demonstrate that C75 resides at the active site, but what information in addition to exogenous base rescue supports the hypothesis that C75 plays more than a structural role in catalysis? A Brønsted analysis (analogous to that conducted for the imidazole rescue reaction) that demonstrates a correlation between the pK_a of the nucleoside, the reaction pK_a (from the pH–rate profile), and the reaction rate would provide the most convincing evidence that C75 mediates proton transfer during the reaction.

However, there exists no straightforward series of cytosine analogues that systematically perturb the nucleobase pK_a within a similar structural context. Moreover, many modifications to the active site cytosine have significant deleterious effects on the reaction. Consequently, a complete Brønsted analysis remains out of reach.

Despite this complication, many lines of evidence support the analogy between the role of C76 and that of imidazole in the rescue reaction described in Section 6.4.1. The reaction of the wild-type (WT) antigenomic ribozyme has similar solvent isotope effects, proton inventory, and pH dependence[35,36] (Figure 6.7C) to those found for the imidazole rescue reaction on the C76Δ mutant of the same ribozyme. Given these similarities, it seems reasonable to draw the same conclusions about the WT reaction: that the chemical step limits the reaction rate, that proton transfer occurs during this step, and that titration of a functional group contributes significantly to the reaction rate.

These features of the cis-acting antigenomic wild-type ribozyme alone do not implicate C76 directly as bearing the functional group that undergoes titration in the pH–rate profile, nor do they implicate C76 directly as the mediator of the proton transfer that occurs in the rate-limiting step. However, the observation that imidazole fulfills these roles in the absence of C76 leads to the suggestion that C76, when present, fulfills these functions. Interestingly, the reaction pK_a for the WT reaction (6.1) is almost two units higher than the solution pK_a of cytosine (4.2).[35,36] Thus, the ribozyme would have to induce a substantial shift in the pK_a of cytosine for the pH–rate profile to reflect titration of C76. NMR studies did not detect a shift of such magnitude in the product form of the ribozyme,[40] suggesting that the pK_a perturbation may depend on the presence of the scissile phosphate.

Evidence that the pH–rate profile does reflect titration of C76 comes from substitution of C76 with adenosine.[35] Although the C76A mutation slows the reaction rate by three orders of magnitude, the pH–rate profile for the adenosine mutants follow the same general titration behavior. The difference in reaction pK_a between the wild-type and C76A ribozymes (6.1 vs. 5.6) nearly matches the difference in pK_a between the free nucleosides (4.2 vs. 3.5),[23,35] supporting the possibility that the pH–rate profile reflects titration of the nucleobase at residue 76. How the C76A ribozyme could retain the same nucleobase pK_a perturbation as the wild-type ribozyme despite the mutation's large deleterious effect on catalysis remains unclear. Alternatively, the correspondence in pK_a shift may arise coincidentally. A series of cytosine and/or adenine analogues with varying pK_a would help to address this question directly.

Whereas Brønsted analysis of the imidazole rescue reaction shows that imidazole participates directly in proton transfer, no such conclusion can be drawn definitively for C76. In addition, even if we assume by analogy to imidazole that the function of C76 involves proton transfer, we face the same ambiguity as to whether the neutral or protonated form of the C76 nucleobase is functional. The same arguments discussed in Section 6.4.1 for the general base or general acid models of imidazole rescue hold true for C76, and both models are equally capable of explaining the pH–rate profile for catalysis by the WT ribozyme.

Nucleotide analogue interference mapping with cytosine analogues having altered N3 pK_as (as compared with cytosine's pK_a of 4.2) has provided further evidence that the pH–rate profile for the HDV ribozyme reflects titration of C75. Oyelere et al.[41] mapped the positions in the ribozyme that exhibited pH-dependent interferences from 5-fluorocytidine (pK_a = 2.3), zebularine (pK_a = 2.5), 6-azacytidine (pK_a = 2.6), and pseudoisocytidine (pK_a = 9.4). Analogues with lower pK_as than cytosine interfered more strongly with ribozyme function at low pH than at high pH, which is consistent with the conclusion that the shape of the pH–rate profile reflects titration of C75. However, these results still cannot distinguish whether the neutral or protonated form of C75 contributes to ribozyme function.

The exocyclic N4 amino group of C75 has been proposed to help position the catalytic base and shift the pK_a of C75 N3 toward neutrality. In the crystal structure of the self-cleaved product, this exocyclic amine participates in a hydrogen bonding network with the 2′-hydroxyl group of U20 and the pro-R$_p$ oxygen of C22,[15] but the corresponding 4-keto group of U75 in the C75U precursor crystal structures does not form these interactions.[13] Hydrogen bond donation by the N4 amino group could favor protonation of N3 at a higher pH than observed for protonation in solution. Biochemical work clearly confirms the importance of the exocyclic amine N4 for catalysis, but its role remains obscure. Exogenous addition of cytosine analogues containing an exocyclic amine at N4 restores cleavage activity of a C75Δ genomic ribozyme, while analogues lacking the exocyclic amine at the proper position fail to support ribozyme-catalyzed cleavage.[36] In addition, direct incorporation into the ribozyme of bases with altered functionality at N4 of C75, including replacement of the amino hydrogens with ethyl groups,[42] replacement of the exocyclic amine with a keto group (uridine,[21,22] 6-azauridine[43]), or removal of functionality at position 4 of the base (zebularine[41]), abolishes cleavage activity.

6.4.3 Resolving the Kinetic Ambiguity

6.4.3.1 Reaction in the Absence of Divalent Cations

Bevilacqua and co-workers excluded divalent cations from the genomic ribozyme reaction and analyzed the pH dependence in the presence of high monovalent cation concentrations.[23,44–46] Under these conditions, the pH–rate profile inverts: the reaction rate maximum occurs at low pH and decreases log–linearly with increasing pH.[23,44] One model to account for this pH–rate profile inversion in the absence of divalent metal ions posits that a second rate-stimulating titration occurs in the presence of divalent metal ions, possibly direct titration of a hydrated magnesium ion itself as suggested by Nakano et al.[23] As discussed in Section 6.4.2, this model is consistent with C75 functioning as a general acid. Although the pH–rate profile in the absence of magnesium ions was later attributed to titration of C41H$^+$ (see Section 6.5.1 for further discussion), it provided researchers in the field with an important clue to the underlying kinetic complexity of the pH–rate profiles, *i.e.*, that the observed profiles may reflect the net effect of multiple titrations.

6.4.3.2 Sulfur Substitution of the Leaving Group

If C75 does play a direct role in HDV ribozyme catalysis by mediating proton transfer, where does the nucleobase transfer the proton? As described in Section 6.4.1 for the imidazole rescue reaction, the nucleobase could transfer a proton to the 5′-leaving group or one of the non-bridging oxygens (general acid), or it could accept a proton from the 2′-hydroxyl nucleophile (general base). To distinguish among these possible models, we must obtain functional evidence that links C75 to one or more of these atoms along the reaction coordinate.

To that end, we used an approach that has been termed chemogenetic suppression, in which we substituted a sulfur atom for the 5′-oxygen leaving group in a trans-acting antigenomic ribozyme[47] (see Section 6.8 for further discussion of trans-cleaving ribozymes). Within RNA the phosphorothiolate linkage undergoes base-catalyzed cleavage about 10 000-fold faster than natural RNA phosphodiester linkages,[48] reflecting the greater leaving ability of the sulfur atom. However, sulfur atoms act as weaker hydrogen bond acceptors than do oxygen atoms, suggesting that phosphorothiolates would exhibit less susceptibility to general acid catalysis than would natural RNA phosphodiesters. Conversely, phosphorothiolate linkages in RNA undergo base-catalyzed cleavage just like natural RNA, suggesting that these analogues retain susceptibility to general base catalysis. Consequently, mutations that disable protonation of the leaving group in the HDV ribozyme reaction should affect cleavage of a substrate containing the 5′ sulfur substitution ($S_{5'S}$) less adversely than cleavage of the native oxygen-containing substrate ($S_{5'O}$). However, perturbation of other catalytic features (general base, active site geometry, electrostatics, *etc.*) should have similar effects on both $S_{5'S}$ and $S_{5'O}$.

We investigated the reactions of $S_{5'S}$ and $S_{5'O}$ in the presence of saturating ribozyme. The reaction of E-$S_{5'O}$ exhibited a bell-shaped pH–rate profile (Figure 6.8). The pH–rate profile for the reaction of E-$S_{5'S}$ closely matches that for E-$S_{5'O}$ from pH 4.5 to 8 both in shape and absolute rate. However, in contrast to the reaction of E-$S_{5'O}$, the reaction of E-$S_{5'S}$ remains pH-independent over the entire basic limb of the profile. In other words, 5′-sulfur substitution suppresses the deleterious effect of alkaline pH. The contrasting behavior of the two substrates might reflect general acid catalysis: deprotonation of the general acid with increasing pH leads to a log–linear decrease in the reaction of E-$S_{5'O}$ but has little effect on the reaction of E-$S_{5'S}$, which experiences little benefit from general acid catalysis.

We next constructed mutations in the catalytic core region: C76U (J4/2), C24U (P1.1), C25U (P1.1), C29A (L3) and U39C (P1). All of these mutations had deleterious effects on the reaction of $S_{5'O}$. The $S_{5'S}$ reaction experiences similar deleterious effects from the mutations with the exception of the C76U ribozyme, which catalyzes the reaction of $S_{5'S}$ far more efficiently than it

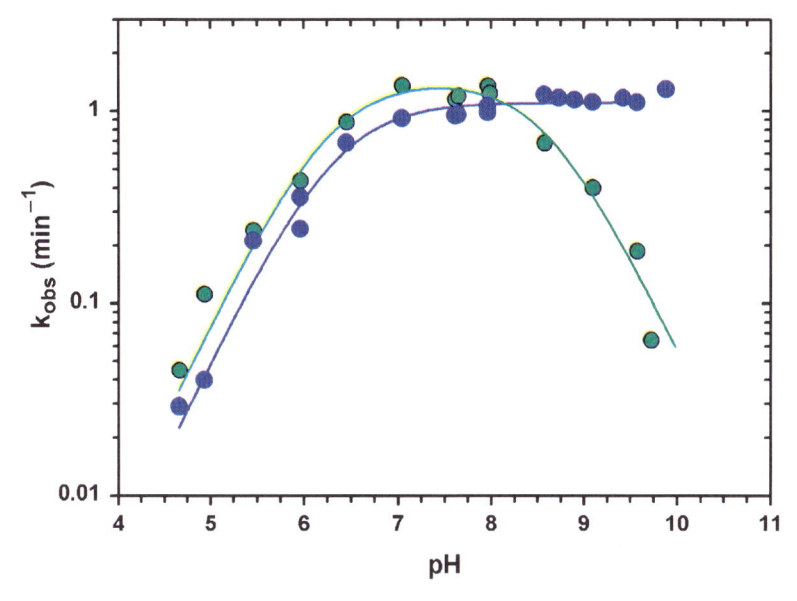

Figure 6.8 pH–rate profiles for a trans-cleaving antigenomic ribozyme. Cleavage of a wild-type substrate ($S_{5'O}$) is shown in green, while cleavage of a substrate containing a 5′-bridging sulfur substitution ($S_{5'S}$) is shown in blue. (Adapted from ref. 47.)

catalyzes the reaction of $S_{5'O}$. In essence, among the mutant constructs, sulfur substitution of the leaving group uniquely suppresses the deleterious effect of C76U. We also attempted to rescue the C76U mutant with imidazole. Imidazole rescues the reaction of $S_{5'O}$ as expected, but not $S_{5'S}$ – consistent with imidazole rescue and sulfur substitution having redundant functions: activation of the leaving group. These observations suggest a possible functional linkage between the active site nucleobase and the leaving group.

Although the C76U and wild-type ribozymes catalyze cleavage of $S_{5'S}$ at similar rates at pH 7.5, comparison of the two ribozymes in the log–linear regime of their pH–rate profiles reveals that, even with the sulfur substrate, the mutation induces a significant deleterious effect on the reaction. Possibly, the nucleobase plays a role beyond that of simply a general acid. As discussed in Section 6.4.2, the crystal structure of the HDV product shows the exocyclic amino group of C75/C76 engaged in a network of hydrogen bonds. Perhaps these interactions contribute to catalysis of the $S_{5'S}$ reaction, and the uracil base disrupts these, accounting for the lack of quantitative suppression with sulfur substitution.

Seeking to disrupt the putative proton transfer capacity of C76 while retaining the exocyclic amino group, we constructed a ribozyme containing at residue 76 3-deazacytidine (c3C), a cytidine analogue in which a CH replaces the N3 imino nitrogen. Perturbation of this one atom in the ribozyme completely abolishes activity against the natural substrate, $S_{5'O}$. However, this

76c3C ribozyme (at increased levels of magnesium) catalyzes cleavage of $S_{5'S}$ only fourfold slower than does the wild-type ribozyme. This near-quantitative suppression upon sulfur substitution functionally links the N3 imino nitrogen of C76 to the leaving group. Notably, as with any suppression experiment, the inferred linkage may reflect direct or indirect interaction between the imino nitrogen and the leaving group. Even with this caveat, functional linkage between the imino nitrogen and the leaving group supports a model of general acid catalysis, in which the imino nitrogen of C76 bears a proton that transfers to the 5′-oxygen leaving group in the transition state.

An analogous chemogenetic approach to identify the general base would complement and strengthen these results. For example, modification of the nucleophile to obviate the need for a general base might provide a strategy to identify the general base and link it to the pH–rate profile. Moreover, a simple prediction would be that the modification would not suppress the deleterious effects of C76 mutations, ruling out the possibility that it functions as a general base while supporting the interpreted linkage of C76 to the general acid function. However, there appear to be no simple chemical modifications to the 2′-hydroxyl group that suppress the need for a general base.

6.5 Metal Ions in the HDV Ribozyme

When RNA molecules adopt compact three-dimensional architectures, the negative charge of the ribose-phosphate backbone generates enormous repulsive energy in the absence of counterions. As a result, cations condense onto all RNA molecules to form a counterion atmosphere that neutralizes this repulsive energy.[49] Within the ion atmosphere, some ions associate more intimately with the RNA, using their hydration shells to bind to specific sites *via* hydrogen bonding (outer-sphere coordination) or shedding water molecules to coordinate directly to a heteroatom from the RNA (inner-sphere coordination). Because the vast majority of cations associated with RNA reside in the diffuse ion atmosphere, we cannot use standard concepts of stoichiometry and ligand–receptor interactions to tease apart the energetics of metal ion–RNA interactions.[49] Consequently, analysis of the role played by individual site-bound metal ions becomes exceedingly difficult. Despite these limitations and complexities, experimentalists have used various approaches to gain insight into the role of divalent metal ions for HDV ribozyme function. Results implicate divalent metal ions in both structural and catalytic roles.

6.5.1 Structural Metal Ions

Initially, the RNA community regarded the HDV ribozyme as an obligate metalloenzyme, requiring divalent ions for folding and catalytic function.[26] Later experiments showed that in the presence of EDTA, which chelates trace amounts of divalent metal ions, high concentrations (1 M) of monovalent salt could support ribozyme activity.[23] Indeed, it was the comparison of the

NaCl-supported reaction to the Mg^{2+}-supported reaction in a cis-acting genomic ribozyme that led Bevilacqua and co-workers to propose specific catalytic roles for C75 (Section 6.4) and a divalent metal ion (Section 6.5.2).[23]

Crystal structures (crystals obtained in low millimolar concentrations of Mg^{2+}) reveal divalent metal ions near the phosphate backbone, but only the precursor structure shows one near the active site. Poisson–Boltzmann analysis[50] of the crystal structures revealed areas of highly negative electrostatic potential, notably near the active site and above P1.1,[44] suggestive of metal ion binding sites. In addition, Mg^{2+} titration of a three-stranded trans-acting HDV ribozyme caused shifts in its NMR spectrum, particularly within P2 and P1 near the cleavage site.[51] These structural studies implicate various sites in the ribozyme as locations where divalent metal ions may bind, but whether or how these metal ions impart catalytic function remains unknown.

To gain deeper insight into how the divalent metal ion(s) impart(s) HDV ribozyme function, Bevilacqua and co-workers performed a series of systematic studies of genomic ribozyme self-cleavage, monitoring thermal unfolding and self-cleavage at different monovalent and divalent salt concentrations to explore the relationship between Mg^{2+} binding and pH.[44-46] They obtained evidence for two classes of divalent metal ions and established conditions to probe each class individually. High concentrations of monovalent cations saturate the ion atmosphere and exclude all but site-bound divalent metal ions from the RNA, allowing the characteristics of these site-bound ions to be studied in more detail. Nakano et al. followed the Mg^{2+} dependence of ribozyme activity at different pHs in the presence of 1 M NaCl. The ribozyme cleavage reaction occurred at the basal NaCl rate at low Mg^{2+} concentrations (10^{-9}–10^{-7} M), increased log–linearly with unit slope at intermediate Mg^{2+} concentrations (10^{-7}–10^{-4} M), and became independent of Mg^{2+} at higher concentrations (10^{-4}–10^{-1} M).

Distinct pH–rate profiles for the intermediate and high Mg^{2+} concentration ranges led to the suggestion that distinct classes of divalent metal ions participate in the reaction. Although intermediate concentrations of Mg^{2+} stimulated the reaction significantly above the basal NaCl rate, the pH dependence at these intermediate Mg^{2+} concentrations remained the same as that of the basal NaCl reaction – an inverse dependence at pH values above 6 (attributed at the time to titration of C75, as described in Section 6.4.3.1). In contrast, for the same pH range, the reaction at high Mg^{2+} concentrations showed no dependence on pH, analogous to the pH profile for the Mg^{2+}-supported reaction at low monovalent salt concentrations. Apparently, the metal ion that binds and stimulates reaction at intermediate Mg^{2+} concentrations cannot offset the rate-inhibiting pH titration, distinguishing this metal ion from the so-called catalytic metal ion. Nakano et al.[44] proposed that this new tighter-binding metal ion fulfilled a structural role in the ribozyme reaction. To account for these and other observations, they adopted a three-channel reaction scheme (Scheme 6.1).

Further analysis of the structural metal ion (under channel 2 conditions) showed that binding correlated inversely with the ionic radius of the divalent metal ion,[46] a biochemical signature for inner-sphere coordination. Consistent

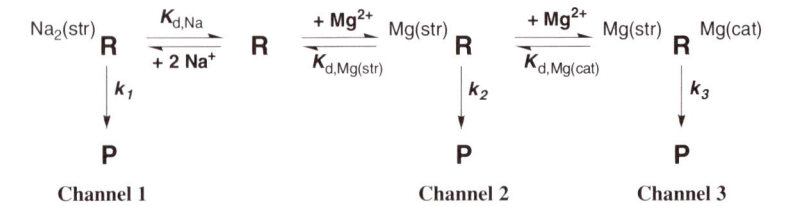

Scheme 6.1 Three-channel mechanism for divalent-metal-independent and -dependent cleavage by the genomic HDV ribozyme. (Adapted with permission from ref. 45).

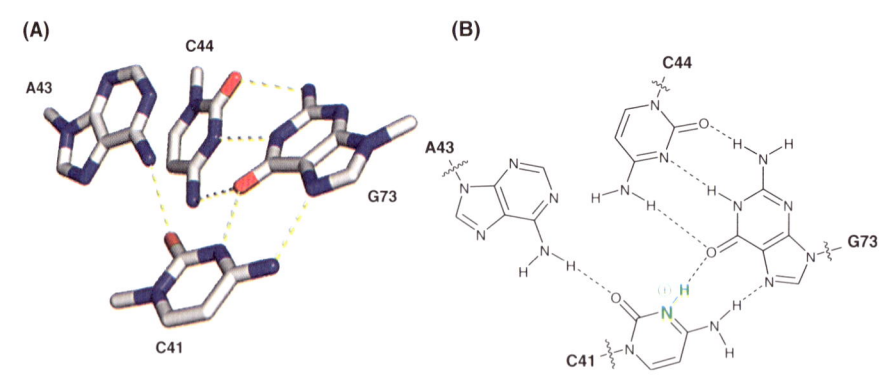

Figure 6.9 C41 participates in a base quadruple interaction. Protonation of N3 of C41 is expected to stabilize this structure. (A) Base quadruple from the product crystal structure. Coloring is by element as in Figure 6.4. (B) Schematic of the base quadruple; the protonated N3 of C41 is shown in green.

with inner-sphere coordination, under channel 2 conditions where Mg^{2+} populates the structural site exclusively (1 M NaCl), increasing concentrations of $Co(NH_3)_6^{3+}$, an exchange-inert complex known to populate binding sites for outer-sphere metal ions in RNA, had no effect on binding of the structural Mg^{2+}.

An additional distinction between the structural and catalytic metal ions, and a clue to the region where a structural metal ion might bind, came from comparative analysis of Mg^{2+} binding *vs.* pH in wild-type and mutant ribozymes.[45] In the wild-type genomic ribozyme, the structural metal ion binds more tightly as the pH decreases (channel 2 conditions), whereas the catalytic metal ion binds more tightly as the pH increases (channel 3 conditions). The former observation suggested that protonation of the RNA contributed to binding of the structural metal ion. Residue C41, located in the J1.1/4 linker region, appeared likely as a candidate site for protonation. Structural and biochemical studies of the genomic ribozyme had already implicated protonated C41 as part of a base quadruple beneath P1.1 that stabilizes the RNA structure[16,52] (Figure 6.9).

To address whether protonation at C41 allows the structural metal ion to bind more tightly, Nakano and Bevilacqua analyzed Mg^{2+} binding vs. pH for a double mutant form of the ribozyme,[45] prepared and analyzed originally by the Been laboratory,[52] that hypothetically forms the base quadruple without the need for C41 protonation. In contrast to the observations for the wild-type ribozyme, binding of the structural metal ion in the double mutant ribozyme exhibited no linkage with pH (pH 6–8). From these and other observations, Nakano and Bevilacqua inferred that formation of the base quadruple supports binding of the structural metal ion and, therefore, that the metal ion may reside in the vicinity of the base quadruple.[45] The decrease in binding affinity with increasing pH observed for the wild-type ribozyme would then reflect disruption of the base quadruple from deprotonation of C41. In contrast, C41 protonation appeared to have no effect on binding of the catalytic metal ion. Other experimental methods have thus far not confirmed the location of this proposed structural divalent metal ion; neither the precursor nor the product crystal structure (obtained at near neutral pH under low millimolar monovalent and divalent salt concentrations) shows evidence of a divalent metal ion near C41.

6.5.2 Catalytic Metal Ions

Structural and functional data suggest that another divalent metal ion, distinct from the structural metal ion defined in Section 6.5.1, contributes directly to catalysis by the HDV ribozyme. Early evidence for an active site metal ion came indirectly from the work of Shih and Been.[53] They investigated ribozyme cleavage of both the native 3′,5′-phosphodiester linkage and a 2′,5′-linkage by a trans-acting antigenomic ribozyme, and demonstrated a link between the scissile phosphodiester bond and the metal ion specificity of the reaction. While both Mg^{2+} and Ca^{2+} support cleavage of the native 3′,5′-phosphodiester linkage equally well, cleavage of the substrate bearing the 2′,5′-linkage, though significantly slower than cleavage of the native linkage, exhibited a 1000-fold preference for Mg^{2+} over Ca^{2+}. Because distinct phosphodiester linkages reacted with distinct metal ion specificity, Shih and Been concluded that a divalent metal ion must sit close to them in the active site.

Bevilacqua and colleagues have proposed that an active site metal ion functions as a general base that enhances the nucleophilicity of the 2′-hydroxyl group.[23,44,46] This reasoning followed from several key observations alluded to in Section 6.5.1: (1) for both the Mg^{2+}-supported reaction (in the absence of high NaCl concentration) and the channel 3 pathway (10^{-4}–10^{-1} M Mg^{2+}, high ionic strength) the reaction rate remains pH-independent at pH > 6. Assuming that under these salt conditions C75 still undergoes deprotonation in this pH-regime, a second titration must occur in the presence of Mg^{2+} that offsets the inhibitory effect of the C75 titration. Titration of $Mg(OH_2)_6^{2+}$ itself provides one attractive model that could account for this behavior. (2) Binding of this Mg^{2+} increases with pH, supporting the possibility that this metal ion sits close to C75. Assuming again that C75 undergoes deprotonation with increasing pH, the loss of positive charge would relieve electrostatic repulsion between C75H$^+$ and the metal ion, allowing the metal ion to bind more tightly. NMR experiments have independently confirmed the linkage between C75 and Mg^{2+}, demonstrating that the pK_a of C75 decreases with increasing Mg^{2+} concentration.[40] Thus, a metal ion in the vicinity of C75, whose presence offsets the negative effect of deprotonating C75, has the kinetic signatures expected of a general base. Direct evidence that links this metal ion to general base function remains elusive.

In contrast to the structural metal ion, the catalytic metal ion appears to interact with the ribozyme through outer-sphere coordination. Co(NH$_3$)$_6^{3+}$ inhibits the channel 3 pathway, suggesting that it competes with the catalytic metal ion for binding at the active site.[23,46] Moreover, binding of the catalytic metal ion does not correlate well with ionic radius as would be expected for an inner-sphere coordinated metal ion.[46] The contribution of this metal ion to catalysis appears to be relatively small. At pH 7.0, the catalytic metal ion apparently contributes only about 25-fold to catalysis, whereas the structural metal ion contributes about 125-fold.[44]

In addition to the functional evidence, NMR,[51] X-ray crystallography,[13] and footprinting analyses with lead[54] and terbium[55] strongly implicate a metal ion near the active site. Although the product crystal structure provides no evidence for this divalent metal ion, the C75U precursor structure does contain a divalent metal ion at the active site, within 5 Å (outer-sphere coordination distance) of the 5′-oxygen leaving group and the 2′-hydroxyl nucleophile[13] (Figure 6.4). Assuming that this is in fact the catalytic metal identified by the functional analysis, further experiments will be required to clarify whether it acts as a general base, general acid, or in some other catalytic role.

6.6 Contributions of Non-active-site Structures to Catalysis

In addition to positioning specific catalytic groups, the three-dimensional architecture of the HDV ribozyme has been implicated in larger-scale contributions to catalysis. Shih and Been[4] used a trans-acting ribozyme with a series of substrates of varying binding affinities to uncover a correlation between substrate binding energy and the activation energy of the reaction. They proposed that sequences upstream of the cleavage site mediate ground-state destabilization of the ribozyme–substrate complex, contributing around 2 kcal mol^{-1} to a total activation-energy reduction (in this system) of 8.5 kcal mol^{-1} compared with the uncatalyzed reaction. This proposal received further support from the precursor crystal structure, in which the RNA backbone adopts an unusual ~180° turn about the scissile phosphate[13] (Figure 6.5). This so-called "U-turn" motif has been compared to a structure found in the hammerhead ribozyme;[56] Walter and co-workers have proposed that, in the HDV ribozyme, this comparatively rigid structure may serve to reduce the entropic barrier to catalysis.

Additional experiments will be needed, however, to establish a functional link between this conformation and any enthalpic or entropic contribution to the reaction.

Another role for the global architecture of the HDV ribozyme in catalysis has emerged from extensive work by the Walter laboratory, which has provided evidence for a large-scale conformational change along the reaction pathway. Fluorescence resonance energy transfer (FRET) evidence,[57,58] coupled with terbium-ion footprinting[55,59] and kinetic solvent isotope effect (KSIE) studies,[39] have all converged on a model in which the global structure of the ribozyme becomes significantly extended upon catalysis. Unfortunately, very little is known about the details of this rearrangement, other than the FRET data that first identified it; according to those results, the distance between the top of P2 and the L4 loop must increase from 52 to 65 Å during catalysis.[57] This conformational change may coincide with the one proposed by Ke et al.,[13] in which the catalytic metal ion is forced out of the active site. However, the crystal structures do not account for the change in distance between P2 and L4, suggesting that conformational change in solution is more dramatic than that inferred from the crystals.

Another possible role for this global conformational change has been proposed from studies of the trefoil turn (Figure 6.6). This distinctive structural feature of J4/2 appears to be intimately linked to the positioning of the catalytic C75, and its conformation has been monitored by installing the fluorescent analogue 2-aminopurine (2AP) in place of G76.[58,60] 2AP fluoresces in solution, but its fluorescence is quenched when it stacks on other nucleotides; as G76 is unstacked in the trefoil turn, the formation of this structure can readily be monitored by 2AP fluorescence. Steady-state fluorescence measurements provide evidence that the trefoil turn may not be formed in the pre-cleavage ribozyme, in contrast to the observations in the precursor crystal structures. However, time-resolved 2AP fluorescence analysis has suggested that the conformational change observed by FRET occurs simultaneously with an increase in the extent of formation of the trefoil turn.[60] Taken together, these observations suggest that the global conformational change during catalysis may be required for the correct positioning of C75 in the active site.

6.7 Dynamics in HDV Function

The time-resolved studies of 2AP fluorescence also highlight the importance of dynamics in the HDV ribozyme. Gondert et al.[60] demonstrated that each of the distinct steady-state levels of 2AP stacking for different product and precursor complexes reflected a combination of at least four distinct conformations in dynamic equilibrium. Thus, the global conformational change discussed in Section 6.6 does not lock the trefoil turn into a single catalytically-active state, but rather biases the existing equilibrium to favor catalysis. Similarly, evidence from the same study suggests that nucleotides 5' to the cleavage site may be able to interact with a conformation of the trefoil turn unfavorable to catalysis, providing a possible rationale for the slight sequence preferences shown by these nucleotides.[4]

It should come as no surprise that dynamics contribute significantly to ribozyme function, considering their rise to prominence in the protein world. One recent study has posited that the enzyme motions necessary for catalysis are encoded in a protein's dynamics, and thus occur naturally whether substrate is present or not.[61] The 2AP results discussed herein provide evidence that the same principle likely governs RNA enzymes as well, making the dynamics of HDV and other ribozymes a promising subject for further study.

In addition to the experimental treatment of HDV dynamics, the Walter laboratory has addressed this subject from a computational perspective, using molecular dynamics simulations to interrogate possible modes of catalysis.[62-64] The current state of the art in molecular dynamics requires that any treatment of RNA be taken with a grain of salt, because of the difficulty in accurately simulating the behavior of the divalent cations so central to RNA structure and function.[63] With this concern in mind, though, we can still gain some useful insight from these simulations. In particular, the simulations highlight regions of the ribozyme that might provide conformational flexibility during the reaction. The J4/2 region we have already discussed, unsurprisingly, displays such dynamic behavior, with C75 showing substantial mobility within the catalytic pocket.

The other major region implicated in these simulations as being highly dynamic is the L3 loop, which forms part of the boundary of the catalytic pocket (Figure 6.10). As this loop interacts directly with C75 in the product crystal structure (through the 2'-OH of U20) and with the active-site metal ion in the precursor structure (through O2 of U20 and a phosphate oxygen of U23), its conformation is likely to have dramatic effects on the local organization of the active site. In addition, phylogenetic studies have identified U23 as among the most conserved nucleotides in the ribozyme, further suggesting the loop's importance.[65]

Functional studies have also identified mutants in and around the L3 loop as having significant effects on reactivity. The loop is interrupted by the P1.1 helix, whose integrity is crucial to ribozyme function.[17] It seems reasonable to speculate that the sensitivity of function to small changes in P1.1 stability might arise, in part, from the shift such changes would likely cause in the dynamics of L3. Another important nucleotide in this loop is G25, which is in a position to extend the P3 helix by base pairing with U20; this interaction is not observed in either the precursor or the product crystal structures, but molecular dynamics simulations indicate that various G·U wobble pairs could be formed at this site.[63] Strikingly, a mutation of G25 to A, allowing for a Watson–Crick U-A pair with U20, causes a 3000-fold reduction in cleavage activity.[22] The severity of this effect suggests that flexibility in L3 is critical to ribozyme function.

Electrostatic potential calculations based on the molecular dynamics simulations suggest that the L3 loop likely has a profound influence on the negative potential well found at the active site, with local unfolding of the loop leading to an appreciable loss of negative potential.[63] This potential well is likely

Catalytic Mechanism of the HDV Ribozyme 115

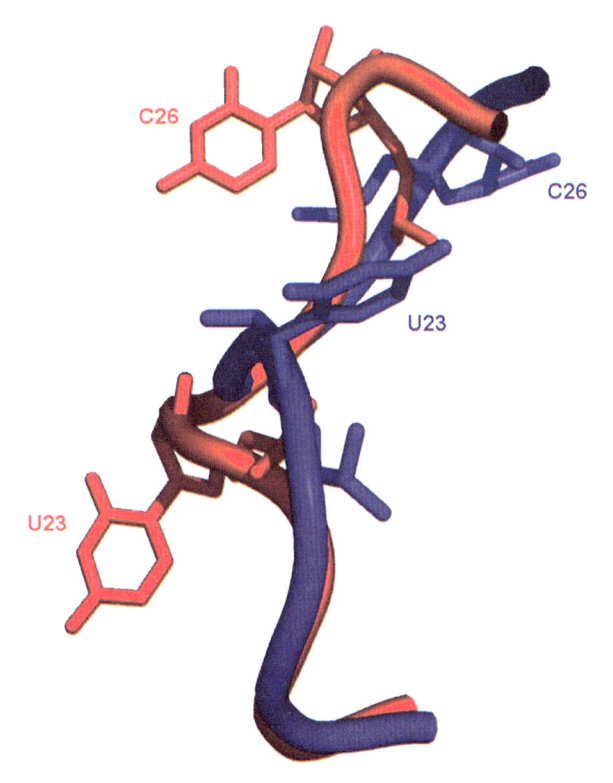

Figure 6.10 Conformational variability of the L3 loop. The L3 loop region of the precursor (blue) and product (red) crystal structures is shown. The extent of differences between the two structures (note the different positions of nucleotides U23 and C26, shown as sticks) suggests that this loop may be particularly flexible, which is consistent with the results of molecular dynamics simulations.

responsible for localizing the catalytic divalent metal ion (regardless of whether its positioning corresponds to that seen in the precursor crystal structures). It is tempting to speculate that the dynamics of the L3 loop may control proper positioning of the metal ion in the same way that the dynamics of J4/2 affect positioning of the catalytic C75. However, L3 has not yet received the same intensity of experimental study that J4/2 has, so several key questions about its role remain unanswered. Among these questions are how L3 dynamics are influenced by the global conformational change during catalysis, and whether factors that alter L3 dynamics affect J4/2 dynamics (or *vice versa*).

6.8 Varieties of Experimental Systems

The experiments discussed in this chapter have profited from the availability of multiple related forms of the ribozyme that have been developed for *in vitro* analysis. At the same time, this proliferation of experimental systems has complicated interpretation of the available data: determining the relevance of results obtained in one system to experiments performed in a different system is a challenging task at best.

One of the major pairs of alternative HDV variants is the naturally-occurring distinction between the genomic and antigenomic ribozymes. Although these two forms of the ribozyme are predicted to adopt very similar structures (Figure 6.1), their sequences are not identical, and only the genomic form has been used in crystallographic studies.

The most notable, and most studied, difference between the genomic and antigenomic ribozymes is the presence of a CAA trinucleotide sequence in the J1.1/4 region of the genomic form (residues C41-A43).[36,45,52,66] This sequence, absent in the antigenomic ribozyme, forms the base quadruple discussed in Section 6.5.1 (Figure 6.9) that appears to be related to the binding of the structural metal ion. Consistent with the expectation that this protonated structure will be stabilized at low pH, the addition of the CAA sequence to an antigenomic ribozyme confers enhanced stability against thermal denaturation at low pH.[52] The extent to which this stabilization reflects the formation of the quadruple itself, as opposed to improved binding of the structural metal ion, has not yet been studied.

While the ability of the C41 base quadruple to stabilize the structure of the HDV ribozyme (whether directly or indirectly) seems clear enough, the importance of this stabilization to catalysis has been more difficult to pin down. Under otherwise wild-type conditions, removal of the CAA sequence from a genomic ribozyme does not appear to have significant effects on reactivity at any pH.[52] However, the same mutation has significant effects when analyzed under certain conditions that inhibit reactivity, such as lack of divalent metal ions. In that case, the C41 base quadruple appears to contribute about two orders of magnitude to reactivity at acidic pH, suggesting that the ribozyme is more sensitive to the loss of this structure when it has already been destabilized for other reasons.

A similar conclusion arises from an examination of the pH–rate profiles for the two forms of the HDV ribozyme.[36] As discussed in Section 6.4.2, the reactivity of the antigenomic ribozyme becomes pH-independent above pH ~7.0 (Figure 6.7C); however, the genomic ribozyme exhibits a second pK_a, decreasing in reactivity in a log–linear fashion above pH 8.5. This result suggests that the genomic and antigenomic ribozymes differ in the number of titratable groups that affect catalytic activity. Perrotta *et al.* have used the C75Δ mutant in conjunction with exogenous cytosine analogues to reveal an effect of mutating the CAA sequence: a C75Δ genomic ribozyme without the C41 base quadruple is no longer inhibited by basic pH during cytosine or imidazole rescue. From this result, it seems likely that the high-pH inhibition of the native genomic ribozyme results from deprotonating C41 and losing the base quadruple. However, it remains unclear why this effect can only be seen during rescue of the C75Δ mutant, and not in the wild-type genomic ribozyme reaction. By the same token, the metal dependence of the antigenomic ribozyme has not been studied in the context of the three-channel mechanism developed for the genomic form, so the effects of the lack of the C41 base quadruple on binding of the structural metal ion to that ribozyme are still unknown.

In addition to the distinction between genomic and antigenomic ribozymes, the other major difference in experimental systems is between cis-cleaving and trans-cleaving ribozymes. As discussed in Section 6.1.1, the ribozyme as found in nature acts in cis, separating the monomers of rolling-circle replication. However, many researchers have modified the HDV sequence to create trans-cleaving ribozymes capable of multiple-turnover reactions. The most common strategy (employed in the trans-cleaving system discussed in Section 6.4.3.2 for analyzing 5′-sulfur substitution) has been to separate the ribozyme at J1/2, creating a substrate strand consisting of the 5′ half of P1, and leaving the rest of the ribozyme intact.[67,68] Even this small modification, however, has its consequences. The break in J1/2 removes one of the two pseudoknots from the tertiary structure; as mentioned in Section 6.2.2, this deletion results in a decrease in both reactivity and stability to denaturation.[20,68,69] Similarly, a trans-cleaving antigenomic ribozyme shows an absolute requirement for two Watson–Crick pairs in P1.1, whereas the corresponding cis-cleaving ribozyme can tolerate mutation of one of these pairs.[17] Thus, it appears that modifying the HDV for trans cleavage weakens enough tertiary contacts to affect the ribozyme's reactivity. This destabilization may be responsible for the unexpected observation that a trans-cleaving antigenomic ribozyme decreases in reactivity at high pH, whereas the cis-cleaving form does not (compare Figures 6.7C and 6.8).

Another approach to constructing a trans-cleaving HDV involves breaking the ribozyme into three segments. The first break is located in J1/2, as in the ribozymes discussed above, while the second is placed in the nonessential L4 loop. Kinetically, these ribozymes behave much like the two-piece versions already discussed, but they facilitate site-specific modifications such as the addition of fluorophores to the RNA. The most-studied three-stranded ribozyme, developed in the Walter laboratory, combines sequence elements from the genomic and antigenomic ribozymes; notably, it lacks the C41 base quadruple.[57]

6.9 Models for HDV Catalysis

This chapter has focused on evidence that supports catalytic roles for C75 and a divalent metal ion in HDV catalysis. In addition to the model where C75 acts as the general acid and a divalent metal ion acts as the general base (Figure 6.11A), Ke et al.[13] have proposed on the basis of structural data a kinetically equivalent model[70] in which C75 acts as the general base (deprotonating the 2′-hydroxyl group) and the divalent metal ion acts as the general acid (protonating the 5′ leaving group) (Figure 6.11B). Support for this model comes from the position of U75 in the C75U precursor crystal structure and from molecular dynamics studies using this crystal structure as a starting point.[62,63] Although N3 of U75 sits roughly equidistant from both the 2′-hydroxyl nucleophile and the

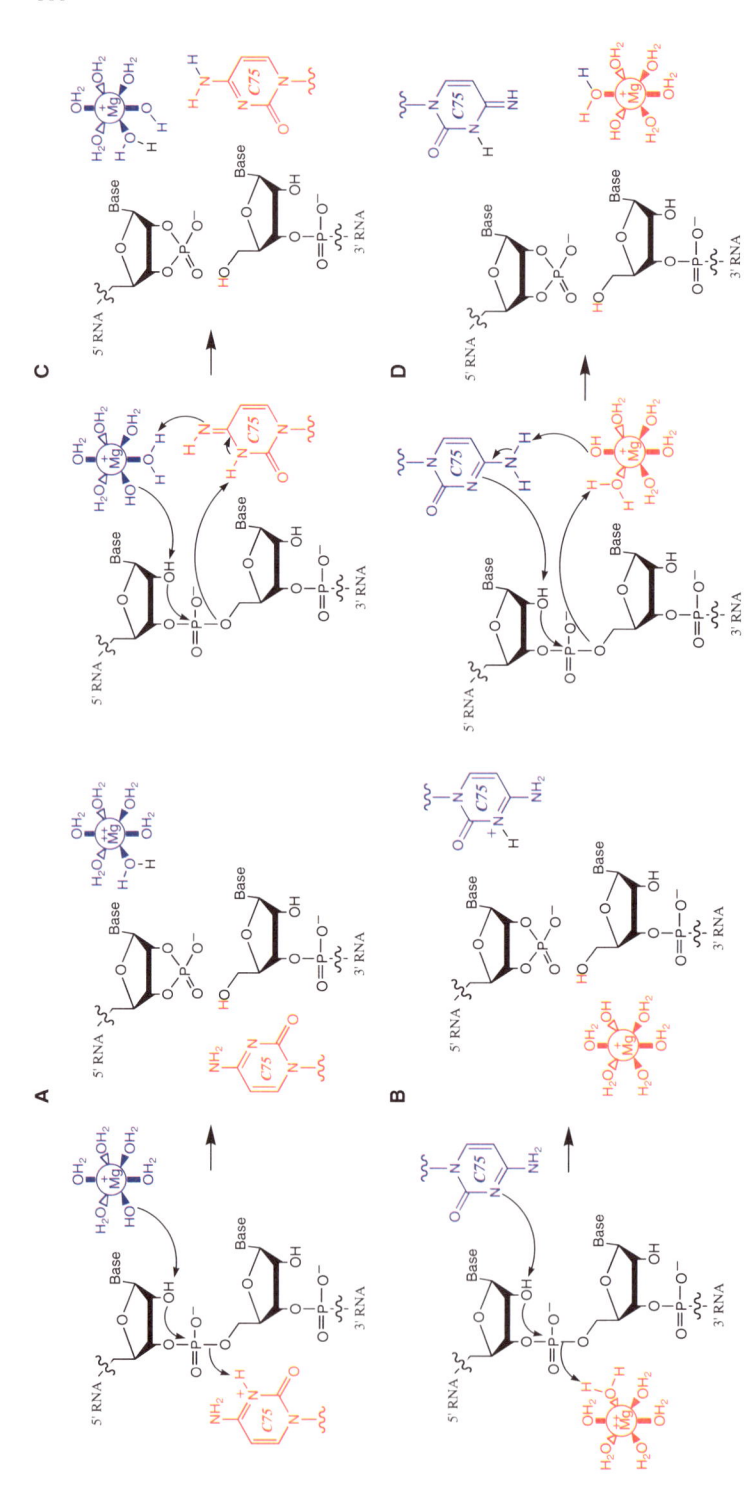

Figure 6.11 Models for general acid–base catalysis by the HDV ribozyme. For clarity, mechanisms depict only the initial and final states of the reaction. The general base is in blue, and the general acid is in red. (*A*) C75 is a general acid, and a divalent metal ion acts as a general base. (*B*) A divalent metal ion is a general acid, and C75 acts as a general base. (*C*) Mechanism accessing a tautomeric state of C75, with C75 as a general acid and a divalent metal ion as a general base. (*D*) Mechanism accessing a tautomeric state of C75, with a divalent metal ion as a general acid and C75 as a general base.[13]

5′-leaving group, simple rotation about the 3′-O–P bond of U-1 brings N3 within hydrogen bonding distance of the 2′-hydroxyl group, which is consistent with the possibility that N3 serves a general base role; bringing the nucleophile within hydrogen bonding distance of the metal ion in the active site results in steric clashes.[13]

As noted in Section 6.3, in addition to protonating the leaving group and/or deprotonating the 2′ nucleophile to effect catalysis, stabilizing negative charge on the non-bridging phosphate oxygens at the cleavage site may also enhance the rate of cleavage. On the basis of density functional theory analysis,[71] Liu et al. have proposed that the non-bridging phosphate oxygens interact with both the exocyclic amine of C75 and a hydrated divalent metal ion. Although the C75 exocyclic amine and the metal ion clearly play significant roles in catalysis, functional evidence for interactions with the non-bridging oxygens is sparse. Phosphorothioate substitution of the *pro*-R$_p$ oxygen at the cleavage site decreases the cleavage rate and Mn^{2+} does not rescue this defect;[72] this is consistent either with steric clashes of the larger sulfur with groups in the active site or with a possible interaction between the *pro*-R$_p$ oxygen and a non-innersphere metal ligand. We are aware of no experimental work on the *pro*-S$_p$ oxygen at the cleavage site.

Another proposed model features a tautomeric state of C75[13] but remains untested. In this mechanism, C75 and a hydrated metal ion interact with each other to shuttle a proton from the 2′-hydroxyl group to the leaving group. Assuming that a divalent metal ion has a general acid role and C75 has a general base role, N3 of C75 accepts the proton from the 2′-hydroxyl nucleophile and donates an exocyclic amine proton to the hydrated divalent metal ion, which then loses a proton to the leaving group (Figure 6.11D). Alternately, if C75 acts as a general acid and the hydrated metal ion acts as a general base, the metal ion accepts the 2′-hydroxyl proton and transfers a proton to the imino nitrogen of C75, which then donates a proton to the leaving group *via* the exocyclic amine to form a C75 tautomer (Figure 6.11C). Both of these intriguing mechanisms involve a net transfer of only one proton because C75 and the hydrated metal ion both accept and donate protons during the reaction. The number of proton transfers that occur during the HDV cleavage reaction remains uncertain, however, as proton inventory experiments on the genomic[73] and antigenomic ribozymes[37] have yielded different answers.

6.10 Conclusion

Powerful, creative, and diverse computational, structural, biochemical, and biophysical approaches have led to a wealth of theoretical and experimental data and penetrating insights into the catalytic mechanism of the HDV ribozyme. Despite these significant strides in mechanistic understanding, great challenges remain. While several important players – C75, structural and catalytic metal ions, the L3 loop – have been identified, in no case have the data converged to define a precise role for one of these components in facilitating phosphoryl transfer. Given the difficulty in interpreting classic analyses such as pH–rate profiles in this system, it seems likely that additional structures and novel experimental approaches will be needed to resolve the existing data into a cohesive model for the function of this catalyst.

Acknowledgements

We thank Michael Been for providing electronic versions of his published figures. We thank Cheryl Small for assistance with the manuscript. S.K. is supported by the Medical Science Training Program (5T32GM007281-32) at the University of Chicago. T.N. was supported by the Predoctoral Training Program at the Interface of Chemistry and Biology (2 T32 GM008720–06) at the University of Chicago. J.A.P. is an Investigator of the Howard Hughes Medical Institute.

References

1. M.M. Lai, *Annu. Rev. Biochem.*, 1995, **64**, 259.
2. J. Filipovska and M.M. Konarska, *RNA*, 2000, **6**, 41.
3. A.T. Perrotta and M.D. Been, *Nature*, 1991, **350**, 434.
4. I. Shih and M.D. Been, *EMBO J.*, 2001, **20**, 4884.
5. D.M. Chadalavada, S.M. Knudsen, S. Nakano and P.C. Bevilacqua, *J. Mol. Biol.*, 2000, **301**, 349.
6. P. Deschenes, D.A. Lafontaine, S. Charland and J.P. Perreault, *Antisense Nucleic Acid Drug Dev.*, 2000, **10**, 53.
7. A.T. Perrotta and M.D. Been, *Nucleic Acids Res.*, 1990, **18**, 6821.
8. K. Salehi-Ashtiani, A. Luptak, A. Litovchick and J.W. Szostak, *Science*, 2006, **313**, 1788.
9. L. Sharmeen, M.Y. Kuo, G. Dinter-Gottlieb and J. Taylor, *J. Virol.*, 1988, **62**, 2674.
10. H.N. Wu, Y.J. Lin, F.P. Lin, S. Makino, M.F. Chang and M.M. Lai, *Proc. Natl. Acad. Sci. U.S.A.*, 1989, **86**, 1831.
11. G. Slim and M.J. Gait, *Nucleic Acids Res.*, 1991, **19**, 1183.
12. H. van Tol, J.M. Buzayan, P.A. Feldstein, F. Eckstein and G. Bruening, *Nucleic Acids Res.*, 1990, **18**, 1971.
13. A. Ke, K. Zhou, F. Ding, J.H. Cate and J.A. Doudna, *Nature*, 2004, **429**, 201.
14. I. Shih and M.D. Been, *Biochemistry*, 2000, **39**, 9055.
15. A.R. Ferre-D'Amare, K. Zhou and J.A. Doudna, *Nature*, 1998, **395**, 567.
16. A.R. Ferre-D'Amare and J.A. Doudna, *J. Mol. Biol.*, 2000, **295**, 541.
17. P. Deschenes, J. Ouellet, J. Perreault and J.P. Perreault, *Nucleic Acids Res.*, 2003, **31**, 2087.
18. T.S. Wadkins, A.T. Perrotta, A.R. Ferre-D'Amare, J.A. Doudna and M.D. Been, *RNA*, 1999, **5**, 720.
19. S.P. Rosenstein and M.D. Been, *Biochemistry*, 1996, **35**, 11403.

20. M.D. Been and G.S. Wickham, *Eur. J. Biochem.*, 1997, **247**, 741.
21. A.T. Perrotta and M.D. Been, *Nucleic Acids Res.*, 1996, **24**, 1314.
22. N.K. Tanner, S. Schaff, G. Thill, E. Petit-Koskas, A.M. Crain-Denoyelle and E. Westhof, *Curr. Biol.*, 1994, **4**, 488.
23. S. Nakano, D.M. Chadalavada and P.C. Bevilacqua, *Science*, 2000, **287**, 1493.
24. A.T. Perrotta and M.D. Been, *Biochemistry*, 2006, **45**, 11357.
25. S.P. Rosenstein and M.D. Been, *Biochemistry*, 1990, **29**, 8011.
26. Y.A. Suh, P.K. Kumar, K. Taira and S. Nishikawa, *Nucleic Acids Res.*, 1993, **21**, 3277.
27. T. Lonnberg and H. Lonnberg, *Curr. Opin. Chem. Biol.*, 2005, **9**, 665.
28. D.M. Perreault and E.V. Anslyn, *Angew. Chem. Int. Ed. Engl.*, 1997, **36**, 432.
29. G.A. Soukup and R.R. Breaker, *RNA*, 1999, **5**, 1308.
30. G.M. Emilsson, S. Nakamura, A. Roth and R.R. Breaker, *RNA*, 2003, **9**, 907.
31. J.D. Ye, N.S. Li, Q. Dai and J.A. Piccirilli, *Angew. Chem. Int. Ed.*, 2007, **46**, 3714.
32. M. Kosonen, K. Hakala and H. Lonnberg, *J. Chem. Soc., Perkin Trans. 2*, 1998, 663.
33. M. Kosonen and H. Lonnberg, *J. Chem. Soc., Perkin Trans. 2*, 1995, 1203.
34. M. Legiewicz, A. Wichlacz, B. Brzezicha and J. Ciesiolka, *Nucleic Acids Res.*, 2006, **34**, 1270.
35. A.T. Perrotta, I. Shih and M.D. Been, *Science*, 1999, **286**, 123.
36. A.T. Perrotta, T.S. Wadkins and M.D. Been, *RNA*, 2006, **12**, 1282.
37. I.H. Shih and M.D. Been, *Proc. Natl. Acad. Sci. U.S.A.*, 2001, **98**, 1489.
38. K.B. Schowen and R.L. Schowen, *Methods Enzymol.*, 1982, **87**, 551.
39. R.A. Tinsley, D.A. Harris and N.G. Walter, *J. Am. Chem. Soc.*, 2003, **125**, 13972.
40. A. Luptak, A.R. Ferre-D'Amare, K. Zhou, K.W. Zilm and J.A. Doudna, *J. Am. Chem. Soc.*, 2001, **123**, 8447.
41. A.K. Oyelere, J.R. Kardon and S.A. Strobel, *Biochemistry*, 2002, **41**, 3667.
42. F. Nishikawa, M. Shirai and S. Nishikawa, *Eur. J. Biochem.*, 2002, **269**, 5792.
43. A.K. Oyelere and S.A. Strobel, *Nucleosides Nucleotides Nucleic Acids*, 2001, **20**, 1851.
44. S. Nakano, D.J. Proctor and P.C. Bevilacqua, *Biochemistry*, 2001, **40**, 12022.
45. S. Nakano and P.C. Bevilacqua, *Biochemistry*, 2007, **46**, 3001.
46. S. Nakano, A.L. Cerrone and P.C. Bevilacqua, *Biochemistry*, 2003, **42**, 2982.
47. S.R. Das and J.A. Piccirilli, *Nat. Chem. Biol.*, 2005, **1**, 45.
48. R.G. Kuimelis and L.W. McLaughlin, *Bioorg. Med. Chem.*, 1997, **5**, 1051.
49. D.E. Draper, D. Grilley and A.M. Soto, *Annu. Rev. Biophys. Biomol. Struct.*, 2005, **34**, 221.
50. K. Chin, K.A. Sharp, B. Honig and A.M. Pyle, *Nat. Struct. Biol.*, 1999, **6**, 1055.

51. Y. Tanaka, M. Tagaya, T. Hori, T. Sakamoto, Y. Kurihara, M. Katahira and S. Uesugi, *Genes Cells*, 2002, **7**, 567.
52. T.S. Wadkins, I. Shih, A.T. Perrotta and M.D. Been, *J. Mol. Biol.*, 2001, **305**, 1045.
53. I.H. Shih and M.D. Been, *RNA*, 1999, **5**, 1140.
54. M. Matysiak, J. Wrzesinski and J. Ciesiolka, *J. Mol. Biol.*, 1999, **291**, 283.
55. D.A. Harris, R.A. Tinsley and N.G. Walter, *J. Mol. Biol.*, 2004, **341**, 389.
56. J. Sefcikova, M.V. Krasovska, J. Sponer and N.G. Walter, *Nucleic Acids Res.*, 2007, **35**, 1933.
57. M.J. Pereira, D.A. Harris, D. Rueda and N.G. Walter, *Biochemistry*, 2002, **41**, 730.
58. D.A. Harris, D. Rueda and N.G. Walter, *Biochemistry*, 2002, **41**, 12051.
59. S. Jeong, J. Sefcikova, R.A. Tinsley, D. Rueda and N.G. Walter, *Biochemistry*, 2003, **42**, 7727.
60. M.E. Gondert, R.A. Tinsley, D. Rueda and N.G. Walter, *Biochemistry*, 2006, **45**, 7563.
61. E.Z. Eisenmesser, O. Millet, W. Labeikovsky, D.M. Korzhnev, M. Wolf-Watz, D.A. Bosco, J.J. Skalicky, L.E. Kay and D. Kern, *Nature*, 2005, **438**, 117.
62. M.V. Krasovska, J. Sefcikova, K. Reblova, B. Schneider, N.G. Walter and J. Sponer, *Biophys. J.*, 2006, **91**, 626.
63. M.V. Krasovska, J. Sefcikova, N. Spackova, J. Sponer and N.G. Walter, *J. Mol. Biol.*, 2005, **351**, 731.
64. J. Sefcikova, M.V. Krasovska, N. Spackova, J. Sponer and N.G. Walter, *Biopolymers*, 2007, **85**, 392.
65. A. Nehdi and J.P. Perreault, *Nucleic Acids Res.*, 2006, **34**, 584.
66. T.S. Wadkins and M.D. Been, *Nucleic Acids Res.*, 1997, **25**, 4085.
67. M. Puttaraju, A.T. Perrotta and M.D. Been, *Nucleic Acids Res.*, 1993, **21**, 4253.
68. A.T. Perrotta and M.D. Been, *Biochemistry*, 1992, **31**, 16.
69. M.D. Been, *Trends Biochem. Sci.*, 1994, **19**, 251.
70. P.C. Bevilacqua, *Biochemistry*, 2003, **42**, 2259.
71. H.N. Liu, J.J. Robinet, S. Ananvoranich and J.W. Gauld, *J. Phys. Chem. B*, 2007, **111**, 439.
72. Y.H. Jeoung, P.K. Kumar, Y.A. Suh, K. Taira and S. Nishikawa, *Nucleic Acids Res.*, 1994, **22**, 3722.
73. S. Nakano and P.C. Bevilacqua, *J. Am. Chem. Soc.*, 2001, **123**, 11333.

CHAPTER 7
Mammalian Self-Cleaving Ribozymes

ANDREJ LUPTÁK AND JACK W. SZOSTAK

Howard Hughes Medical Institute, Center for Computational and Integrative Biology, and Department of Molecular Biology, Massachusetts General Hospital, 185 Cambridge Street, Boston, MA 02114, USA

7.1 Introduction

The fact that there are catalytic RNAs in mammals should no longer come as any surprise. RNase P, the ribosome and the spliceosome catalyze hydrolytic RNA cleavage, peptide formation and splicing, respectively, most likely without direct involvement of proteins in the catalyzed transformations. However, to find self-cleaving ribozymes encoded in mammalian genomes is perhaps somewhat unexpected. Until recently this group of catalytic RNAs had been associated solely with subviral pathogens and DNA satellite sequences where they are responsible for the production of unit-length RNA transcripts. In the last three years several reports have provided the first evidence for chromosomal self-cleaving RNAs: cofactor-dependent ribozymes in bacteria[1] and primates[2] (although the *in vitro* self-cleavage activity of this primate "CoTC" ribozyme failed to be confirmed[3]), a plant genomic hammerhead ribozyme in *Arabidopsis*[4] and a group I intron-like ribozyme in the pre-rRNA of some slime molds.[5] These reports were followed by the discovery of four self-cleaving ribozymes isolated from the human genome,[6] further suggesting that self-cleaving ribozymes play a bigger role in biology than was previously thought. However, outside of satellite transcripts, self-cleaving ribozymes have not been conceptually integrated into the field of molecular biology. The broad questions about these catalytic RNAs are: How widespread are they in the genomes of higher eukaryotes? What is their biological function? How are they regulated? This chapter discusses our current understanding of the mammalian self-cleaving RNAs, particularly the *CPEB3* ribozyme.

7.2 General Features of Small Self-cleaving Sequences

Naturally occurring small ($\sim$ 50–150 nt) self-cleaving ribozymes catalyze sequence-specific scission of their own RNA strand. This reaction typically starts with the nucleophilic attack of a 2′ hydroxyl on the adjacent scissile phosphate and proceeds *via* a pentavalent phosphorane transition state or intermediate. This transformation is therefore innate to RNA and happens spontaneously at a slow rate at any position, but it can be vastly accelerated by ribozymes and other catalysts. Among the catalytic strategies that have been proposed for ribozymes are: (1) alignment of the nucleophile and the scissile phosphate for an in-line attack, (2) deprotonation of the nucleophile by a base, (3) stabilization of the phosphorane transition state or intermediate by a Lewis or Brønsted–Lowry acid, and (4) stabilization and protonation of the oxyanion leaving group.[7,8] All small ribozymes have been shown to use more than one of these strategies to accelerate RNA scission. The multitude of catalytic strategies that can be employed by ribozymes allows for a large variation in the cleavage rate. In at least one case – the hammerhead ribozyme – the catalytic mechanism may differ in similar, but structurally distinct, constructs of the ribozyme.[9,10]

7.3 Genome-wide Selection of Self-cleaving Ribozymes

Most ribozymes have been discovered by careful analysis of individual gene transcripts. In contrast, *in vitro* selection methodology allows isolation of hitherto unknown functional molecules from a mix of informational molecules and it has been used with great success to select aptamers and ribozymes from random RNA sequences transcribed from synthetic DNA templates.[11] When the starting library is derived from genomic DNA (or cDNA), *in vitro* selection can be used to search for genomically encoded functional RNAs.[12–14]

While selection of aptamers and some ribozymes is straightforward, the isolation and amplification of self-cleaving ribozymes, whether from synthetic or genomic DNA, requires an extra trick in the experimental design. This is because self-cleaved molecules are no longer covalently intact and therefore can not be amplified in a straightforward manner unless the cleavage site sequence is known *a priori*. To circumvent these issues, the RNA library can be circularized so that self-cleavage at any position yields a linear molecule that can be enzymatically religated and then amplified.[15,16] Alternatively, the RNA transcript can be produced from a circular DNA template in a rolling circle fashion[6,17–19] to yield a concatemeric transcript that contains multiple covalently linked copies of the encoded sequence on a single RNA chain (Figure 7.1). If the RNA harbors self-cleavage activity, it will yield products corresponding to progressively shorter multimers of the library. All but the terminal fractions will contain the entire encoded sequence, including the promoter and constant regions (Figure 7.1). However, a transcript cleaved down to monomer-length has the primer-specific constant regions adjacent to each other, preventing it from being amplified. Dimer and longer multimers,

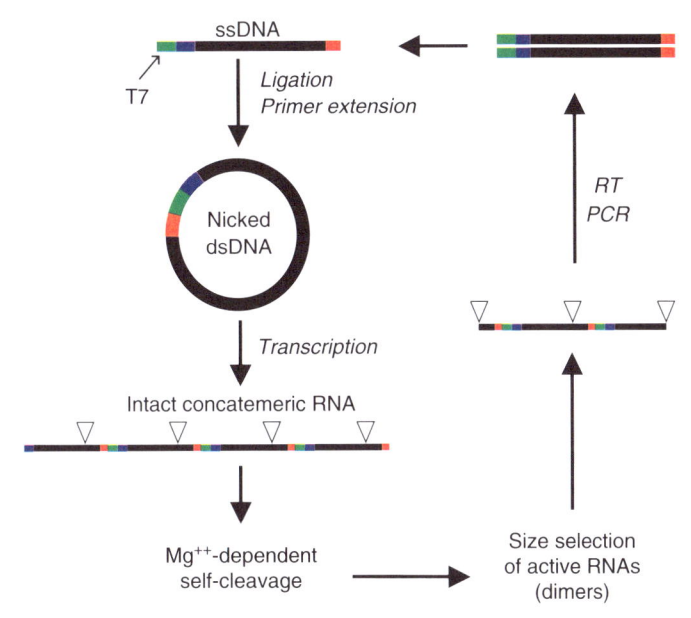

Figure 7.1 *In vitro* selection scheme for sequence-independent isolation of self-cleaving ribozymes. The DNA library is converted into single-stranded form and subsequently circularized by splint-ligation. Primer extension of the splint oligodeoxynucleotide produces a nicked double-stranded circular form of each library sequence that is transcribed into long concatemeric RNA. If the transcript harbors self-cleavage activity (indicated by open triangles), it will yield shorter forms of the transcript upon incubation under selection conditions (in this case in the presence of Mg^{2+}). Dimers and longer concatemers retain all the elements of the library in the same order as the starting library and can be amplified directly by RT-PCR.

though, contain all of the elements of the sequence in the same order as on the DNA template, therefore DNA reverse-transcribed from these multimers can be amplified and the circular template regenerated for the next round of selection (Figure 7.1).

When the rolling-circle transcription-selection procedure was applied to a library of short (~150 nt) human genomic fragments, the selection yielded several sequences (with multiple independent isolates) that exhibited self-cleavage activity.[6] These sequences mapped to the following sites in the human genome: an intron of the *CPEB3* gene, the 5′ UTR of a LINE 1 retroposon, the antisense strand of a monoexonic olfactory receptor gene *OR4K15*, and the antisense strand of the insulin-like growth factor 1 receptor precursor (*IGF1R*) intron 2.

7.4 *CPEB3* Ribozyme

Of the newly discovered human self-cleaving RNAs, only the *CPEB3* ribozyme has so far been analyzed in detail.[6] The ribozyme maps to the second intron of the mammalian cytoplasmic polyadenylation element binding protein 3 locus (the intron is located between the first and the second coding exons of the gene; however, the 5′ UTRs are also spliced, therefore we refer to the ribozyme-containing intron as the second intron of the gene). Like the *CPEB3* gene, this ribozyme is highly conserved among mammals but does not have orthologs in other eukaryotes that possess genes from the *CPEB* family. CPEBs are RNA-binding proteins that regulate cytoplasmic mRNA translation and have been implicated in *Xenopus* oogenesis and activity-dependent local translational regulation in rat neurons.[20] The mouse CPEB3 is expressed in various tissues and is upregulated in mouse hippocampus upon kainate-induced seizure.[21] The CPEB3 protein binds an RNA sequence that is different from the canonical cytoplasmic polyadenylation elements bound by CPEB1 and may therefore regulate translation of mRNAs (*e.g.*, the mRNA of GluR2, an AMPA receptor) differing from the targets of other members of the CPEB family.[22] The *CPEB3* ribozyme is located from ~10 to 25 kbp upstream of the third exon, depending on the species. At this point it is not clear how it affects the transcription or translation of the *CPEB3* gene, given that its self-cleavage rate is rather slow (in the order of ~1 h^{-1} in physiological conditions *in vitro*). Thus in the absence of other factors that may upregulate its activity, or significant downstream transcriptional pausing, cleavage occurs too slowly to affect splicing of exons 2 and 3. This is because the Pol II RNA polymerase transcribes at ~1000 nt min^{-1} and would synthesize the remainder of the intron in 10–15 min. Given that the next splice site would be marked for splicing almost immediately,[3,23] the ribozyme is not likely to affect the splicing of the *CPEB3* mRNA. However, if the ribozyme cleavage rate is accelerated (*e.g.*, by the binding of a trans acting factor) or if the RNA polymerase pauses for a significant period of time, the self-cleavage could affect the integrity of the *CPEB3* pre-mRNA. Whether the cleaved RNA is further processed by exo-nucleases or whether the cleavage products are stable and further utilized for some cellular function is currently under investigation.

7.4.1 Expression of the *CPEB3* Ribozyme

Several expressed sequence tags (ESTs) corresponding to the self-cleaved products of the ribozyme have been detected in human, mouse and opossum tissues, suggesting that the *CPEB3* ribozyme is expressed and active *in vivo*. All of the ESTs share a common 5′ terminus that is identical to the cleavage site mapped *in vitro*, suggesting that the ribozyme – the sequence immediately downstream of the cleavage site – is not rapidly digested by a 5′→3′ exo-nuclease and may be stable enough to protect the 5′ end of the transcript. The cellular fate of the transcript after ribozyme self-cleavage is not known. 5′ RACE (rapid amplification of 5′ cDNA ends) analysis of total RNA samples from various human and murine tissues confirmed that the ribozyme is expressed and self-cleaved prior to RNA isolation, suggesting that self-cleavage occurs *in vivo*.[6] Self-cleaved ribozyme was detected in human testis and placenta and in murine brain, testis and spleen tissues. In addition, initial analysis of the ribozyme cleavage state in various tissues using RT-PCR amplicons that either span or neighbor the cleavage site suggested that self-cleavage is tissue specific. Subsequent quantitative RT-PCR (RT-qPCR) experiments showed that in mouse the ribozyme expression (*i.e.*, steady-state levels of ribozyme-containing RNA) varies over about two orders of magnitude, with the highest expression occurring in ovaries and kidneys (A.L. and J.W.S., unpublished). Skeletal muscle exhibits the lowest ribozyme levels among the tissues we have tested while the full-length *CPEB3* mRNA expression is high, suggesting the ribozyme-containing intron is not constitutively retained.

To further probe the cellular status of the ribozyme, RT-qPCR was used to measure the relative levels of uncleaved ribozyme and the downstream product of self-cleavage. In most murine tissues where it was detected, the ribozyme was detected in self-cleaved form (*i.e.*, the levels of the cleavage products were significantly higher than the levels of uncleaved precursor; A.L. and J.W.S., unpublished) while in bone marrow, skeletal muscle and lung tissues the ribozyme was found almost completely uncleaved. Since the role of this ribozyme in the regulation of the *CPEB3* gene is not known, the implications of these preliminary data remain uncertain; however, they do suggest tissue-specific regulation of the ribozyme self-cleavage rate.

7.4.2 Structural Features of the *CPEB3* and HDV Ribozymes

Initial biochemical characterization of the *CPEB3* ribozyme showed that it requires divalent metal ions for activity (low millimolar Mg^{2+}, Mn^{2+}, Ca^{2+} and Co^{2+} support self-cleavage; however, 3 M Li^+ does not) and is active in 4 M urea, suggesting that it folds into a stable structure. In addition, near neutral pH the ribozyme has a flat kinetic pH profile that drops off at low and high pH and the cleavage rate exhibits a solvent kinetic isotope effect. These observations suggest that the rate-limiting step most likely involves a transfer of at least one proton. Finally, the minimum sequence required for self-cleavage maps downstream of the cleavage site. Taken together, these experiments revealed that the ribozyme's biochemical properties closely resemble those of the ribozymes isolated from the human Hepatitis Delta Virus (HDV, Figure 7.2A and B) (Chapter 6). Threading the *CPEB3* sequence through the secondary structure of the HDV ribozymes revealed that the *CPEB3* ribozyme can form the same double pseudoknot structure as the HDV ribozymes (Figure 7.2C). Mapping all of the known variants of the ribozyme from different mammalian sequences and from *in vitro* selected isolates onto the proposed secondary structure revealed that none of the known mutations significantly perturb the HDV-derived structure. The ribozyme from the dog genome (*Canis familiaris*, *Cf*) is the least like the human sequence, but its mutations affect only four base-pairs (out of twenty five) in the proposed structure, converting one G-C pair into an A-U pair, two base-pairs into wobble pairs and one wobble-pair into a C-U mispair (Figure 7.2C). Most other variants of

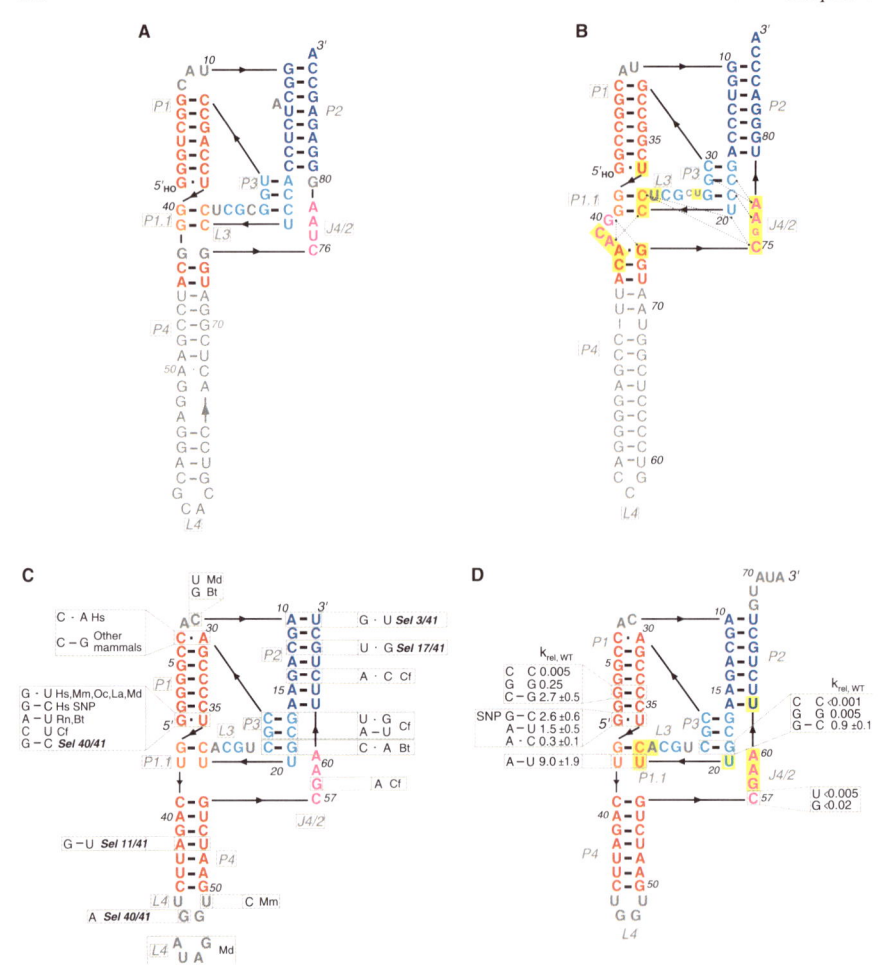

Figure 7.2 Secondary structures of the self-cleaved HDV antigenomic (*A*), genomic (*B*) and mammalian *CPEB3* (*C, D*) ribozymes; 5' indicates the cleavage site. Paired regions (P1–P4) that are found stacked in the crystal structures of the genomic HDV ribozyme[24,25] are aligned above each other and colored with similar colors. Positions of phosphorothioate interference in the genomic HDV[33] and the human *CPEB3*[6] ribozymes are indicated by yellow boxes (*B, D*). Known *CPEB3* mammalian variants and mutants of the human sequence isolated by the *in vitro* selection are shown in (*C*). Rates of self-cleavage, relative to the wild-type human sequence, of select mutants are shown in (*D*).

the ribozyme support the secondary structure with most mutations retaining base-pairing interactions or affecting non-essential single-stranded elements. Kinetic analysis of mutants prepared to test the predicted secondary structure further confirmed that the *CPEB3* ribozyme forms the same base-pairing interactions as do the HDV ribozymes (Figure 7.2D). The HDV ribozymes fold to form an intricate tertiary structure in which two co-axially stacked helical regions (orange-red and blue in Figure 7.2) are stabilized by four cross-over strands (Chapter 6). The crystal structures of the precursor and product forms of the genomic (viral "plus" strand) HDV ribozyme show that the overall fold changes only slightly upon self-cleavage and that the ribozyme active site contains an essential cytosine provided by the J4/2 strand.[24,25] This active-site cytosine (C75), which in the product form of the ribozyme has a somewhat elevated pK_a compared with a base-paired or free cytosine,[26] is thought to be directly involved in proton shuttling during the transesterification reaction.[27–31] The *CPEB3* ribozyme has a cytosine (C57) at the analogous position. Mutation of the cytosine to guanosine or uridine abolishes the ribozyme self-cleavage activity, suggesting that it has the same role in catalysis as does C75 in the genomic HDV ribozyme. This suggestion is further supported by interference experiments, which identify those phosphates in the RNA backbone that are sensitive to sulfur substitution at the *pro*-R$_p$ position.[32] The phosphorothioate-sensitive positions map to the L3 and J4/2 regions of both the *CPEB3* ribozyme (Figure 7.2D) and a short version of the genomic HDV ribozyme[33] and overlap with positions that form the solvent-inaccessible core.[24,25,34]

7.4.3 Linkage of HDV to the Human Transcriptome

The complex structure of the HDV ribozyme implies that its secondary structure elements provide spacing and register for the interhelical cross-over elements, so that they can participate in the correct tertiary interactions. Thus both the structural complexity and the phylogenetic conservation imply an informational complexity of the sequence, *i.e.*, sequences that can fold into the HDV-like fold are rare compared with other, simpler self-cleaving ribozymes. Two recent *in vitro* selection experiments designed to probe the informational complexity of the antigenomic HDV ribozyme have confirmed that the core elements of the ribozyme are surprisingly resistant to mutation.[35,36] In addition, when we searched a library of self-cleaving RNAs selected from a random starting library[37] for any sequence that can fold into an HDV-like structure, we found none. Likewise when we searched the conserved elements of the human genome using a secondary structure motif descriptor (RNABOB) we found none that could fit the HDV secondary structure and conserved sequences of the joining strands, supporting the hypothesis that the HDV-like fold is rare and that HDV-like ribozymes are unlikely to have arisen independently multiple times. This is in contrast to the hammerhead ribozyme, which has been independently isolated from random sequences several times[37] and occurs in many subviral plant pathogens and transcripts of satellite DNAs of amphibians, schistosomes and crickets, as well as in the genome of *Arabidopsis thaliana*.[4,38–42]

Given the high informational complexity of the HDV and *CPEB3* ribozymes, it is possible that they are evolutionarily linked. Although the base-paired sequences of these ribozymes are highly divergent – even between the genomic and antigenomic HDV ribozymes – the fact that their secondary structure and single stranded regions of the ribozyme core are so conserved suggests that they may have a common ancestor. Rapid evolution of sequences that satisfy secondary and tertiary structure constraints among various structured RNA orthologs shows that in catalytic RNAs the primary sequence does not have to be conserved among phylogenetically related molecules, except at a few critical positions. The direct implication of the common-ancestor hypothesis is that either the mammalian lineages acquired a self-cleaving ribozyme from HDV (or its ancestor) or that the HDV ribozymes were acquired from the mammalian transcriptome. We favor the scenario in which the HDV single-stranded RNA genome arose from the human transcriptome, because the virus is unique to humans with no known similar pathogens isolated from other animals. HDV is a satellite of the human Hepatitis B Virus (HBV), which it requires for infection because HDV uses the HBV coat proteins.[43] While HBV is infectious by itself and is common among many animals, even non-mammals, HDV has only been isolated from human HBV-infected hepatocytes. Although we cannot rule out the possibility that an HDV-like virus does exist in other animals or that it once existed in non-human mammals, current data support the hypothesis that HDV is a uniquely human pathogen.

The HDV virus codes for three known elements: two self-cleaving ribozymes and one protein, the delta antigen. The delta antigen is required for viral replication and has been suggested to be similar to, and to interact with a host protein, the vertebrate-conserved delta antigen-interacting protein A (DIPA).[44] Thus all three HDV elements have related molecules in the human genome that are highly conserved in all mammals. If HDV is unique to humans, then the less likely scenario is one in which the *CPEB3* ribozyme was acquired from an HDV ancestor by early mammals and DIPA by early vertebrates, but that the virus remains only in humans.

7.5 Possible Biological Roles of Self-cleaving Ribozymes

What are the biological roles of self-cleaving ribozymes? Cleavage of the CoTC element – whether caused by the RNA itself or by another factor – induces transcriptional termination.[45] Similar roles could in theory be played by self-cleaving sequences encoded by other transcripts – whether a LINE1 element or the *CPEB3* pre-mRNA. However, these two ribozymes, together with self-cleaving ribozymes located antisense to known genes, do not map to the 3′ UTRs of the genes, and therefore they cannot have as simple a role as the CoTC element.

In the *CPEB3* gene, the ribozyme may affect the fate of the *CPEB3* pre-mRNA by generating products of self-cleavage that may be further utilized by the cell. If the *CPEB3* mRNA is split due to ribozyme self-cleavage but correctly spliced, exported and polyadenylated, then it may end up being translated into shorter versions of the *CPEB3* protein. For this to occur, new termini produced by self-cleavage would have to be either processed – the upstream product polyadenylated and/or the downstream product capped, both perhaps following nucleolytic digestion – or recognized as *bona fide* mRNA termini that can be used directly in translation. This last option is particularly attractive given that the HDV ribozyme fold is very stable: the structure is sufficiently robust to enable it to function in high molar urea or formamide and to retain half of the secondary structure in 18 M formamide.[46–48] In addition, the HDV ribozymes have been shown to self-cleave at elevated temperatures. The structural stability may be used in the *CPEB3* gene to create a stable, exonuclease-inaccessible 5′ end of the self-cleaved pre-mRNA. For the same reason, such a 5′ terminus would most likely not be capped; therefore, if it were to be translated, it would have to act through a mechanism akin to cap-independent internal ribosomal entry.

Another potential role of self-cleaving ribozymes may be trans cleavage. Trans cleavage – scission of an RNA strand other than or far removed from the one containing the ribozyme sequence – has been described for all small self-cleaving ribozymes, in some cases even if they contained the "substrate" strand.[18,19] If such trans cleavage can be performed by the mammalian ribozymes, it may represent a novel network of gene regulation. In addition, trans cleavage may link the expression and self-cleavage of a ribozyme encoded on one strand to scission of the opposite strand, if the "substrate" strand is either palindromic or recognized by base pairing alone (as is the case in the P1 region of HDV/*CPEB3* ribozymes). Such mechanism may allow self-cleaving ribozymes that are anti-sense to the human *OR4K15* and *IGF1R* genes to affect the sense RNA expression. Recent evidence suggests that the mammalian transcriptome contains many "non-coding" RNAs that are expressed from intergenic regions and strands that are antisense to protein-coding genes.[49–51] Thus it is possible that ribozymes oriented antisense to protein-coding genes are expressed and active *in vivo*.

7.6 Closing Remarks

The discovery of new non-coding RNAs has often been followed by the description of fascinating new cellular roles for these RNAs. Whatever the biological roles of self-cleaving ribozymes may be, their conservation and tissue-specific expression hints at their importance and possible roles in gene expression regulation. At this point we do not know how many ribozymes are present in the human genome, but *in vitro* selections performed under different conditions than were used by Salehi *et al.*[6] and in the presence of co-factors should let us estimate whether the ribozymes discussed above are unique or whether they represent just the tip of an (RNA) iceberg.

References

1. W.C. Winkler, A. Nahvi, A. Roth, J.A. Collins and R.R. Breaker, *Nature*, 2004, **428**, 281–286.

2. A. Teixeira, A. Tahiri-Alaoui, S. West, B. Thomas, A. Ramadass, I. Martianov, M. Dye, W. James, N.J. Proudfoot and A. Akoulitchev, *Nature*, 2004, **432**, 526–530.
3. M.J. Dye, N. Gromak and N.J. Proudfoot, *Mol. Cell*, 2006, **21**, 849–859.
4. R. Przybilski, S. Graf, A. Lescoute, W. Nellen, E. Westhof, G. Steger and C. Hammann, *Plant Cell*, 2005, **17**, 1877–1885.
5. H. Nielsen, E. Westhof and S. Johansen, *Science*, 2005, **309**, 1584–1587.
6. K. Salehi-Ashtiani, A. Lupták, A. Litovchick and J.W. Szostak, *Science*, 2006, **313**, 1788–1792.
7. D.M. Perreault and E.V. Anslyn, *Angew. Chem. Int. Ed.*, 1997, **36**, 432–450.
8. G.M. Emilsson, S. Nakamura, A. Roth and R.R. Breaker, *RNA*, 2003, **9**, 907–918.
9. A. Khvorova, A. Lescoute, E. Westhof and S.D. Jayasena, *Nat. Struct. Biol.*, 2003, **10**, 708–712.
10. M. Martick and W.G. Scott, *Cell*, 2006, **126**, 309–320.
11. D.S. Wilson and J.W. Szostak, *Annu. Rev. Biochem.*, 1999, **68**, 611–647.
12. B.S. Singer, T. Shtatland, D. Brown and L. Gold, *Nucleic Acids Res.*, 1997, **25**, 781–786.
13. T. Shtatland, S.C. Gill, B.E. Javornik, H.E. Johansson, B.S. Singer, O.C. Uhlenbeck, D.A. Zichi and L. Gold, *Nucleic Acids Res.*, 2000, **28**, E93.
14. S. Kim, H. Shi, D.K. Lee and J.T. Lis, *Nucleic Acids Res.*, 2003, **31**, 1955–1961.
15. T. Pan and O.C. Uhlenbeck, *Biochemistry*, 1992, **31**, 3887–3895.
16. V.K. Jayasena and L. Gold, *Proc. Natl. Acad. Sci. U.S.A.*, 1997, **94**, 10612–10617.
17. S.L. Daubendiek and E.T. Kool, *Nat. Biotechnol.*, 1997, **15**, 273–277.
18. A.M. Diegelman and E.T. Kool, *Nucleic Acids Res.*, 1998, **26**, 3235–3241.
19. A.M. Diegelman and E.T. Kool, *Chem. Biol.*, 1999, **6**, 569–576.
20. J.D. Richter, *Proc. Natl. Acad. Sci. U.S.A.*, 2001, **98**, 7069–7071.
21. M. Theis, K. Si and E.R. Kandel, *Proc. Natl. Acad. Sci. U.S.A.*, 2003, **100**, 9602–9607.
22. Y.S. Huang, M.C. Kan, C.L. Lin and J.D. Richter, *EMBO J.*, 2006, **25**, 4865–4876.
23. S.A. Lacadie, D.F. Tardiff, S. Kadener and M. Rosbash, *Genes Dev.*, 2006, **20**, 2055–2066.
24. A.R. Ferré-D'Amaré, K. Zhou and J.A. Doudna, *Nature*, 1998, **395**, 567–574.
25. A. Ke, K. Zhou, F. Ding, J.H. Cate and J.A. Doudna, *Nature*, 2004, **429**, 201–205.
26. A. Lupták, A.R. Ferre-D'Amare, K. Zhou, K.W. Zilm and J.A. Doudna, *J. Am. Chem. Soc.*, 2001, **123**, 8447–8452.
27. A.T. Perrotta, I. Shih and M.D. Been, *Science*, 1999, **286**, 123–126.
28. S. Nakano, D.M. Chadalavada and P.C. Bevilacqua, *Science*, 2000, **287**, 1493–1497.

29. S. Nakano and P.C. Bevilacqua, *J. Am. Chem. Soc.*, 2001, **123**, 11333–11334.
30. I. Shih and M.D. Been, *Proc. Natl. Acad. Sci. U.S.A.*, 2001, **98**, 1489–1494.
31. S.R. Das and J.A. Piccirilli, *Nat. Chem. Biol.*, 2005, **1**, 45–52.
32. E.L. Christian and M. Yarus, *J. Mol. Biol.*, 1992, **228**, 743–758.
33. N.S. Prabhu, G. Dinter-Gottlieb and P.A. Gottlieb, *Nucleic Acids Res.*, 1997, **25**, 5119–5124.
34. S.P. Rosenstein and M.D. Been, *Biochemistry*, 1996, **35**, 11403–11413.
35. A. Nehdi and J.P. Perreault, *Nucleic Acids Res.*, 2006, **34**, 584–592.
36. M. Legiewicz, A. Wichlacz, B. Brzezicha and J. Ciesiolka, *Nucleic Acids Res.*, 2006, **34**, 1270–1280.
37. K. Salehi-Ashtiani and J.W. Szostak, *Nature*, 2001, **414**, 82–84.
38. G.A. Prody, J.T. Bakos, J.M. Buzayan, I.R. Scheneider and G. Bruining, *Science*, 1986, **231**, 1577–1580.
39. L.M. Pabon-Pena, Y. Zhang and L.M. Epstein, *Mol. Cell Biol.*, 1991, **11**, 6109–6115.
40. Y. Zhang and L.M. Epstein, *Gene*, 1996, **172**, 183–190.
41. G. Ferbeyre, J.M. Smith and R. Cedergren, *Mol. Cell Biol.*, 1998, **18**, 3880–3888.
42. A.A. Rojas, A. Vazquez-Tello, G. Ferbeyre, F. Venanzetti, L. Bachmann, B. Paquin, V. Sbordoni and R. Cedergren, *Nucleic Acids Res.*, 2000, **28**, 4037–4043.
43. M.M. Lai, *Annu. Rev. Biochem.*, 1995, **64**, 259–286.
44. R. Brazas and D. Ganem, *Science*, 1996, **274**, 90–94.
45. S. West, N. Gromak and N.J. Proudfoot, *Nature*, 2004, **432**, 522–525.
46. S.P. Rosenstein and M.D. Been, *Biochemistry*, 1990, **29**, 8011–8016.
47. J.B. Smith and G. Dinter-Gottlieb, *Nucleic Acids Res.*, 1991, **19**, 1285–1289.
48. J. Duhamel, D.M. Liu, C. Evilia, N. Fleysh, G. Dinter-Gottlieb and P. Lu, *Nucleic Acids Res.*, 1996, **24**, 3911–3917.
49. P. Kapranov, S.E. Cawley, J. Drenkow, S. Bekiranov, R.L. Strausberg, S.P. Fodor and T.R. Gingeras, *Science*, 2002, **296**, 916–919.
50. E.M. Reis, H.I. Nakaya, R. Louro, F.C. Canavez, A.V. Flatschart, G.T. Almeida, C.M. Egidio, A.C. Paquola, A.A. Machado, F. Festa, D. Yamamoto, R. Alvarenga, C.C. da Silva, G.C. Brito, S.D. Simon, C.A. Moreira-Filho, K.R. Leite, L.H. Camara-Lopes, F.S. Campos, E. Gimba, G.M. Vignal, H. El-Dorry, M.C. Sogayar, M.A. Barcinski, A.M. da Silva and S. Verjovski-Almeida, *Oncogene*, 2004, **23**, 6684–6692.
51. S. Katayama, Y. Tomaru, T. Kasukawa, K. Waki, M. Nakanishi, M. Nakamura, H. Nishida, C.C. Yap, M. Suzuki, J. Kawai, H. Suzuki, P. Carninci, Y. Hayashizaki, C. Wells, M. Frith, T. Ravasi, K.C. Pang, J. Hallinan, J. Mattick, D.A. Hume, L. Lipovich, S. Batalov, P.G. Engstrom, Y. Mizuno, M.A. Faghihi, A. Sandelin, A.M. Chalk, S. Mottagui-Tabar, Z. Liang, B. Lenhard and C. Wahlestedt, *Science*, 2005, **309**, 1564–1566.

CHAPTER 8
The Structure and Action of glmS Ribozymes

KRISTIAN H. LINK[a] AND RONALD R. BREAKER[a,b,c]

[a] Howard Hughes Medical Institute, Yale University, New Haven, Connecticut 06520-8103, USA; [b] Department of Molecular, Cellular and Developmental Biology, Yale University, New Haven, Connecticut 06520-8103, USA; [c] Department of Molecular Biophysics and Biochemistry, Yale University, New Haven, Connecticut 06520-8103, USA

8.1 Introduction

Riboswitches are structured RNA elements that most commonly reside within the 5′ untranslated regions (UTRs) of bacterial mRNAs where they control gene expression *via* direct binding of small molecules.[1–3] These RNA-based regulatory motifs have been shown to specifically bind small molecule metabolites, including coenzymes, amino acids, and nucleobases. Riboswitches are typically composed of two different functional domains: an aptamer domain, which binds the small-molecule metabolite, and an expression platform, which interfaces with factors that are responsible for expressing genes. Ligand binding typically stabilizes one structural state of a riboswitch aptamer, and subsequently alters the folding pathway taken by the expression platform located immediately downstream of (or sometimes overlapping) the aptamer.

Aptamer domains are frequently used to classify riboswitches because metabolite recognition requires a structural organization that is well conserved phylogenetically, whereas expression platforms can vary considerably in both sequence and structure. In some eukaryotes, the characteristic consensus sequence and structure of a riboswitch aptamer is flanked on each side by nucleotides that serve as split expression platform components. For example, in the fungus *Neurospora crassa* this divided expression platform controls access to intron nucleotides that are critical for mRNA splicing, and metabolite binding alters the structures near these key nucleotides and thereby controls alternative splicing.[4]

In bacteria, the vast majority of riboswitches appear to control gene expression by modulating transcription termination or translation initiation. Computer-aided nucleotide sequence analyses[5-10] have shown that many riboswitch aptamers reside upstream of intrinsic transcription terminators,[11,12] and some riboswitches have been experimentally demonstrated to modulate transcription termination upon ligand binding *in vitro*.[13-21] Additionally, bioinformatics[6,8,9,22] and biochemical[13,23,24] studies have demonstrated that the binding status of the aptamer can influence the accessibility of ribosome binding sites within some mRNAs.

One of the most distinct riboswitch classes known is represented by RNA control elements termed *glmS* ribozymes (Figure 8.1A). Representative *glmS* ribozymes are always located within the 5′ UTRs of *glmS* mRNAs, as reported for 18 Gram-positive bacteria.[23,25] *glmS* ribozymes undergo self cleavage by internal phosphoester transfer in a process that is activated by glucosamine-6-phosphate (GlcN6P, **1**) (Figure 8.1B). GlcN6P is a metabolic product of the enzyme encoded by *glmS* mRNAs, which is L-glutamine:fructose 6-phosphate amidotransferase.[26] GlcN6P production is the first step in the biosynthesis of UDP-GlcNAc. This nucleotide-modified glucosamine compound is used in bacterial cell wall biosynthesis and is essential for bacterial cell survival.[26]

As noted above, *glmS* ribozymes have some characteristics that are unique among riboswitches that suggest they control gene expression using a distinct regulatory pathway. The most prominent difference between typical riboswitches and *glmS* ribozymes is that *glmS* ribozymes undergo efficient site-specific self-cleavage that is accelerated by at least several orders of magnitude by GlcN6P.[23,25,27,28] The rate constant for self-cleavage of the *Bacillus subtilis glmS* ribozyme is $\sim 3 \text{ min}^{-1}$ in the presence of saturating amounts of GlcN6P and divalent metal ions.[23,27] *glmS* ribozymes from some other species exhibit equal or even greater rate constants for self-cleavage.[28,29] Therefore, the rate enhancements generated upon GlcN6P binding are at least 10 000-fold over that measured in the absence of ligand[27,28] and at least 10-million-fold over the uncatalyzed rate constant measured for spontaneous RNA transesterification.[30] Furthermore, *glmS* ribozymes respond linearly to increases in GlcN6P concentrations, which is consistent with a 1:1 complex between ribozyme and GlcN6P.[23]

Self-cleavage occurs *via* a phosphoester transfer reaction in which the 2′ oxygen of the labile linkage functions as a nucleophile and attacks the phosphorus center at the cleavage site. The products of the cleavage reaction possess a 5′-hydroxyl group and a 2′,3′-cyclic phosphate group on the 3′ and 5′ RNA fragments, respectively. Since *glmS* ribozymes reside within the 5′ UTRs of *glmS* mRNAs, self-cleavage cannot directly deactivate translation by disrupting the continuity of the open reading frame (ORF). Although the precise mechanism for gene control has yet to be reported, it has been demonstrated that ribozyme activity is essential for gene repression.[23]

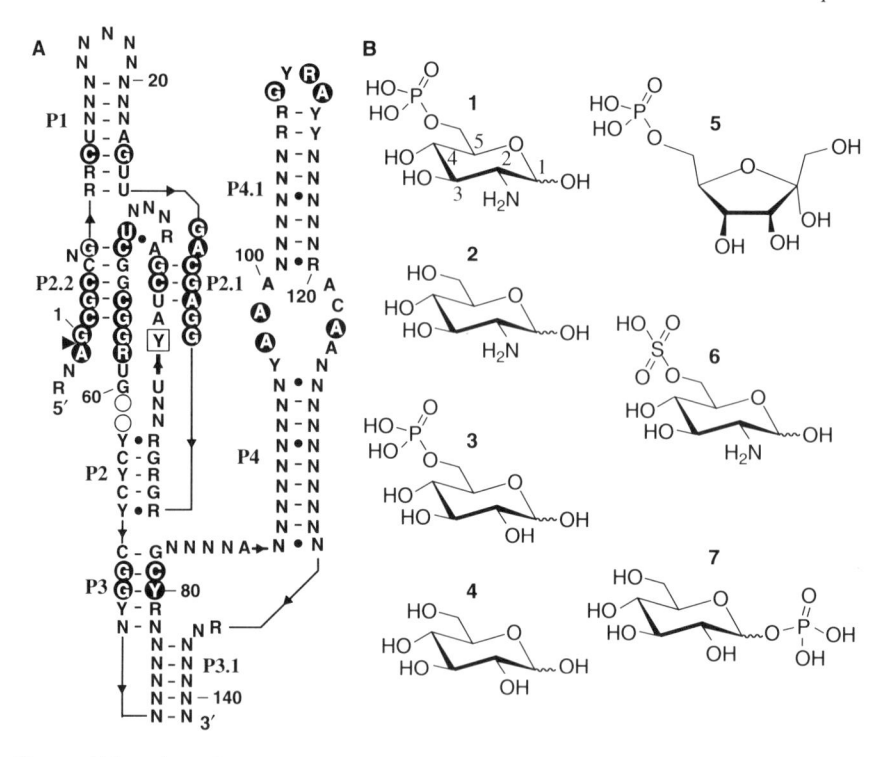

Figure 8.1 *glmS* ribozymes and molecular recognition. (*A*) Consensus sequence and secondary structure model for the most common *glmS* ribozyme type. Nucleotide numbers conform to those used previously for the *Bacillus cereus* RNA.[45] The boxed nucleotide is absent from the *B. cereus* sequence, and open circles represent the two extra nucleotides present in *B. cereus* and *B. anthracis* ribozymes that are not present in all *glmS* ribozyme representatives. Nucleotides in black circles are conserved in at least 97% of *glmS* ribozyme representatives known. N represents any nucleotide identity, where some stretches are variable in length. Nucleotides denoted Y are either C or U, and nucleotides denoted R are either G or A. The arrowhead identifies the site of ribozyme cleavage. This figure is adapted from Ref. 45 with permission. (*B*) Chemical structures of various compounds used initially[23] to probe the molecular recognition characteristics of the *B. subtilis glmS* ribozyme.

8.2 Biochemical Characteristics of *glmS* Ribozymes

8.2.1 Divalent Metal Ions Support Structure and Not Chemistry

Most highly-structured RNAs require divalent metal ions to support their complex-folded structures and their resulting biological activities. The structures and actions of *glmS* ribozymes are no exception. However, *glmS* ribozymes display relaxed divalent metal ion specificity,[23,27] as a diverse collection of divalent metals can activate the ribozyme (*i.e.*, Mn^{2+}, Ca^{2+}, Co^{2+} and Sr^{2+}).

Relaxed divalent ion specificity has also been observed for other natural self-cleaving ribozymes, including the hairpin, hammerhead and VS ribozymes.[31–34] These ribozymes do not rely on direct interactions with divalent metal ions to support catalysis. Instead, they require the presence of a positive charge in the form of metal complexes, organic polycations or monovalent ions to support proper structure formation.[35] Similarly, K^+ ions can support self-cleavage of glmS ribozymes,[27] but to a lesser extent than Mg^{2+}. The ability of monovalent ions to support self-cleavage of the glmS ribozyme indicates that metal ions are required for structure and are not directly involved in chemistry.

To further probe the role metal ions play in glmS ribozyme self-cleavage, reactions were carried out in the presence of $Co(NH_3)_6^{3+}$. It was found[25] that the glmS ribozyme from *Bacillus cereus* is activated to near maximal levels in the presence of $Co(NH_3)_6^{3+}$. This result suggests that glmS ribozymes do not require inner-sphere divalent metal ion coordination for either catalysis or folding. Direct coordination of divalent metal ions with non-bridging phosphate oxygen (NBPO) atoms also can be assessed by using thiophosphate substitutions. This was carried out at the site of cleavage within the *B. cereus* glmS ribozyme.[27] Substantial changes in the rate constant resulting from the substitutions are expected to reflect major changes in the affinity of the ribozyme for different divalent metal ions.[36,37] For example, if either of the NPBOs is directly coordinated by divalent metal ions, thiophosphate substitution should cause a reduced affinity for hard metals such as Mg^{2+} compared with the unmodified phosphate, which should result in reduced ribozyme activity. Furthermore, soft metals such as Mn^{2+} should restore much of the lost activity observed when Mg^{2+} is used with the thiophosphate-containing ribozyme, owing to the higher affinity that Mn^{2+} has for sulfur.

The results of thiophosphate substitution reveals that the glmS ribozyme displays only about a one-third reduction in the rate constant for self-cleavage compared with the wild type when 10 mM Mg^{2+} is present. This same reduction in the rate constant for self-cleavage was also observed when substituted ribozymes were allowed to self-cleave in the presence of 10 mM Mn^{2+}. The modest reduction in rate constant upon thiophosphate introduction, and the inability of Mn^{2+} to fully rescue the activity of these ribozymes indicates that divalent metal ions are not coordinated to the NBPOs during catalysis. Thus, the characteristics of metal ion function with glmS ribozymes are consistent with structural and not catalytic roles. Furthermore, glmS ribozymes are unlikely to employ catalytic strategies[38] that utilize divalent metal ions as Lewis acids that directly facilitate deprotonation or that neutralize negative charges on the labile internucleotide linkage.

8.2.2 Ligand Specificity of *glmS* Ribozymes

A striking feature of glmS ribozymes is their ability to specifically recognize GlcN6P and exclude even closely-related analogs from activating ribozyme function. For example, the *B. subtilis* glmS ribozyme is selectively activated by

GlcN6P (**1**), but discriminates by at least two orders of magnitude against glucosamine (**2**) and glucosamine-6-sulfate (**6**)[23,39] (Figure 8.1B). Therefore, the RNA is one of only a few natural metabolite-sensing RNAs that make productive interactions with phosphate moieties of their ligands.[13,40]

Some compounds that are completely ineffective at activating *glmS* ribozymes,[23] such as glucose (**4**), fructose-6-phosphate (**5**), and glucose-1-phosphate (**7**), are very different from the target ligand. However, glucose-6-phosphate (**3**), which differs from GlcN6P only by the replacement of an amine group with a hydroxyl group, also results in no ribozyme activity. Both functional groups can serve as hydrogen bond donors, so the chemical distinction made by the ribozyme must involve characteristics that are different from the formation of a single hydrogen bond. Indeed, biochemical evidence indicates that **3** can inhibit ribozyme self-cleavage when present in millimolar concentrations,[28] suggesting that **3** might be a competitive inhibitor. This conclusion is supported by X-ray crystallography data,[41,42] which show both **3** and GlcN6P can bind in the same pocket and contact the RNA linkage to be cleaved by the ribozyme. Although the affinity that the ribozyme exhibits for **3** is somewhat less than the $\sim 200\,\mu\text{M}$ K_D demonstrated for GlcN6P,[23] the inactivity of **3** must be due to something other than poor molecular recognition of the analog.

The ligand specificity of *glmS* ribozymes have for GlcN6P was further established by a structure–activity relationship (SAR) analysis using the *B. cereus glmS* ribozyme.[43] Of the twelve GlcN6P analogs used in this study, only compounds **10**, **14**, **15**, **18** and **19** (Figure 8.2A) activate ribozyme function to a detectable level. The findings generally indicate that the ribozyme forms a restricted pocket for GlcN6P that likely involves hydrogen bonding interactions with most of the hydroxyl groups and the phosphate group of the ligand. For example, the reduced ability of the phosphorothioate analog **10** to activate ribozyme function indicates that a phosphate oxygen atom is important for specific recognition by the ribozyme. However, since inner-sphere coordination by Mg^{2+} had previously been ruled out,[27] the NBPOs most likely participate in hydrogen-bonding interactions.

The effects of the changes to the amine group were most intriguing. For example, compound **15** carries a methyl group on the amine, but retains the ability to activate the ribozyme. In contrast, **16**, which carries the quaternary amine modification, is rendered inactive as a ribozyme ligand. Perhaps most striking is the action of **18**, which differs from GlcN6P in its stereochemical configuration at position 3 of the ring. Compound **18** can activate the ribozyme to within 1/100th the activity induced by GlcN6P despite the fact that the amine at position 3 is critical for ribozyme activation. The critical role that this amine plays in ribozyme activation either does not require this group to be in a single precise position in space, or the ligand binding pocket is sufficiently flexible to maintain the position of the amine regardless of stereochemical differences. The GlcN6P molecular recognition determinants were predicted (Figure 8.2B) based on the cumulative SAR data[43] and both confirmed and expanded by examining X-ray crystallography data[41,42] (see below).

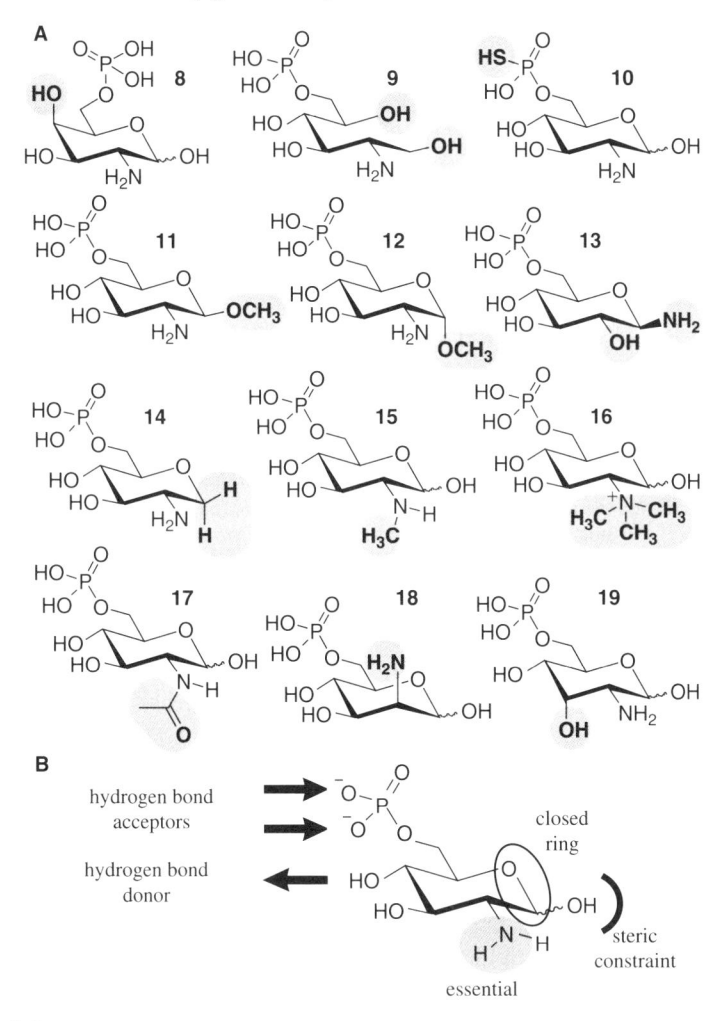

Figure 8.2 GlcN6P analogs and molecular recognition determinants. (*A*) Chemical structures of GlcN6P analogs used for structure–activity relationship (SAR) analysis of the *Bacillus cereus glmS* ribozyme.[43] Shaded portions denote chemical differences of the various analogs from GlcN6P. (*B*) Molecular recognition characteristics of *glmS* ribozymes predicted previously based upon SAR data.[43] Figure adapted from Ref. 43 with permission.

8.2.3 Evidence for a Coenzyme Role for GlcN6P

Although *glmS* ribozymes are highly selective for GlcN6P recognition, it was found that the commonly used buffer tris(hydroxymethyl)aminomethane buffer (Tris; compound **20**, Figure 8.3) is also capable of activating *glmS* ribozymes.[27,28] However, Tris is not nearly as potent an activator as GlcN6P. It was noted initially[23] that saturating GlcN6P concentrations increase the rate

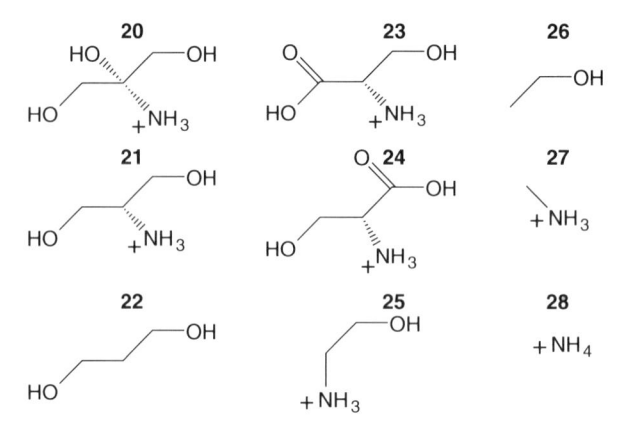

Figure 8.3 Chemical structures of compounds used to probe the importance of functional groups on Tris (**20**) in ribozyme activation. The most potent activator compound in this series is serinol (**21**), which is a closer mimic of the portion of GlcN6P that carries the 2-amino group than is Tris.

constant for *B. subtilis glmS* ribozyme self-cleavage by 1000 fold compared to the rate constant measured without ligand. However, these measurements were conducted in the presence of 50 mM Tris-HCl, which is a concentration providing a more than two orders of magnitude increase in rate constant[28] over that measured for the uncatalyzed transesterification of RNA under similar conditions.[30] Therefore, at least some *glmS* ribozymes are activated by more than 100 000-fold when bound to their target ligand.

One possible explanation for the activity of Tris is that GlcN6P is not a true allosteric effector that activates the ribozyme by inducing structural changes. In contrast, the ligand might serve as a coenzyme by supplying a key functional group that participates in the chemical step of RNA cleavage. Indeed, several observations are consistent with this hypothesis. For example, the fact that glucose-6-phosphate (or any other compound lacking an amine group) cannot support ribozyme self-cleavage suggests the amine is essential for chemical activation. As noted above, glucose-6-phosphate functions as an inhibitor of ribozyme function, indicating that it might bind to the same site as GlcN6P but remains inactive when bound. Furthermore, there is little or no modulation of RNA folding when GlcN6P binds to *glmS* ribozymes,[23,44] which suggest that the ligand is a ribozyme cofactor and not an allosteric effector.

If GlcN6P is a cofactor for *glmS* ribozymes, then the activity of Tris can be rationalized because its structure mimics a part of the GlcN6P ring that carries the essential amine group. This hypothesis was explored by using a series of small molecules similar to Tris to assess the ability of compounds with various configurations of amine and hydroxyl groups (Figure 8.3) to activate *glmS* ribozymes.[28] It was observed that several other amine-containing compounds also activate *B. cereus glmS* ribozymes to varying extents. With the exception of D-serine (**24**), all of the compounds that contained at least one hydroxyl group and an amine group supported detectable levels of ribozyme activity.[28]

Serinol (**21**) was a more potent activator than Tris, which is consistent with the fact that serinol is a closer mimic of the amine-carrying portion of GlcN6P. The inactivity of D-serine might be due to the different stereochemical orientations of the functional groups relative to L-serine and to GlcN6P. Most likely, the hydroxyl groups (and the phosphate moiety) provide molecular recognition contacts that increase ligand affinity and aid in the positioning of the amine group, which directly participates in the chemical step of RNA self-cleavage.

Additional evidence for coenzyme function of GlcN6P comes from an *in vitro* selection study that was performed on the *glmS* ribozyme from *Bacillus cereus*.[45] The nucleotide sequence of the parental ribozyme was mutagenized such that each nucleotide had a 3% chance of carrying a mutation. The population was then subjected to a selection for activation with GlcN6P or GlcN6P analogs (**1–4**, **6**, **9**, **11**, **12**). Although numerous variants were isolated that are activated by GlcN6P and by GlcN6S (**6**), the population of variant RNAs did not exhibit any activation by analogs that cannot activate the parental ribozyme. If GlcN6P is being used as an allosteric effector, it might have been relatively easy for the ribozyme to accrue mutations that permit other effectors to trigger the proper folding change in the ribozyme. In contrast, if GlcN6P is being used as a coenzyme, the simultaneous demands of molecular recognition and accurate coenzyme positioning within the binding site should make it exceedingly uncommon to encounter variants that display the ability to self-cleave with GlcN6P analogs.

8.3 Atomic-resolution Structure of *glmS* Ribozymes

8.3.1 Secondary and Tertiary Structures of *glmS* Ribozymes

Four forms of the *Thermoanaerobacter tengcongensis glmS* ribozyme have been solved by X-ray crystallography.[41] These are (1) ribozyme with synthetic substrate containing a 2′-deoxy modification at the cleavage site, (2) ribozyme with synthetic substrate containing 2′-amino modification at the cleavage site, (3) ribozyme with synthetic substrate containing 2′-amino modification at the cleavage site with the inhibitor glucose-6-phosphate bound, and (4) the cleaved ribozyme product. The chemical modifications carried in the first three constructs prevent cleavage of the otherwise labile linkage because the 2′ hydroxyl nucleophile has been removed. Despite the differences in these constructs, all four structures are essentially identical in their global architecture, which implies that both the active site and the metabolite-binding site are rigid and preformed. A similar structure model has also been proposed using X-ray crystallography data derived from the *glmS* ribozyme from *B. anthracis*.[42]

The ribozyme forms three helical regions that packed side by side (Figure 8.4). The longest helical region is composed of a coaxial stack of P1, P2.2, P2, P3, and P3.1 structures that together span approximately 100 Å. This stack and the stack composed of P4 and P4.1 sandwich a centrally-located P2.1 structure. The P2.1 and P3.1 elements are pseudoknots that help form the

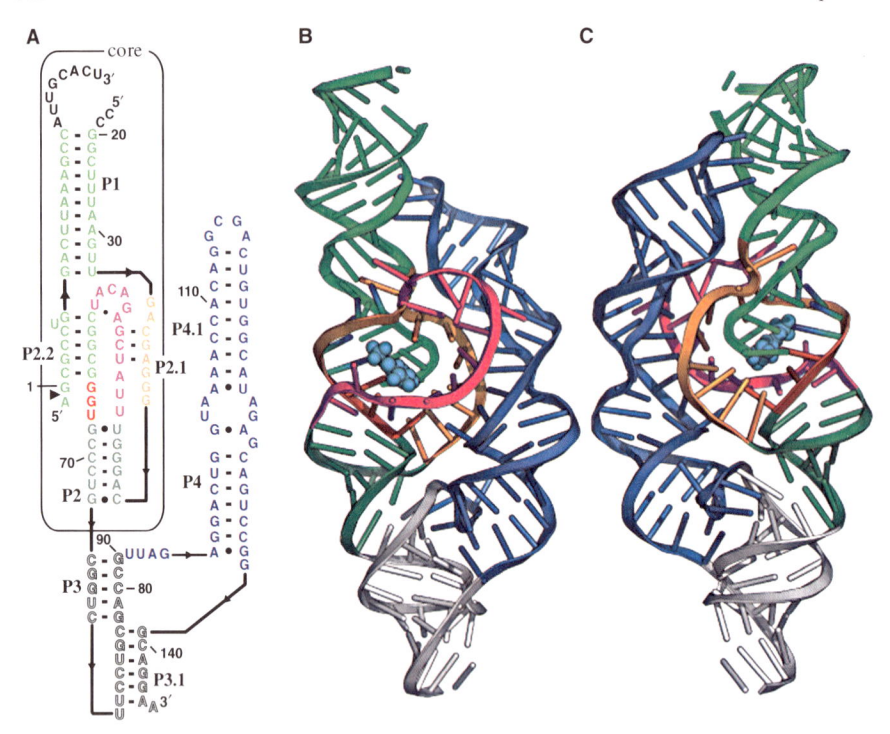

Figure 8.4 Secondary and tertiary structure models for the *T. tengcongensis glmS* ribozyme. (*A*) Secondary structure.[41] P1 nucleotides shown in black are not found in the wild-type ribozyme and are included in the construct engineered for crystallization. (*B*) Ribbon diagram of the *T. tengcongensis glmS* ribozyme bound to glucosamine-6-phosphate (space-filling model). Colored regions of the ribozyme correspond to the colored nucleotides in (*A*). The image was prepared with data from the protein data bank (PDB) file 2HOZ using PyMOL. (*C*) Image as in (*B*) rotated ∼180°.

active site and that tether two substructures together, respectively. Both these pseudoknots were not evident from the initial sequence analyses.[23] However, the P2.1 pseudoknot was nearly predicted[46] and the P3.1 pseudoknot was demonstrated biochemically[27,29] before being confirmed by crystallographic analyses.[41,42]

This history of *glmS* ribozyme secondary-structure prediction and revisions reveals some of the challenges that researchers have when modeling new-found RNAs. Originally, a P2a stem was predicted to form due to base-pairing potential present in all known representatives of this ribozyme class. Despite the apparent thermodynamic favorability of this proposed interaction, the RNA actually folds to form the P2.1 and P2.2 stems (Figure 8.1A) that preclude formation of P2a. In addition, the P3.1 pseudoknot interaction does not require specific nucleotide identities to form this structure, and therefore the lack of sequence conservation made it difficult to identify by comparative sequence analysis. Thus, a synthesis of data from several approaches, including comparative sequence analysis, biochemical probing, and X-ray crystallography, was required to produce an accurate representation of the full *glmS* ribozyme architecture.

As initially noted,[23] the *glmS* ribozyme from *B. subtilis* retains much of its ligand-binding and self-cleavage activities even when the P3 through P4 stems are deleted. Furthermore, active variants of this ribozyme isolated by *in vitro* selection[45] that retain their GlcN6P-triggered activity carry most mutations outside the region encompassing stems P1 through P2. These findings suggested that conserved nucleotides and structures that reside outside the P1 through P2 core participate as accessory components required for the core to exhibit maximal rate enhancement. If true, then it seemed possible that alternative accessory structures might exist naturally that also support the core structure and function. As expected, a distinct variant of the P3 through P4 region was identified in the *glmS* ribozyme of *Deinococcus radiodurans*.[27] This variant ribozyme retains the molecular recognition and chemical characteristics of the typical core, but likely supports this core with an alternate accessory domain architecture.

These observations are explained by the atomic-resolution structure that reveals the P1 through P2 core binds the ligand. The nucleotides at the roof of the active site are highly conserved and participate in base triples that bracket the three-way junction between P1, P2.2 and P2.1. The local structure of the P1 through P2 core appears to be supported by the particular fold of the P3 through P4 stems. Thus, although the core of the ribozyme exhibits ligand-mediated activation, the presence of the conserved P3 through P4 region assists the core in cleaving RNA under low Mg^{2+} concentrations that more closely mimic those expected inside cells.[27]

8.3.2 Metabolite Recognition by *glmS* Ribozymes

Shortly after the structure of the *T. tengcongensis glmS* ribozyme bound to glucose-6-phosphate was reported,[41] the structure of the *B. anthracis glmS* ribozyme was solved with GlcN6P bound.[42] The structures reveal that both compounds bind at the same site using similar molecular recognition contacts. For example, GlcN6P is bound by the *B. anthracis* ribozyme by the formation of contacts to the both the phosphate and hydroxyl groups of the ligand (Figure 8.5). The phosphate moiety is recognized by two fully hydrated Mg^{2+} ions. This observation is consistent with biochemical and SAR data that suggested the RNA interacts with Mg^{2+} through outer-sphere interactions[27] and that the phosphate was an important moiety for molecular recognition.[43] Interestingly, thiamine pyrophosphate (TPP) riboswitches also recognize phosphate moieties of their ligand *via* hexahydrated Mg^{2+}.[47–50] The use of divalent metal ions to bridge between RNA and phosphate moieties might be a general way that negatively-charged ligands can be bound by RNA, despite the anionic character of these polymers.

144	Chapter 8

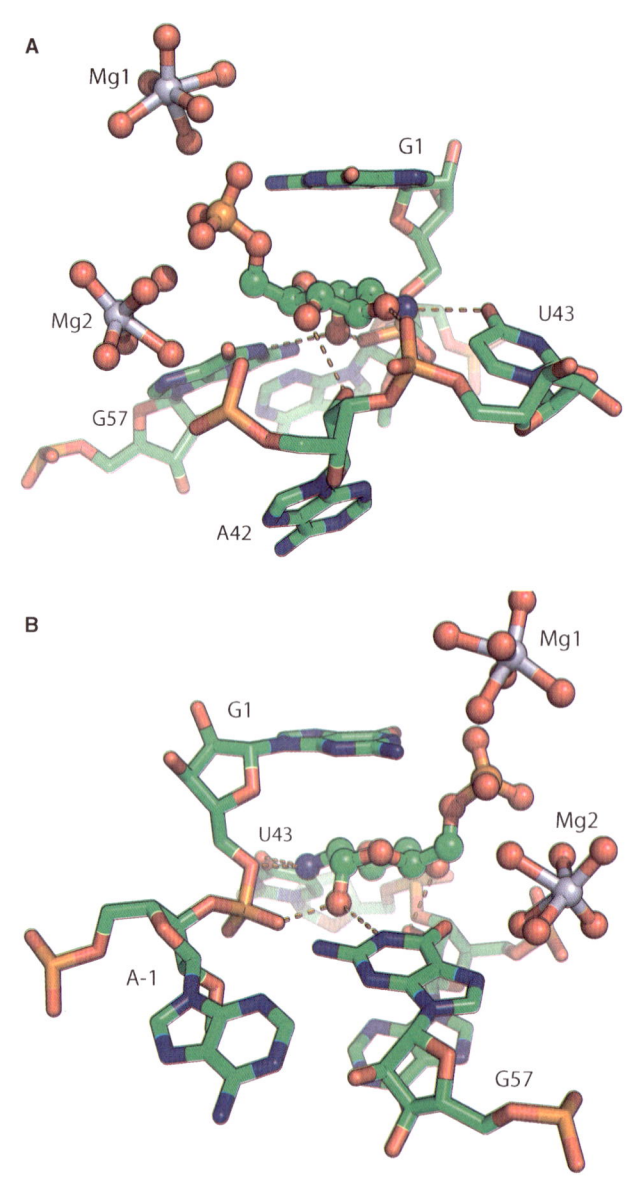

Figure 8.5 Recognition of GlcN6P by hexahydrated Mg^{2+} ions and functional groups of nucleotides of the *B. anthracis glmS* ribozyme. (*A*) Model of a portion of the *glmS* ribozyme binding pocket with potential hydrogen-bonding interactions depicted as dashed lines. Carbon atoms are green, nitrogen atoms are blue, oxygen atoms are red and phosphorus atoms are orange. The image was prepared with data from PDB file 2NZ4 using PyMOL. (*B*) Ligand binding pocket rotated ~180° from the view in (*A*).

The ribozyme also interacts with the sugar *via* an extensive set of hydrogen bonding interactions that recognize all four of the exocyclic functional groups of the glucosamine ring. The hydroxyl groups at the C1, C3 and C4 position of the ring all appear to be forming hydrogen bonds to various parts of the ribozyme. For example, the C1-hydroxyl group resides close to the N1 position of G57 of the *B. anthracis* RNA, and similarly is close to the *pro*-S$_p$ nonbridging oxygen of the scissile phosphate moiety. This restricts the space near this position, which explains why methylation of the C1-hydroxyl group strongly disrupts activity.[43] The C3-OH and C4-OH of GlcN6P are involved in hydrogen bonding interactions with the *pro*-R$_p$ oxygen of U43 and the 2′-OH of A42, respectively. However, altering the stereochemical configuration of the C3-OH group causes only a modest decrease in the activity of the ligand, suggesting that the ribozyme can accommodate this change without substantial loss of ligand affinity or disruption of the active site.

The most intriguing functional group positioning is that observed for the C2-amine group. In the *B. anthracis* structure, the amine of GlcN6P is within hydrogen-bonding distance of the O4 of U43 and the 5′-oxygen atom of nucleotide G1, which is the leaving group for the reaction catalyzed by the ribozyme. Since the amine group of GlcN6P is essential for ribozyme activity, unsurprisingly, the amine group is one of only two direct contacts made between the ligand and the atoms of the labile linkage. Both removal of the C1-hydroxyl group and altering the stereochemical configuration of the C2-amine group cause substantial losses of ribozyme activity.[43] The roles these groups might play in the chemical step of ribozyme action are discussed in more detail below.

8.4 Mechanism of *glmS* Ribozyme Self-cleavage

There are four major catalytic strategies an enzyme can employ to accelerate the breaking of phosphodiester bonds.[38,51] These are (1) in-line positioning of nucleophile and leaving group, (2) deprotonation of the 2′-oxygen nucleophile, (3) neutralization of negative charge on the non-bridging oxygen atoms of the scissile phosphate, and (4) protonation of the 5′-oxygen leaving group. Divalent metal ions could directly participate in these strategies, but biochemical data[27] indicate that divalent metal ions are dispensable, and structural data[41,42] indicate that no metal ions make contact with the labile linkage. Only the GlcN6P ligand is making contact to the cleavage site phosphodiester linkage. If GlcN6P is acting as a cofactor, it is expected to be held in a position to perform one or more of these strategies.

The structure of the *glmS* ribozyme bound to GlcN6P reveals that the nucleophile and leaving group are held in close proximity and are preorganized for chemistry (Figure 8.6). Moreover, the angle formed between the 2′-oxygen nucleophile, the phosphorus, and the 5′-oxygen leaving group is 165°, which is very close to the optimal angle of 180° for in-line attack.[52] The closest functional group to the A-1 2′-oxygen atom that could potentially act as a

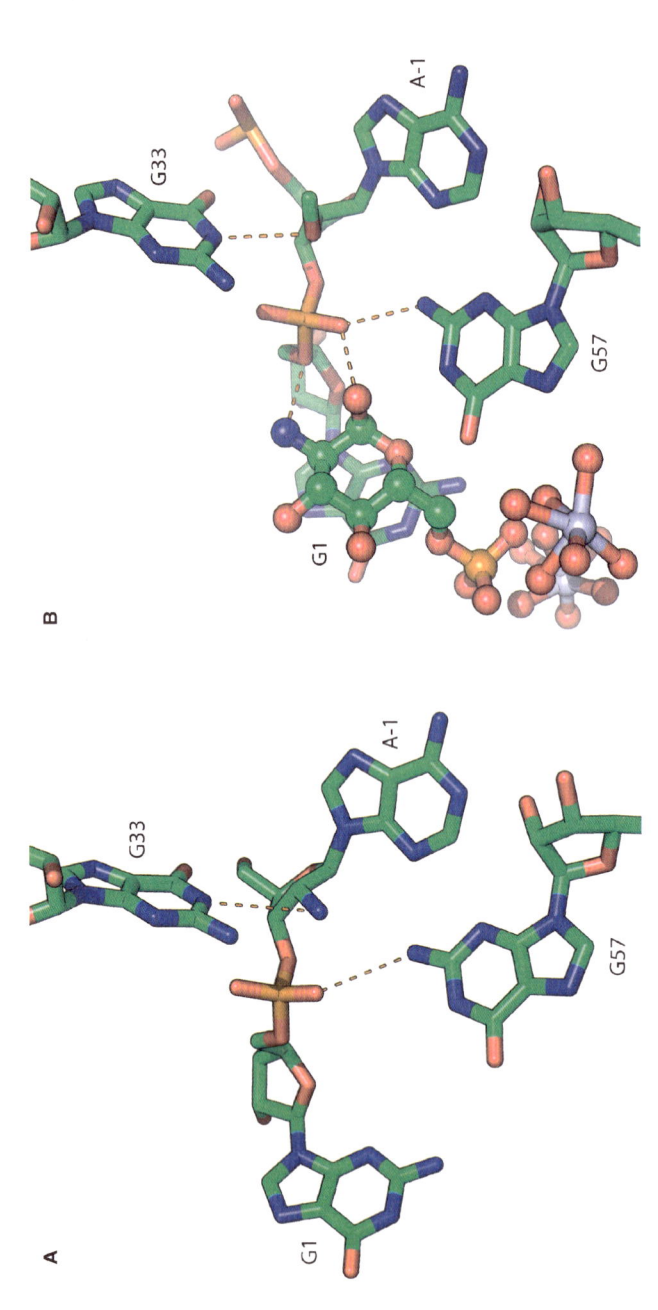

Figure 8.6 Comparison of *glmS* ribozyme active site structures in the absence and in the presence of GlcN6P. (*A*) Structure of the *T. tengcongensis glmS* ribozyme active site[41] in the absence of ligand reveals G33 in a position that could assist in deprotonation of the 2′-oxygen nucleophile. In addition, the exocyclic amine of G57 is positioned favorably for neutralization of the negative charge on the *pro*-S$_p$ NBPO. Atoms are colored as described in the legend to Figure 8.5(A) and the image was prepared as described in the legend to Figure 8.4(B). Note that the nucleophilic 2′ hydroxyl group is replaced with an amine group in this structure. (*B*) Structure of the *B. anthracis glmS* ribozyme active site[42] in the presence of GlcN6P. The analogous G33 and G57 nucleotides are positioned similarly to that for the ligand-free structure depicted in (*A*). In addition, the C1 hydroxyl group of GlcN6P is positioned where it could neutralize a negative charge on the *pro*-S$_p$ NBPO, and the C2-amino group of GlcN6P is positioned where it could protonate the 5′-oxygen leaving group. Atoms are colored as described in the legend to Figure 8.5(A) and the image was prepared as described in the legend to Figure 8.5(B). Note that the nucleophilic 2′ hydroxyl group is replaced with 2′-methoxy group in this structure.

general base and deprotonate the 2′-hydroxyl group in the wild-type ribozyme is the N1 of the strictly conserved G33 nucleotide. Mutation of G33 to any of the other 3 nucleotides reduces the rate of the reaction by at least 1000-fold compared with the wild-type ribozyme,[42] indicating it has a critical role in supporting ribozyme function.

Stabilization of negative charge on the scissile phosphate could potentially be achieved *via* interactions with both GlcN6P and nucleobases. The exocyclic amine of G57 and the C1-hydroxyl group of GlcN6P are within hydrogen bonding distance of the *pro*-S$_p$ NBPO on this phosphate. The importance of the exocyclic amine of G57 is evident from the interference observed with inosine and *N*-methylguanosine substitutions,[53] and the importance of the C1-hydroxyl group of GlcN6P or its analogs also has been demonstrated.[28,43] Therefore, the exocyclic amine of G57 and the C1-hydroxyl group of GlcN6P could accelerate self-cleavage by neutralizing the negative charge on the *pro*-S$_p$ NBPO.

The best candidate for protonation of the leaving group is the C2-amine group of GlcN6P. This functional group (and its replacement in glucose-6-phosphate) is positioned within hydrogen bonding distance of the 5′-oxygen leaving group.[41,42] Such a critical role for the amine group would explain why it is absolutely required for ribozyme action, as protonation of the leaving group could contribute many orders of magnitude in rate acceleration.[38] However, the precise role of the amine group needs clarification by additional experimentation. The rate constant for ribozyme cleavage increases log–linearly with decreasing proton concentrations,[23,27,28,54] and this fact was used to support speculation that the amine might be involved in deprotonating the 2′-hydroxyl group at the cleavage site.[43]

However, the location of the amine group in the atomic-resolution structure of the *B. anthracis* RNA rules out a role for the amine as a general base in this structural model. It has been suggested[42] that the increase in rate constant with increasing pH is not due to a change in the chemical step of ribozyme function, but results because the ribozyme binds the deprotonated (–NH$_2$) form of GlcN6P. If true, then ribozymes will more easily become bound to GlcN6P, but presumably would need to have the bound ligand revert to the protonated (–NH$_3^+$) form to function as a general acid and protonate the 5′-oxyanion leaving group.

Phosphoryl transfer reactions could employ either associative or dissociative mechanisms.[55,56] An associative mechanism involves the initial attack of the 2′-oxygen nucleophile, which results in the formation of a pentavalent transition state (or intermediate),[55] followed by subsequent elimination of the 5′-oxygen leaving group. In contrast, the dissociative mechanism involves elimination of the 5′-oxygen leaving group, resulting in formation of an unstable metaphosphate intermediate, followed by nucleophilic attack of the 2′-oxygen atom. One way to discriminate between associative or dissociative mechanisms is to determine the stereochemistry of the 2′,3′-cyclic phosphate product when using thiophosphate-substituted substrates. A dissociative mechanism should produce racemic products, whereas an associative mechanism should cause inversion of stereochemistry. Although representatives of several self-cleaving ribozyme classes have been shown to function *via* an associative or S$_N$2-like mechanism,[57,58] the *glmS* ribozyme might employ GlcN6P as a general acid to protonate 5′-oxygen leaving group and therefore might operate *via* a dissociative or S$_N$1-like mechanism.

The initial report on *glmS* ribozymes[23] noted that its dual role as riboswitch and ribozyme might be due to some unique role that this RNA has compared with other non-reactive riboswitches. We now speculate that the molecular recognition challenge faced by the *glmS* ribozyme is made exceedingly difficult by the presence of both GlcN6P and glucose-6-phosphate in cells. Although *glmS* ribozymes discriminate modestly against glucose-6-phosphate, they might rely on the chemical role of the amine group for self-cleavage to create an even greater level of discrimination against natural GlcN6P analogs. In other words, the cofactor role of GlcN6P could be exploited by cells to ensure that compounds like glucose-6-phosphate do not mistakenly cause repression of GlcN6P production.

8.5 Can *glmS* Ribozymes be Drug Targets?

The increasing number of antibiotic resistant strains of microbial pathogens,[59] coupled with the small number of new chemical scaffolds for novel antibiotics,[60] seriously threatens our ability to control microbial infections. Riboswitches might be attractive new targets for the development of novel antibiotics because nearly all riboswitch classes discovered to date control the expression of genes whose protein products participate in the biosynthesis or import of essential metabolites. It is noteworthy that some compounds known for decades to inhibit bacterial or fungal growth have been found to target riboswitches that bind TPP[61] or lysine.[62,63] Furthermore, riboswitches are unique RNA targets for drug development because they represent nearly all of the RNAs known to have evolved to form structured receptors that purposefully bind small organic compounds.[64]

glmS ribozymes might also be good targets for a new class of antibiotics because they are found in several microbial pathogens, including *B. anthracis* and *Staphylococcus aureus*.[23,25] Each *glmS* ribozyme represses the expression of the adjoining *glmS* gene whose product is essential for the growth of *B. subtilis*.[65] Therefore, an analog of GlcN6P that triggers ribozyme action without serving as a GlcN6P supplement might deactivate *glmS* gene expression to a level that is deleterious to the pathogen. The high level of sequence conservation surrounding the active site of *glmS* ribozymes indicates that a single compound is likely to hit all ribozymes in this structural class.

The identification of analogs that trigger ribozyme activity should be facilitated by the available atomic-resolution structures of *glmS* ribozymes from two different species of bacteria. These structures provide opportunities for designing compounds that could efficiently trigger ribozyme self-cleavage using rational design strategies such as computer-aided modeling. Moreover, two high-throughput screening assays[39,66] have recently been developed that can be used to screen for compounds in large chemical libraries that trigger *glmS* ribozyme activity. Although research and development efforts that target RNA structures with small molecules are not as common as those intent on targeting proteins, the characteristics of riboswitches should allow the use of similar strategies and techniques to those used to develop other small-molecule drugs.

8.6 Conclusions

glmS ribozymes represent a unique type of "ribozyme-riboswitch" that reside within the 5′ UTRs of *glmS* genes in various Gram-positive bacteria. Unlike other classes of riboswitches, *glmS* ribozymes undergo efficient self-cleavage in the presence of GlcN6P, which is the metabolite product of the enzyme encoded by *glmS* mRNA. Biochemical and crystallographic data support structural (and not catalytic) roles for bound divalent metal ions. In contrast, these data demonstrate that the RNA specifically recognizes GlcN6P though nucleobase- and Mg^{2+} ion-mediated contacts to sugar hydroxyl groups or the phosphate

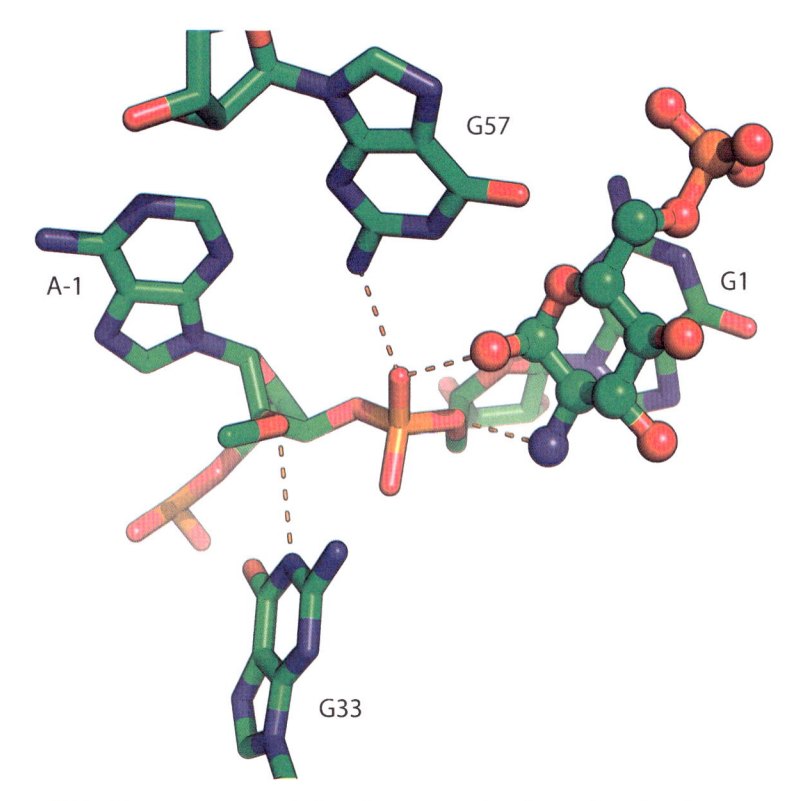

Figure 8.7 Key contacts made by ribozyme and cofactor functional groups to the labile phosphodiester linkage of *glmS* ribozymes. Dashed lines identify possible binding partners whose interactions could facilitate RNA cleavage.

moiety, respectively. Structural data support a mechanism of self-cleavage wherein a bound GlcN6P molecule functions as a coenzyme to protonate the 5′-oxyanion leaving group (Figure 8.7), although additional biochemical data are required to confirm mechanistic details. Furthermore, since *glmS* ribozymes control essential genes within several microbial pathogens, these RNAs might serve as targets for a new class of antibiotics that bind to *glmS* ribozymes and trigger self-cleavage even when microbes are starved for GlcN6P.

References

1. W.C. Winkler and R.R. Breaker, *Annu. Rev. Microbiol.*, 2005, **59**, 487.
2. B.J. Tucker and R.R. Breaker, *Curr. Opin. Struct. Biol.*, 2005, **15**, 342.
3. M. Mandal and R.R. Breaker, *Nat. Rev. Mol. Cell Biol.*, 2004, **5**, 451.
4. M.T. Cheah, A. Wachter, N. Sudarsan and R.R. Breaker, *Nature*, 2007, **447**, 497.
5. W.C. Winkler, A. Nahvi, N. Sudarsan, J.E. Barrick and R.R. Breaker, *Nat. Struct. Biol.*, 2003, **10**, 701.
6. D.A. Rodionov, A.G. Vitreschak, A.A. Mironov and M.S. Gelfand, *Nucleic Acids Res.*, 2003, **31**, 6748.
7. A.G. Vitreschak, D.A. Rodionov, A.A. Mironov and M.S. Gelfand, *RNA*, 2003, **9**, 1084.
8. D.A. Rodionov, A.G. Vitreschak, A.A. Mironov and M.S. Gelfand, *J. Biol. Chem.*, 2002, **277**, 48949.
9. A.G. Vitreschak, D.A. Rodionov, A.A. Mironov and M.S. Gelfand, *Nucleic Acids Res.*, 2002, **30**, 3141.
10. A. Nahvi, J.E. Barrick and R.R. Breaker, *Nucleic Acids Res.*, 2004, **32**, 143.
11. I. Gusarov and E. Nudler, *Mol. Cell*, 1999, **3**, 495.
12. W.S. Yarnell and J.W. Roberts, *Science*, 1999, **284**, 611.
13. W.C. Winkler, S. Cohen-Chalamish and R.R. Breaker, *Proc. Natl. Acad. Sci. U.S.A.*, 2002, **99**, 15908.
14. A.S. Mironov, I. Gusarov, R. Rafikov, L.E. Lopez, K. Shatalin, R.A. Kreneva, D.A. Perumov and E. Nudler, *Cell*, 2002, **111**, 747.
15. B.A.M. McDaniel, F.J. Grundy, I. Artsimovitch and T.M. Henkin, *Proc. Natl. Acad. Sci. U.S.A.*, 2003, **100**, 3083.
16. V. Epshtein, A.S. Mironov and E. Nudler, *Proc. Natl. Acad. Sci. U.S.A.*, 2003, **100**, 5052.
17. N. Sudarsan, J.K. Wickiser, S. Nakamura, M.S. Ebert and R.R. Breaker, *Genes Dev.*, 2003, **17**, 2688.
18. M. Mandal and R.R. Breaker, *Nat. Struct. Mol. Biol.*, 2004, **11**, 29.
19. M. Mandal, M. Lee, J.E. Barrick, Z. Weinberg, G.M. Emilsson, W.L. Ruzzo and R.R. Breaker, *Science*, 2004, **306**, 275.
20. J.K. Wickiser, W.C. Winkler, R.R. Breaker and D.M. Crothers, *Mol. Cell*, 2005, **18**, 49.
21. F.J. Grundy, S.C. Lehman and T.M. Henkin, *Proc. Natl. Acad. Sci. U.S.A.*, 2003, **100**, 12057.

22. D.A. Rodionov, A.G. Vitreschak, A.A. Mironov and M.S. Gelfand, *J. Biol. Chem.*, 2003, **278**, 41148.
23. W.C. Winkler, A. Nahvi, A. Roth, J.A. Collins and R.R. Breaker, *Nature*, 2004, **428**, 281.
24. X.W. Nou and R.J. Kadner, *Proc. Natl. Acad. Sci. U.S.A.*, 2000, **97**, 7190.
25. J.E. Barrick, K.A. Corbino, W.C. Winkler, A. Nahvi, M. Mandal, J. Collins, M. Lee, A. Roth, N. Sudarsan, I. Jona, J.K. Wickiser and R.R. Breaker, *Proc. Natl. Acad. Sci. U.S.A.*, 2004, **101**, 6421.
26. S. Milewski, *Biochim. Biophys. Acta*, 2002, **1597**, 173.
27. A. Roth, A. Nahvi, M. Lee, I. Jona and R.R. Breaker, *RNA*, 2006, **12**, 607.
28. T.J. McCarthy, M.A. Plog, S.A. Floy, J.A. Jansen, J.K. Soukup and G.A. Soukup, *Chem. Biol.*, 2005, **12**, 1221.
29. S.R. Wilkinson and M.D. Been, *RNA*, 2005, **11**, 1788.
30. Y.F. Li and R.R. Breaker, *J. Am. Chem. Soc.*, 1999, **121**, 5364.
31. H.C.T. Guo, D.M. Deabreu, E.R.M. Tillier, B.J. Saville, J.E. Olive and R.A. Collins, *J. Mol. Biol.*, 1993, **232**, 351.
32. Y.A. Suh, P.K.R. Kumar, K. Taira and S. Nishikawa, *Nucleic Acids Res.*, 1993, **21**, 3277.
33. B.M. Chowrira, A. Berzalherranz and J.M. Burke, *Biochemistry*, 1993 **32**, 1088.
34. S.C. Dahm and O.C. Uhlenbeck, *Biochemistry*, 1991, **30**, 9464.
35. M.J. Fedor, *Curr. Opin. Struct. Biol.*, 2002, **12**, 289.
36. V.L. Pecoraro, J.D. Hermes and W.W. Cleland, *Biochemistry*, 1984 **23**, 5262.
37. E.K. Jaffe and M. Cohn, *J. Biol. Chem.*, 1979, **254**, 839.
38. G.M. Emilsson, S. Nakamura, A. Roth and R.R. Breaker, *RNA*, 2003 **9**, 907.
39. K. Blount, I. Puskarz, R. Penchovsky and R.R. Breaker, *RNA Biol.*, 2006, **3**, 77.
40. W. Winkler, A. Nahvi and R.R. Breaker, *Nature*, 2002, **419**, 952.
41. D.J. Klein and A.R. Ferre-D'Amare, *Science*, 2006, **313**, 1752.
42. J.C. Cochrane, S.V. Lipchock and S.A. Strobel, *Chem. Biol.*, 2007, **14**, 97.
43. J. Lim, B.C. Grove, A. Roth and R.R. Breaker, *Angew. Chem. Int. Ed.*, 2006, **45**, 6689.
44. K.J. Hampel and M.M. Tinsley, *Biochemistry*, 2006, **45**, 7861.
45. K.H. Link, L.X. Guo and R.R. Breaker, *Nucleic Acids Res.*, 2006, **34**, 4968.
46. G.A. Soukup, *Nucleic Acids Res.*, 2006, **34**, 968.
47. S. Thore, M. Leibundgut and N.N. Ban, *Science*, 2006, **312**, 1208.
48. A. Serganov, A. Polonskaia, A.T. Phan, R.R. Breaker and D.J. Patel, *Nature*, 2006, **441**, 1167.
49. J. Noeske, C. Richter, E. Stirnal, H. Schwalbe and J. Wohnert, *ChemBioChem.*, 2006, **7**, 1451.
50. T.E. Edwards and A.R. Ferre-D'Amare, *Structure*, 2006, **14**, 1459.
51. R.R. Breaker, G.M. Emilsson, D. Lazarev, S. Nakamura, I.J. Puskarz, A. Roth and N. Sudarsan, *RNA*, 2003, **9**, 949.
52. G.A. Soukup and R.R. Breaker, *RNA*, 1999, **5**, 1308.

53. J.A. Jansen, T.J. McCarthy, G.A. Soukup and J.K. Soukup, *Nat. Struct. Mol. Biol.*, 2006, **13**, 517.
54. R.A. Tinsley, J.R.W. Furchak and N.G. Walter, *RNA*, 2007, **13**, 468.
55. A. Fersht, *Enzyme Structure and Mechanism*, 2nd edn. W. H. Freeman and Co., New York, 1985.
56. S.J. Admiraal and D. Herschlag, *Chem. Biol.*, 1995, **2**, 729.
57. H. van Tol, J.M. Buzayan, P.A. Feldstein, F. Eckstein and G. Bruening, *Nucleic Acids Res.*, 1990, **18**, 1971.
58. G. Slim and M.J. Gait, *Nucleic Acids Res.*, 1991, **19**, 1183.
59. W. Wolfson, *Chem. Biol.*, 2006, **13**, 1.
60. U. Theuretzbacher and J.H. Toney, *Curr. Opin. Investig. Drugs*, 2006, **7**, 158.
61. N. Sudarsan, S. Cohen-Chalamish, S. Nakamura, G.M. Emilsson and R.R. Breaker, *Chem. Biol.*, 2005, **12**, 1325.
62. N. Sudarsan, J.K. Wickiser, S. Nakamura, M.S. Ebert and R.R. Breaker, *Genes Dev.*, 2003, **17**, 2688.
63. K.F. Blount, J.X. Wang, J. Lim, N. Sudarsan and R.R. Breaker, *Nat. Chem. Biol.*, 2007, **3**, 44.
64. K.F. Blount and R.R. Breaker, *Nat. Biotechnol.*, 2006, **24**, 1558.
65. K. Kobayashi, S.D. Ehrlich, A. Albertini and G. Amati *et al.*, *Proc. Natl. Acad. Sci. U.S.A.*, 2003, **100**, 4678.
66. G. Mayer and M. Famulok, *ChemBioChem.*, 2006, **7**, 602.

CHAPTER 9
A Structural Analysis of Ribonuclease P

STEVEN M. MARQUEZ, DONALD EVANS, ALEXEI V. KAZANTSEV AND NORMAN R. PACE

Department of Molecular, Cellular and Developmental Biology, University of Colorado at Boulder, Boulder, Colorado 80309-0347, USA

9.1 Introduction

The endonucleolytic cleavage of precursor sequences from the 5′-end of transfer RNA (tRNA) precursors to form the mature 5′-end is catalyzed by a remarkably complex and unusual enzyme, ribonuclease P (RNase P) (Figure 9.1).[1–4] RNase P is ubiquitous, found in cells from all three domains of life: Bacteria, Eucarya and Archaea. RNase P is a ribonucleoprotein particle, composed of a single RNA and at least one protein component. One of the distinctive characteristics of RNase P is that the RNA component of RNase P can catalyze 5′-end tRNA maturation *in vitro* in the absence of proteins.[5–7] The RNA from these organisms is the catalytic subunit; thus RNase P is an RNA-based enzyme, or "ribozyme".

In contrast to most other natural ribozymes in this volume, (group I and group II introns, hammerhead and hairpin ribozymes, *etc.*), RNase P functions in trans and conducts multiple-turnovers. The others catalyze one-time, sometimes reversible structural rearrangements in RNA that involve the making and breaking of phosphodiester bonds. Another distinctive feature of RNase P is its ability to correctly process various different substrates. Besides all of the different pre-tRNAs (in some cells over a hundred[8]), RNase P processes 4.5S RNA (signal recognition particle or SRP RNA);[9] bacteriophage φ80-induced RNA,[10] the 3′-terminal structure of the turnip yellow mosaic virus genomic RNA,[11] tmRNA (transfer-messenger RNA, SsrA or 10Sa RNA);[12] the mRNA from the polycistronic *his*-operon,[13] the pre-C4 repressor RNA from bacteriophages P1 and P7[14] and some transient structures adopted by riboswitches.[15] While not all RNase P substrates have been well characterized, they likely contain similar structural features that mimic tRNAs.

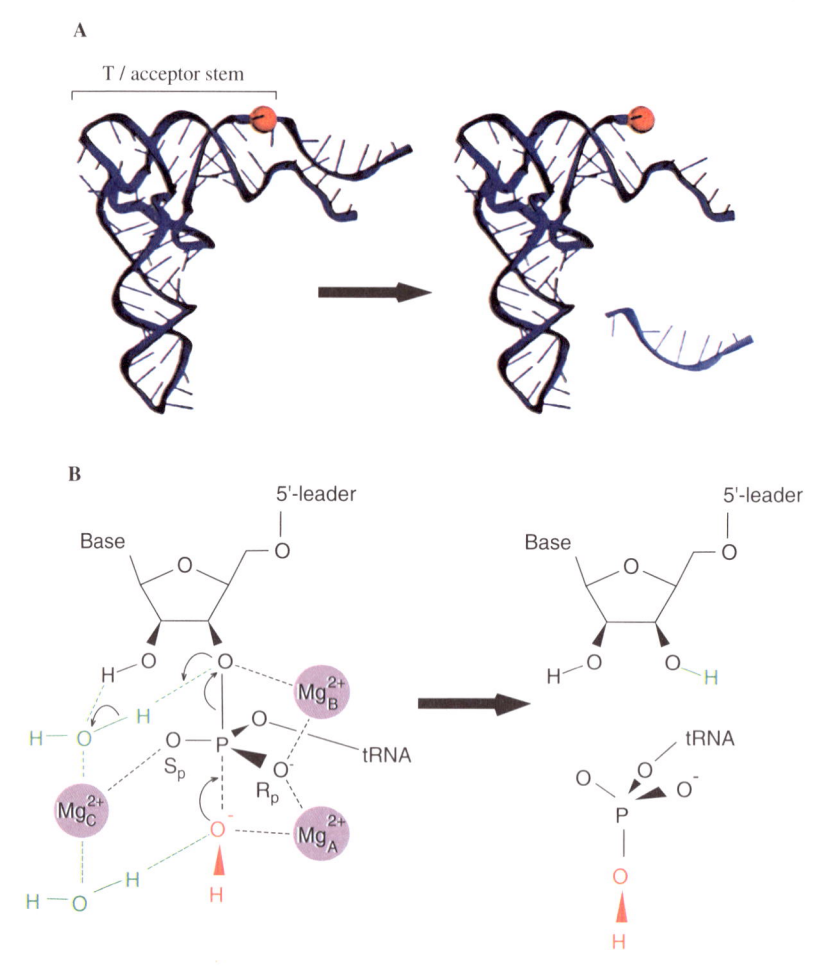

Figure 9.1 RNase P catalyzes maturation of the 5′-end of tRNA. (*A*) Cartoon representation of the RNase P reaction substrate on the left and the products tRNA and 5′-precursor on the right. The red sphere indicates the substrate phosphodiester. (*B*) Mechanism of the reaction proposed on the basis of biochemical studies. The putative structure of the transition state is shown on the left. Electron flow is indicated by arrows. The *pro*-R$_p$ oxygen of the substrate phosphate coordinates two magnesium ions, Mg$^{2+}_A$ and Mg$^{2+}_B$. Mg$^{2+}_A$ serves to activate and position the nucleophile (OH$^-$, red). Mg$^{2+}_B$ is involved in electrostatic stabilization of the transition state and activation of the leaving group (3′-oxygen) by neutralizing a developing negative charge. The third ion, Mg$^{2+}_C$, is expected to make an outer sphere contact with the 2′-OH of the preceding nucleotide (N$_{-1}$) and to contribute to activation of the leaving group by lowering the pK_a of the coordinated water (green).[2]

The protein composition of the RNase P holoenzyme differs dramatically between bacterial (one small basic protein), archaeal (≥ 4 different protein subunits) and eucaryal (≥ 9 different protein components) versions of the enzyme.[16] While the RNA is sufficient for catalysis *in vitro*, both RNA and protein components of RNase P are essential *in vivo*.[17-22]

The discovery of the catalytic activity of the bacterial RNase P RNA[5] triggered numerous mechanistic and biochemical studies. Most biochemical, mutational and enzymatic investigations have focused on the bacterial version of the enzyme due to its relatively simple composition and the high level of catalytic activity by the RNA alone. Missing from the picture is a detailed structural perspective on the ternary complex and interactions that comprise the chemical mechanism. An understanding of the structural relationships is necessary to explain the results of biochemical studies and to facilitate the design of experiments that address the catalytic and substrate-recognition properties of the enzyme at the atomic level. Until recently it was not even clear which nucleotides in the bacterial ribozyme would constitute the active site. In recent years, structures of individual components of RNase P have been determined and a structural perspective on function has begun to emerge. In addition, knowledge of the compositions of archaeal and eucaryal RNase P has made it possible to highlight the similarities and differences with the bacterial holoenzyme. This chapter summarizes the extensive biochemical and phylogenetic information about RNase P in the context of recent structural studies.

9.2 Chemistry of RNase P RNA

9.2.1 Universal

RNase P enzymes from organisms as diverse as *Homo sapiens* and *Escherichia coli* all contain a homologous RNA subunit. The common evolutionary origin of these various RNase P RNAs is evidenced by the conservation of a core RNase P RNA structure (Figure 9.2). This universal core structure contains most of the determinants shown to be essential for function of the bacterial RNase P RNA, which suggests that the RNA subunit performs an essential function in the archaeal and eucaryal RNase P. In fact, RNA-alone activity has been reported for representative RNAs from each of the three main lines of descent, although the archaeal and eucaryal RNAs are much less active than their bacterial counterparts.[6,7] Sulfur substitution of the *pro*-R_p oxygen atom at the scissile phosphodiester resulted in a similar inhibition of cleavage by both bacterial and eucaryal RNase P enzymes. This observation is consistent with a common RNA-based mechanism.[23]

9.2.2 S_N2-type Reaction

RNase P catalyzes hydrolysis of a phosphodiester bond, to generate a phosphate at the 5'-end of mature tRNA[24] and a precursor sequence with a

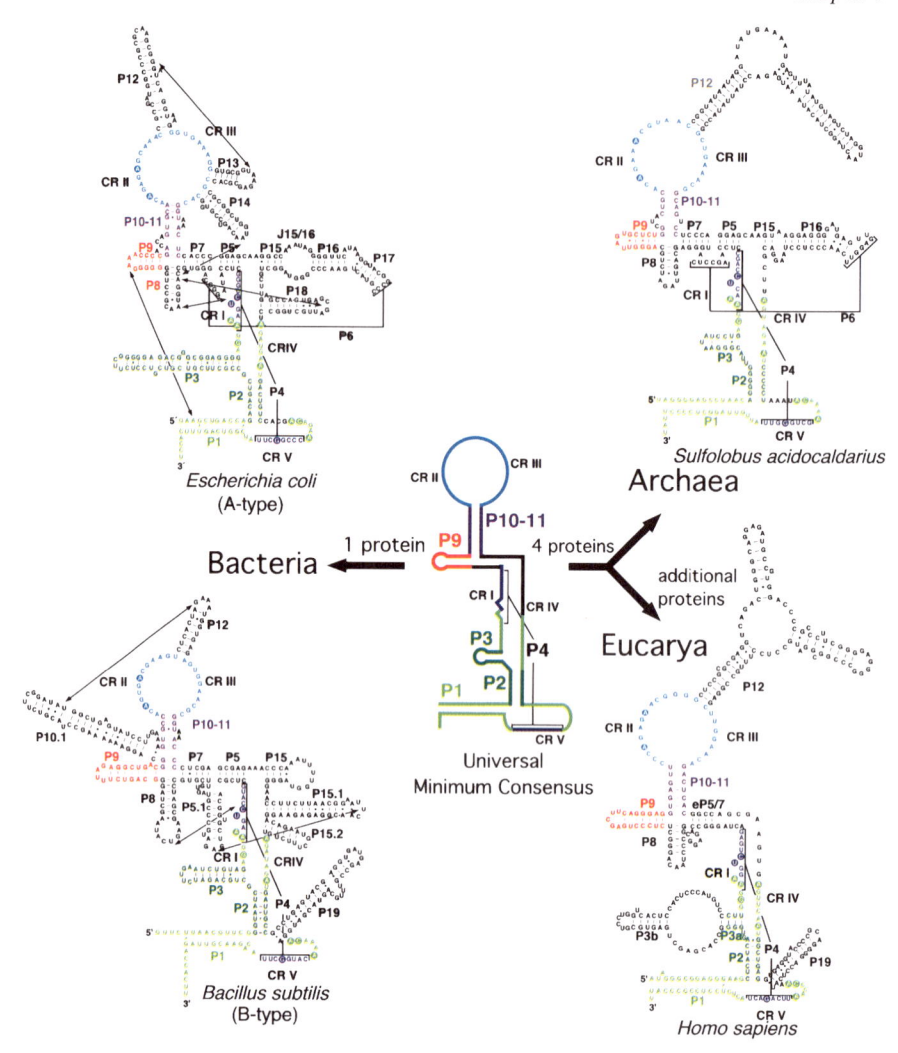

Figure 9.2 Diversity and conservation among RNase P RNA secondary structures. Representative RNase P RNA secondary structures from the three domains of life, Bacteria, Archaea and Eucarya are shown along with the universal minimum consensus structure. The universal minimum consensus structure includes only the structural elements conserved in all known RNase P RNAs. The secondary structures are color coded to represent homologous structural elements of the representative RNAs. Structural elements unique to a specific phylogenetic group are colored black. Structural elements in the RNAs are labeled in the order of their occurrence from 5′ to 3′: P for paired region, L for loop and J for joining region. Base pairs represented by dots are noncanonical. Long-range tertiary interactions are shown as brackets and lines. The five universally conserved sequence regions are labeled CRI–V (ref.1).

3′ hydroxyl (Figure 9.1B). The mechanistic pathway of RNase P does not involve formation of a 2′,3′-cyclic phosphate or a covalent RNase P RNA:tRNA intermediate. Biochemical data support a concerted S_N2-like nucleophilic substitution mechanism.[25–28]

9.2.3 pH-Dependence of the Reaction: Hydroxide Ion as the Nucleophile

Under single-turnover conditions the rate of the chemical step of catalysis increases with pH and is essentially first-order throughout the pH range 6.0–8.0.[28] This observation is consistent with either a hydroxide ion attacking a nucleophile or a general base-catalyzed hydrolysis mechanism. Pre-steady-state kinetic analysis at pH 6.0 reduces the rate of the chemical step sufficiently (2.3 min^{-1}) to allow measurement using conventional techniques.[28] With rapid quench-flow techniques the rate of the chemical step could be determined at the more physiologically relevant pH 8.0 (6 s^{-1}).[25] The slow turnover rate of RNase P when compared with enzymes that operate at high catalytic efficiency suggests that this enzyme has evolved for optimal cleavage site selection and recognition of multiple substrates rather than for a high rate of activity.[25]

9.2.4 Metal Ions in Catalysis

Divalent metal ions are required for catalysis by RNase P. While Mg^{2+} is the natural metal, other divalent cations, including Mn^{2+}, Ca^{2+}, Zn^{2+} and Pb^{2+}, have been shown to support some catalytic activity, although in some cases Ba^{2+}, Sr^{2+} or $Co(NH_3)_6^{3+}$ are additionally required.[27,29–32] Stabilization of the transition state by at least three Mg^{2+} ions is supported by Hill analysis of the dependence of the chemical rate on the concentration of Mg^{2+}.[28] However, interpretation of this result is complicated by the additional requirement for divalent metal ions for both the folding of the enzyme[33–35] and for the binding of substrate.[36] Additionally, applications of mass action law do not adequately describe metal binding to RNA undergoing metal-dependent structural transitions.[37] The metal ion specificity of RNase P can be switched from Mg^{2+} to more thiophilic metals such as Cd^{2+} and Mn^{2+} by substituting a phosphorothioate for the substrate phosphodiester.[38,39] These results provide additional support for the role of Mg^{2+} in the chemical mechanism and confirm the metalloenzyme nature of RNase P.

The RNase P catalyzed reaction appears to be a variation on the two-metal ion mechanism[40] utilized by many phosphoryl transfer enzymes, both RNA (group I and group II introns) and protein (DNA and RNA polymerases, RNase H, restriction endonucleases, *etc.*; reviewed in ref. 41). However, the RNase P mechanism appears also to utilize the substrate 2′-OH adjacent to the site of cleavage to coordinate a third catalytic metal ion (Figure 9.1B). This mechanistic difference was suggested by the experimental observation that substitutions of the 2′-OH adjacent to the cleaved phosphodiester bond results in a substantial reduction in the rate of cleavage and lowered catalytic metal ion cooperativity.[28,42] The mechanism of RNase P catalysis is modeled as formation of a trigonal bipyramidal transition state stabilized by Mg^{2+} hydrate complexes, with protonation of the leaving 3′ oxygen facilitated by the precursor 2′-OH (Figure 9.1B).

9.3 Phylogenetic Variation and Structure of RNase P RNA

The diversity of RNase P substrates and the ability of the RNase P RNA to perform catalysis in the absence of protein made determination of the structure an obvious scientific goal. Perspective on the structure of RNase P RNA has accumulated incrementally, beginning with the evolutionary approach of phylogenetic comparative sequence analysis.[43] Evolutionary conservation of any particular RNA helical structure is achieved by maintenance of base paired interactions between nucleotides that form helices, structural barrels. In a phylogenetic comparative analysis of RNAs, an alignment of homologous (of common ancestry) sequences is scrutinized to identify bases that systematically co-vary in some way, usually to maintain complementarity, and thereby to identify potentially base-paired helical elements of secondary structure.[44,45] About two-thirds of any RNase P RNA is proven by covariation analysis to be involved in the formation of anti-parallel, double-stranded helices of a predictable structure, an A-form helix. As an RNA secondary structure is refined by multiple iterations of such covariation analysis, with a large number of diverse sequences, further structural information can be extracted by identification of known RNA motifs and of long-range interactions such as base triples,[46] tetraloops and tetraloop receptors,[47] *etc.*

Comparative analysis of several hundred RNase P RNA sequences has resulted in refined secondary structure models (Figure 9.2).[46,48–50] Despite some significant differences in peripheral substructures, all RNase P RNAs share a set of highly conserved structural elements represented by the phylogenetic minimal structure and thought to represent the structural and catalytic core of the enzyme (Figure 9.2). Five distinct, universally Conserved Regions (CRI–V), found in the structural core, contain many highly conserved nucleotides, implying their functional importance.[51] Portions of CRI and CRV pair to form helix P4, one of the most conserved substructures found in all RNase P RNAs. CRII and CRIII are located in a universally conserved loop between helices P10-P11 and P12. Universal conservation suggests that this minimum structural core must have been present in earlier life.

Phylogenetically-variable structural elements are attached to the structural core and participate in the formation of long-range docking interactions that are expected to stabilize the overall conformation of the ribozyme. The bacterial RNase P RNAs, typically 330–400 nt, are seen to fall into two distinct secondary structure classes, A-type (Ancestral-type; *e.g.*, that of *Escherichia coli*) and B-type (*Bacillus*-type; *e.g.*, from *Bacillus subtilis*) (Figure 9.2),[52] due to differences in their non-core phylogenetic variable elements. Examples of stabilizing helices in the bacterial RNase P RNAs are P6, P14 and P18 for A-type RNAs and P5.1 and P15.1 for B-type RNAs. The discovery of an RNase P RNA from *Aquificales*, which lacks the long-range P18-P8 interaction, and yet remains thermostable suggests that this loss may be compensated for by strengthening the remaining interactions (*e.g.*, P1-P9).[53,54] An artificial, simplified RNA consisting of only the structural core of the bacterial RNase P RNA is catalytically active *in vitro*, although far less so than the native RNA, due to structural instability.[55,56]

Comparative analysis shows that the archaeal and eucaryal RNase P RNAs are homologs of the bacterial version and contain most of the elements shown to be essential for activity in the bacterial RNAs (Figure 9.2).[48,49] A low level of activity could be detected with some archaeal RNase P RNAs *in vitro* under conditions of extraordinarily high concentrations of monovalent and divalent ions, well above those required for bacterial RNase P RNAs.[6] The archaeal RNAs tend to be similar in size to the bacterial RNAs and they either lack, or have different versions of, structurally stabilizing helices found in bacterial RNAs. Recently, RNA-alone activity was also demonstrated for eucaryal RNase P RNAs, although the activity was down approximately six orders of magnitude when compared with the bacterial RNA.[7] The typical eucaryotic RNase P RNA is roughly two-thirds the sequence length of the typical bacterial RNase P RNA and lacks stabilizing helices found in bacterial RNAs (Figure 9.2). While eucaryal RNAs fold into similar structures as their bacterial homologs, the eucaryal RNA structures are less conformationally stable, as assayed by temperature gradient gel electrophoresis.[57] Thus the RNA-alone activity of RNase P RNAs appears to correlate with the presence of structural elements likely to contribute to structural stability. More stable RNAs are more active than less stable ones. Another major difference between eucaryal and bacterial RNase P RNAs is the presence, in most eucaryal RNAs, of a bulge-loop in helix P3.[48] Crosslinking experiments place the eucaryal specific loop in the P3 stem in a location similar to that of P15/L15 of the bacterial RNase P RNA, an element known to be important in binding the 3′ CCA of tRNA.[57,58]

9.4 Early Studies of the RNase P RNA Structure

Insight into the overall structural organization of the bacterial RNase P RNA was first achieved by determination of the orientations of helical barrels identified by comparative analysis, based on long-range distance constraints. These distance constraints were obtained from both phylogenetic information (long-range covariations) and biochemical photoaffinity crosslinking.[59-62] The resulting models showed that, despite significant structural differences, A and B-type RNase P RNAs fold into similar global structures.[63,64] However, these tertiary structure models could provide no detail on inter-helical sequences, which contain many highly conserved nucleotides. Only a high-resolution crystal structure of the RNA could reveal the spatial organization of these sequence elements, which were expected to be critical to both structure and function.

9.5 Crystallographic Studies of Bacterial RNase P RNAs

Our perspective of the bacterial RNase P RNA structure has been significantly advanced by several structures solved by X-ray crystallography. Crystal structures of bacterial RNase P RNA emerged initially from analysis of an independently folding domain containing CRII/CRIII[35,65] from *Thermus thermophilus* (A-type);[66] and *Bacillus subtilis* (B-type).[67] More recently the crystal structures of full-length RNase P RNAs from *Thermotoga maritima* (A-type),[68] and *Bacillus stearothermophilus* (B-type),[69] were solved (Figure 9.3).

The crystal structure of the full length (338 nt) RNA subunit from *T. maritima* (A-type) was solved to 3.85 Å resolution, which allows the general structure (ordering of helices) of the entire RNA to be traced, but not the detail of the highly conserved interhelical regions. The *B. stearothermophilus* structure (B-type; 417 nt), could be resolved to 3.3 Å, which revealed much structural detail, especially of the conserved interhelical elements that could not be seen in the *T. maritima* structure. Data obtained from phylogenetic comparative analysis and photoaffinity crosslinking are, in general, consistent with the crystal structures.

Together, these crystal structures represent considerable progress in understanding the structure of RNase P RNA. However, as each individual structure possesses obvious structural flaws as a result of crystallization, comparative interpretation is necessary. Each of the independently solved CRII/CRIII-domain structures is distorted by the absence of the "catalytic" domain of RNase P RNA and any anchor points and contacts associated with it. Dimerization of the RNA from *T. maritima*, through the A-type specific helical element P6 between two symmetry related molecules in the crystal, is likely an artifact of crystallization. This formation of the dimer in the crystal structure disrupts a long-range stabilizing interaction consisting of P6 and P17, which force a misfolded conformation of the catalytic core. Since RNase P RNA is active as a monomer *in vitro*,[70,71] such dimerization is expected to be biologically irrelevant. A major disadvantage of the *B. stearothermophilus* structure is disorder of a large part of the CRII/CRIII-domain. The lack of crystallographic data makes inference of the structure of that part of the molecule problematic. The full-size B-type RNase P RNA structure can be reconstructed by superimposition of the overlapping, independently solved structure of the CRII/CRIII-domain from *B. subtilis*,[67] but it results in different positioning of the CRII/CRIII-domain in the full-length A- and B-type RNAs (Figure 9.3).

Comparison of the two full-length structures identifies some general features of RNase P. Both A- and B-type RNase P RNA structures are formed by coaxially stacked helical domains, joined together by long-range docking interactions. The spatial arrangement of these elements of the catalytic domain

A Structural Analysis of Ribonuclease P

B. stearothermophilus   **T. maritima**

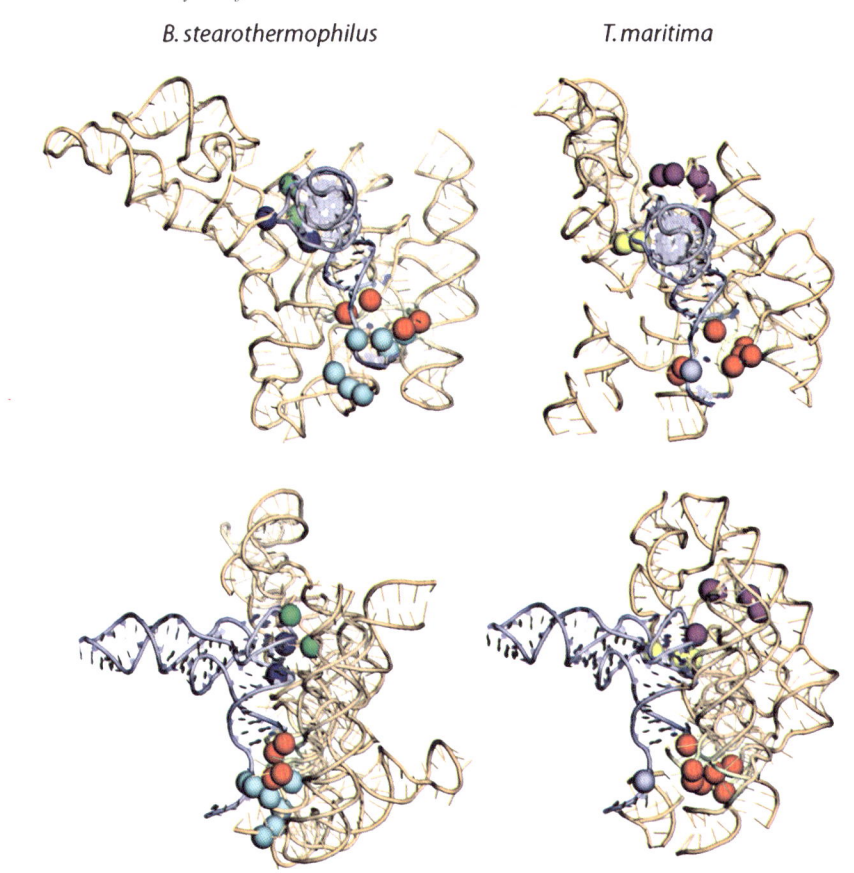

Figure 9.3 Comparison of the *in silico* reconstructions of A- (*Thermotoga maritima*) and B-type (*Bacillus stearothermophilus*) RNase P RNAs. Front-view (top row) and side-view (bottom row) reconstructions of *B. stearothermophilus* (left) and *T. maritima* (right) RNase P RNA complexed with tRNA. In the side-view reconstruction the coaxial stack of P19, P2 and P3 is in the forefront. In the front-view reconstruction the anticodon loop of tRNA is in the forefront. The sites of crosslinking from tRNA to RNase P RNA are shown as spheres, color-coded by the site of photoagent attachment in the tRNA. Crosslinks from the product 5′-phosphate in tRNA are in red.

is highly similar between A- and B-type RNAs (Figure 9.3). In both cases the structure results in a compactly folded RNA with a flat surface, which has been shown experimentally to present the pre-tRNA binding site. Phylogenetically variable structural elements are located on the surface of the core structure, away from the substrate-binding face, and tend to participate in long-range docking interactions that contribute to global structural stability. Conserved nucleotides, which are located primarily in five regions throughout the secondary structure are found in two clusters in the crystal structure: the domain consisting of CRII and CRIII; and the catalytic core (CRI, CRIV, CRV and a few other highly conserved nucleotides).[68,69] The face of the RNA containing these conserved clusters molds the substrate-binding patch in the RNase P RNA. These observations are all consistent with biochemical and phylogenetic results, confirming that the structures solved generally are biologically relevant.

Consistent with phylogenetic predictions,[72] the differences in the peripheral docking elements between A and B-type RNAs may explain some gross differences between the two structures. These long-range docking interactions can be phylogenetically volatile. For example, in the A-type RNA structure, long-range bridging of the A-type specific element P14 (L14) to the universally present helix P8 results in a well ordered CRII/CRIII-domain. An equivalent long-range contact is not present in the B-type RNA, which may explain why this domain is disordered in the B-type crystal structure. In addition, different and unrelated structures can perform equivalent stabilizing tasks in different structural types of the RNA. For instance, docking between P12 and P13 in the A-type RNA is replaced by a docking interaction of P10.1 to P12 in the B-type RNA.[66] Another example is the P16-P17-P6 structure in the A-type RNase P RNA, not present in the B-type RNA. Instead, this structurally stabilizing strut is replaced in the B-type RNA by the unrelated elements P5.1 and P15.1, which dock with each other and occupy the same space in the overall structure.[69]

Proteins generally are stabilized internally by a hydrophobic core. In contrast, a general principle of RNA packing highlighted by these structures is the stabilization of global structure by helical struts or supports. In folded RNAs, helices are ordered by docking interactions that occur on the surface of the RNAs and involve a growing number of RNA structural motifs.[73-75] Many of the structural motifs involved in long-range docking interactions in the bacterial RNase P RNA structures have been observed previously in other RNA structures. These motifs include tetraloops and tetraloop receptor;[76,77] ribose zippers,[78] A-minor interactions,[79] and dinucleotide platforms.[80,81] However, two highly conserved structural features in RNase P RNAs have not been observed in any other RNA structure. The first, consisting of conserved regions CRII and CRIII, forms two, intertwined T-loop-like structures.[67] The second, is formed by the universally conserved helix P4 and five adjacent, highly conserved, inter-helical joining regions, including phylogenetically absolutely conserved sequences in CRI, CRIV and CRV.[69] Phylogenetic and biochemical data implicate this complex structure as the catalytic core of RNase P.

While crystal structures at the current resolution provide a new perspective on RNase P RNA, specific details of the catalytic mechanism will be known only when the structure of the ternary complex, including both the protein and substrate pre-tRNA, becomes available. Nonetheless, the individual structures of tRNAs and bacterial P proteins are known and can be modeled onto the RNase P RNA crystal structure using phylogenetic and biochemical criteria.

9.6 Modeling an RNase P RNA:tRNA Complex

The crystal structures of the RNase P RNAs alone offer no specific information on the structural aspects of the recognition of the substrate pre-tRNA.

However, the structures provide a framework with which to correlate a wealth of biochemical results in order to approximate those interactions. Intermolecular photoaffinity crosslinking data[62,63] provide distance constraints between locations in the tRNA and the RNase P RNA. In these crosslinking experiments an arylazide photoagent was attached to various sites throughout the tRNA structure. Upon UV-irradiation of the pre-formed complex between such derivatized tRNAs and RNase P RNA, the photoagent forms crosslinks with the nucleotides that are within 9 Å of the site of its attachment. Crosslinked complexes containing both tRNA and P RNA are then purified by gel electrophoresis and the sites of crosslinking in the RNase P RNA are identified by primer-extension analysis.[62] The tRNA binding site on the RNase P RNA crystal structure was identified by crosslinks from variously positioned agents in tRNA to RNase P RNA. By combining these intermolecular crosslinking distance constraints with the independently solved crystal structures of the RNase P RNA and tRNA, models of the complex between RNase P RNA and product tRNA can be constructed (Figure 9.3). These models, similar between A- and B-types of RNase P RNA, are generally consistent with the location of known substrate-binding elements in the RNase P structure. However, not all biochemically imposed distance constraints are satisfied in these models, which may suggest conformational rearrangements in the RNA upon substrate binding. For instance, a highly conserved nucleotide on the surface of the independently folding CRII/CRIII domain is protected by tRNA from chemical modification,[82,83] but is separated by ~5 Å (A-type complex) to ~11 Å (B-type complex) from the nearest atom of the tRNA in the *in silico* reconstructions.

The structural organization of the chemically active site is revealed by combining the bacterial RNase P RNA crystal structure with crosslinking data in which the product phosphate is modified with the photoagent.[84] The highly conserved core structure adjacent to helix P4 contains these crosslinked nucleotides (Figure 9.3). Previously, the major groove of helix P4 was proposed to constitute the active site of RNase P due to its high conservation and considerable biochemical and biophysical evidence implicating helix P4 in binding divalent metal ions important for some aspect of catalysis.[85–91] However, the structure-based analysis of the crosslinking data places the product 5′-phosphate of tRNA in the cleft formed by inter-helical joining regions that contain the highly conserved sequence elements CRIV and CRV.

9.7 Modeling the Bacterial RNase P Holoenzyme

The bacterial holoenzyme can be modeled by combining the structures of the individual RNase P RNAs and bacterial P protein subunits, using biochemical results.[92] The structures of diverse bacterial RNase P proteins have been determined by X-ray crystallography[93,94] and NMR.[95] The protein components of both A and B-type RNase P are structurally highly similar, which is not surprising since heterologous holoenzymes with A and B-type components function both *in vitro* and *in vivo*.[96] These homologous proteins adopt an RNP-like fold of a α-β sandwich and have a highly conserved electrostatic surface. This high level of structural and functional conservation makes it likely that bacterial RNase P protein binds to a region that is conserved among all bacterial RNase P RNAs, part of the structural core. One peculiar structural feature of the bacterial RNase P protein is an unusual left-handed β-α-β crossover, seen also among related RNA-binding motifs of the ribosomal protein S5 superfamily.[93] This structure contains the highly conserved "RNR" sequence motif that contributes to formation of a highly positively charged surface on one side of the protein.[97] This face of the protein has been implicated in binding to the RNase P RNA by both biochemical and mutational data.[98–101] In addition mutational,[98] photo-affinity crosslinking[102] and NMR studies[95] indicate that the 5'-leader sequence interacts with a cleft formed by an N-terminal helix α-1 and the central β-sheet.

Significant efforts have been made over the years to use footprint studies to identify the protein-binding site on the bacterial RNase P RNA, but results have not been uniform and were sometimes contradictory. Factors such as nonspecific binding of excess RNase P protein, conformational heterogeneity in the RNA and aggregation of the holoenzyme possibly account for inconsistent results obtained by different research groups.[103–106] Chemical footprinting of RNAs and RNA-protein complexes following their resolution by native gel electrophoresis eliminated these caveats.[92] In this assay non-bridging phosphorothioate nucleotides were randomly incorporated into the RNase P RNA by *in vitro* transcription. Radioactively end-labeled native RNase P RNAs or holoenzymes were then separated from misfolded or aggregate forms by native polyacrylamide gel electrophoresis. In-gel treatment with iodine results in breaks of the RNA chain at the phosphorothioate bond unless it is somehow protected and the sizes of fragments reveal the break points. A patch of protection common to both A- and B-type RNase P RNAs in the presence of bacterial P protein was observed that is distributed in the sequence, but drawn together in the tertiary structure.

These biochemical data, coupled with the crystal structures of the bacterial RNase P RNAs and proteins, were used to reconstruct the overall organization of the bacterial RNase P holoenzyme.[92] The protected residues occupy an area roughly equivalent to the size of the protein as shown in Figure 9.4. One presumption of the model is that no significant conformational rearrangements occur upon formation of the holoenzyme. The location of the identified protein-binding site on RNase P RNA is adjacent to, not immediately at, the proposed chemically active site. This position is consistent with crosslinking and affinity cleavage data between the protein and RNA subunit.[107] The protein locale is compatible with biochemical data that implicate the protein component in binding of the 5'-leader sequence,[102] decreasing the requirement for Mg^{2+},[108] and allowing the uniform processing of various substrates.[109] This model is also consistent with the general belief that the bacterial RNase P protein plays no direct role in catalysis.

Figure 9.4 Reconstructed ternary complex between product tRNA (gray), *B. stearothermophilus* RNase P RNA (brown) and *B. subtilis* RNase P protein (purple). The product phosphate at the tRNA 5′-end is shown as a red sphere.

9.8 Substrate Recognition

RNase P binds pre-tRNAs along the coaxially stacked helical domain consisting of the T-stem-loop and the acceptor helix of tRNA substrates (Figure 9.1).[110–112] Model substrates that lack anticodon and D-stem-loops are processed efficiently, suggesting that these elements are not important for binding RNase P. RNase P appears to recognize substrates by contributions from a few enzyme–substrate interactions that are scattered on the binding faces and contribute cooperatively to substrate affinity and the correct selection of the site of hydrolysis. The T-loop, the T-stem and a few nucleotides proximal to the cleavage site, including the 3′-CCA of the tRNA and the 5′-leader sequence, have all been shown to be important for interaction with RNase P. In addition, the RNase P-tRNA model based on crystal structures suggests that the backbone of the acceptor stem also may contribute to substrate recognition. While the atomic details of these interactions are not known, biochemical studies provide some insight. For instance, a reduction in processing by RNase P RNA observed upon conversion of a few 2′-hydroxyls in the T-loop, T-stem and the 3′-CCA of pre-tRNA into 2′,3′-cyclic phosphates confirms the importance of 2′-OH contacts.[113] In addition, two highly conserved adenosines that bulge from RNase P RNA (A130 in the base of helix P9 and A230 in the base of helix P11 in *B. subtilis*) were suggested to participate in docking with the T-stem-loop.[114,115] The results of photoaffinity crosslinking experiments corroborate an interaction between the CRII/CRIII-domain of RNase P RNA and the T-stem-loop of the pre-tRNA,[63] although the molecular details of this interaction are unknown. The reconstructed RNase P-tRNA model described above would require a conformational change in the RNase P RNA to accommodate this particular constraint.

An important recognition determinant for many pre-tRNAs is the 3′-terminal CCA.[116–118] However, 3′-CCA is often not encoded in pre-tRNA genes and in these cases the 5′-end is thought to be processed before 3′-end maturation occurs via an enzymatic pathway involving RNase Z and terminal nucleotidyl transferase.[119–121] In pre-tRNAs that contain the 3′-CCA, this sequence base pairs with a sequence in the highly conserved loop L15 of the RNase P RNA,[122–125] an interaction critical for the correct selection of the site of processing and Mg^{2+} binding.[124,126,127]

The 5′-leader sequence of pre-tRNA contains additional determinants important for processing. An absolutely conserved adenosine in the J5-15 inter-helical junction of RNase P RNA interacts with the N_{-1} nucleotide of the 5′-leader sequence.[128,129] The N_{-2} nucleotide of the 5′-leader sequence also may be important in binding,[104,130,131] although the details of the interaction with RNase P are currently unknown. 5′-Precursor sequences longer than four nucleotides interact with the protein component of the RNase P holoenzyme;[102] perhaps causing the 5′-precursor sequence to adopt an extended conformation.[132]

The importance of helix P4 in RNase P RNA is evidenced by the high degree of phylogenetic conservation. Before crystallization of the bacterial RNase P RNAs and the identification of the actual active site, helix P4 was thought to be part of the chemically active site. Biochemical and physical data had suggested that the major groove of helix P4 is involved in loading divalent metal ions into the active site of the ribozyme.[85–91] While the crystallographic[69] and photoaffinity crosslinking studies[133] indicate that helix P4 must be important for correct binding of the substrate, the structural data summarized above locate helix P4 between the proposed chemically active site and the T-loop-binding elements in the RNase P RNA. This localization places the metal-binding sites in the major groove of the helix P4 at least 15 Å away from the chemically active site. This distance suggests that Mg^{2+} bound to helix P4 is not directly involved in catalysis and that the primary function of these metal-binding sites is to accommodate and correctly position the negatively charged backbone of the substrate acceptor stem.[69,133] The close proximity of the highly negatively charged RNA chains of the enzyme and substrate creates an intense repulsion that must be screened by divalent cations for proper orientation to occur. Conservation of helix P4 is likely explained by the fidelity of processing that results from the interaction of this stem and the accompanying Mg^{2+} with the acceptor stem of the substrate.

9.9 Archaeal and Eucaryal Holoenzymes – More Proteins

Eucaryal RNase P holoenzymes have been purified and components identified from both human and yeast.[20,134] The protein composition of these enzymes is much greater than seen in the bacterial enzymes. The protein content of the bacterial RNase P holoenzyme makes up approximately 10% of the total mass, whereas the eucaryotic holoenzyme is composed of at least 70% protein. Figure 9.5 summarizes homologous sets of the proteins from different

A Structural Analysis of Ribonuclease P 167

	BsuP protein	Pfu Pop5	Pho Rpp1	Pho Rpr2	Afu Pop4	Pho Pop3			
Bacteria	Eco	RNA (121)	RnpA (14)							
Eucarya	Sce	RNA (118)		Pop5 (20)	Rpp1 (32)	Rpr2 (16)	Pop4 (33)	Pop3 (23)	Pop1 (101)	Pop7 (16)	Pop6 (18)	Pop8 (16)
	Hsa	RNA (109)		hPop5 (19)	Rpp30 (29)	Rpp21* (18)	Rpp29* (25)	Rpp38 (32)	hPop1(115)	Rpp20 (16)	Rpp25 (21)	Rpp14 (14)	Rpp40 (35)
Archaea	Pfu	RNA (106)		PF1378* (14)	PF1914* (25)	PF1613* (14)	PF1816* (15)					
	Pho	RNA (106)		PH1481** (14)	PH1877** (25)	PH1601* (15)	PH1771 (15)	PH1496* (13)				
	Mth	RNA (94)		MTH687(15)	MTH688 (28)	MTH1618 (17)	MTH11 (11)					

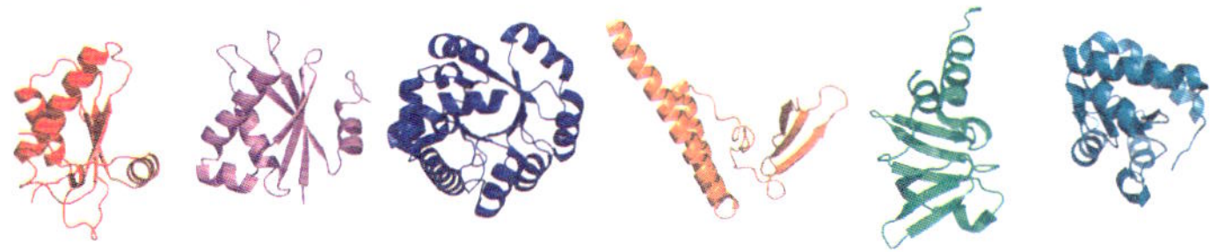

Figure 9.5 Summary of the composition and in vitro reconstitution requirements of phylogenetically diverse RNase P holoenzymes. RNase P holoenzyme components are listed for Bacteria: *Escherichia coli* (Eco); Eucarya: *Saccharomyces cerevisiae* (Sce), *Homo sapiens* (Hsa); and Archaea: *Pyrococcus furiosus* (Pfu), *Pyrococcus horikoshii* (Pho) and *Methanothermobacter thermoautotrophicus* (Mth). The approximate molecular weight of each subunit in kDa is shown in parentheses. The universal RNA subunit, essential *in vivo* and both necessary and sufficient *in vitro*, is enclosed within the red box. The composition of the *E. coli* holoenzyme, a single RNA subunit with the small basic protein RnpA, is within the blue box. The minimal holoenzyme components present before the divergence of Archaea and Eucarya, the RNA subunit and at least four of the proteins, are within the green box. Asterisks label the protein subunits that are the minimum required, along with the RNA subunit, for *in vitro* reconstitution of archaeal[150,152] and eucaryal RNase P holoenzyme activity;[148] for Pfu, either Pop5 and Rpp1 (green asterisks) or Rpr2 and Pop4 (pink asterisks); for Pho, Pop5, Rpp1 and either Rpr2 (red asterisks) or Pop3 (blue asterisks). On the bottom, the bacterial and archaeal protein structures are compared. The ribbon structure of the *B. subtilis* bacterial RNase P protein (red; PDB 1A6F) is shown with *P. furiosus* Pop5 (magenta; PDB 2AV5); *P. horikoshii* Rpp1 (blue; PDB 1V77; *P. horikoshii* Rpr2 (orange; PDB 1X0T); *A. fulgidus* Pop4 (green; PDB ID1TS9); and *P. horikoshii* Pop3 (cyan; PDB 2CZW). Figure created using MacPyMol (DeLano Scientific; http://www.pymol.org).

organisms. The nomenclature of homologous proteins between organisms is not uniformly standardized. Therefore, to aid the following discussion the terminology for the *Saccharomyces cerevisiae* proteins will be used as a reference for the homologous protein from other organisms. At least seven homologous proteins are common to the yeast and human RNase P. A recent bioinformatics survey identified additional homologies between Pop3 and Rpp14 as well as Pop6 and Rpp25, suggesting human and yeast holoenzymes are even more similar than previously known.[135] None of these proteins share obvious sequence similarity to the bacterial RNase P proteins. The different proteins vary greatly in size, ranging from 15.5 to 100.5 kDa in the yeast holoenzyme, with similar size variation in the human holoenzyme (Figure 9.5). Genetic knockout experiments have shown that each of the nine yeast RNase P protein subunits, in addition to the RNA, is essential for viability.[20,136]

The completion of several archaeal genome sequences enabled database searches for genes encoding RNase P protein subunits. Although no genes encoding proteins similar to the bacterial RNase P protein were discovered, open reading frames with homology to four eucaryal RNase P proteins (Rpr2, Pop4, Pop5 and Rpp1; Figure 9.5) were identified and subsequently shown to be associated with RNase P activity in *Methanothermobacter thermoautotrophicus*.[137]

The complexity of the archaeal and eucaryal RNase P holoenzymes compared to the bacterial holoenzyme reflects different evolutionary pressures on this ribonucleoprotein among different domains of life. General trends include smaller size and complexity of the RNA structure compared to the bacterial version and more protein cofactors in archaeal and eucaryal holoenzymes compared to the bacterial RNase P. Such considerable complexity of a ribonucleoprotein particle thought to perform a relatively simple function in the cell raises intriguing questions regarding the specific functions of the multiple protein components. Since reduction in complexity in archaeal and eucaryal RNase P RNAs appears to be achieved at the expense of elements important for structural stabilization of the bacterial RNA, it is thought that at least some protein components in archaeal and eucaryal holoenzymes perform structural roles, functionally substituting for the missing RNA substructures. Other proposed roles for which there is some experimental evidence are the biogenesis and localization of the RNase P holoenzyme. The ability of RNase P proteins to target a reporter protein to the nucleolus has been documented.[138] The binding of several eucaryotic proteins to the eucaryote-specific bulge-loop in the P3 stem of the RNase P RNA was shown to be essential for both the proper localization and maturation of the RNase P RNA.[139–141] Eucaryal RNase P proteins may still function by contributing to the formation of the chemically active site.[142]

Binding to the RNase P RNA *in vivo* has been demonstrated for human and yeast protein subunits using a yeast three-hybrid genetic screen.[143,144] Potential protein–protein interactions within yeast and human RNase P holoenzymes have been investigated by directed yeast two-hybrid genetic screens.[143,144] Maps of *Methanothermobacter thermoautotrophicus* and *Pyrococcus horikoshii* RNase

P protein interactions were generated using the yeast two-hybrid screen.[145,146] The strongest protein–protein interactions observed in both archaeal RNase Ps (Pop4 with Rpr2 and Pop5 with Rpp1) are generally consistent with results from the eucaryal two-hybrid screens. Details of the Pop5 interaction with Rpp1 have been provided by a recent crystal structure of that complex.[147] Such experiments are beginning to offer insight into the structural organization of the eucaryotic and archaeal RNase P holoenzymes.

Functional *in vitro* reconstitution of eucaryal and archaeal holoenzymes from individual components has proven difficult, but some successes have been reported recently. The reconstitution of some activity was demonstrated utilizing human RNase P RNA and proteins Pop4 and Rpr2.[148] However, results from *S. cerevisiae* are inconsistent with an important role for the Rpr2 protein in pre-tRNA processing, because an *S. cerevisiae* RNase P precursor holoenzyme lacking Rpr2 retains *in vitro* activity.[149] RNase P activity of the archeaon *Pyrococcus furiosus* could also be reconstituted with its RNA subunit and the Pop4 and Rpr2 proteins. However, a more active *P. furiosus* holoenzyme was obtained with a distinct complex containing the RNA and the Pop5 and Rpp1 proteins.[150] The importance of these complexes (Pop4 and Rpr2; Pop5 and Rpp1) is supported by the two-hybrid analysis. In addition, RNA footprinting suggests that the *P. furiosus* Pop5 and Rpp30 complex protects regions of the *P. furiosus* RNase P RNA in a manner similar to that of the bacterial RNase P protein on the bacterial RNase P RNAs.[150] Reconstitution of RNase P activity from the closely related archaeon *Pyrococcus horikoshii* required a minimum of three protein subunits – no activity was observed with the two protein minimal complexes that activated the *P. furiosus* RNA.[151,152] In addition, a fifth gene was identified in *P. horikoshii* with similarity to the protein human Pop3, which when added to *in vitro* processing reactions increased temperature optima.[153] A comparative approach with versions of RNase P from more diverse organisms is needed to provide a clearer picture of the importance of each protein subunit in the eucaryal and archaeal holoenzymes.

The eucaryal and archaeal RNase P holoenzymes lack an obvious homolog of the bacterial RNase P protein.[16] However, some archaeal RNase P RNAs are weakly activated by the bacterial RNase P protein[6,154] and conversely the human Pop4 protein reportedly can activate to some extent the *E. coli* RNase P RNA under low salt conditions.[148,155] These results suggest that a common shape and charge distribution is shared between these proteins or, alternatively, more than a single structural solution is possible for obtaining the function necessary.

The archaeal RNase P protein subunits have proven to be useful targets for structural studies by virtue of their solubility and conformational stability. The structures of archaeal RNase P proteins (Pop4, Rpp1, Rpr2, Pop5 and Pop3) are shown with a bacterial protein for comparison in Figure 9.5 (bottom). The Pop4 proteins from three archaea (*M. thermoautotrophicus*, *P. horikoshii*, and *Archaeoglobus fulgidus*) have been solved by NMR and/or crystallography[156–159] (Figure 9.5). This protein adopts an oligonucleotide or oligosaccharide-binding fold (OB-fold) or Sm-like fold, a structure present in many RNA-binding proteins, consistent with its role in binding RNase P RNA in the eucaryal holoenzymes, as indicated by three-hybrid analysis. The crystal structure of the *P. horikoshii* Rpr2 homolog consists of two long α-helices that interact through hydrophobic amino acids at the N terminus, a central domain consisting of an unstructured loop, and a C-terminal zinc ribbon[160] (Figure 9.5). The *P. horikoshii* Rpp1 homolog, also determined by X-ray crystallography, adopts a αβ barrel that is similar to a structure found in the metallo-dependent hydrolase superfamily[161] (Figure 9.5).

Structural studies of the *P. furiosus* and *P. horikoshii* Pop5 homologs show that it adopts an α-β sandwich fold similar to the single stranded RNA binding RRM (RNA recognition motif), or RNP domain, seen in the bacterial RNase P protein (Figure 9.5).[147,162] Although the *P. furiosus* Pop5 protein resembles, to some extent, the bacterial P protein in overall shape, these proteins have no obvious sequence similarity, suggesting that they evolved independently. The fact that a homologous RNA is found in all three domains of life, while the proteins subunits appear to have evolved separately in bacteria when compared to the proteins found in archaea and eucarya, may suggest that the primordial RNase P was composed solely of RNA. While this protein resembles the bacterial RNase P protein in overall shape, the functions of the eucaryal proteins are likely to be more complex as a single protein is unable to activate the RNA. In addition, database searches suggest that the gene encoding the Pop5 protein is not universally conserved in archaea and eucaryotes,[16,135] indicating that in its absence other protein subunits can substitute, a finding consistent with the reconstitution of *P. furiosus* RNase P activity. Because of the homologies of both protein and RNA elements of the archaeal and eucaryal holoenzymes, the archaeal system is expected to reflect the essence of the eucaryotic version.

9.10 Concluding Remarks

Our understanding of the structure and evolution of RNase P has been broadened by the recent advances described above. It appears that the ancestral RNase P possessed a core RNA structure that persists in contemporary versions of the enzyme. The bacterial RNase P RNAs arrived at an intramolecular solution to the problem of global structural stability by use of RNA struts, whereas archaeal and eucaryal RNase P RNAs followed a different evolutionary path to stability, *i.e.*, acquisition of more proteins for RNA–protein and protein–protein interactions. A putative example of a protein enzyme that catalyzes the same reaction as catalyzed by the ribozyme RNase P occurs in the spinach chloroplast,[163] which begs the question: Why does the RNA subunit, responsible for catalysis, persist in all of life? Perhaps the RNA subunit is retained because it confers unique qualities to the holoenzyme, such as flexibility in substrate selection. Such a role may be supported by RNase MRP, an enzyme apparently evolved from eucaryal RNase P. In the case of this enzyme, a duplication of the RNA subunit has allowed the repertoire of substrates to expand to pre-ribosomal RNAs and mRNAs important for cell cycle progression.[164]

The future of RNase P lies in several avenues of study. The roles of individual protein components in the holoenzymes are largely unknown. Reconstitution experiments are rudimentary and some results are conflicting. We expect that results of the reconstitution experiments will reflect the homologies of the various protein and RNA components. Crystal structures of ternary complexes will be needed to shed light on the mechanism of RNase P and the role of the RNase P RNA in catalysis.

Acknowledgements

Support for this work was provided by a National Institutes of Health grant to N.R. Pace.

References

1. D. Evans, S.M. Marquez and N.R. Pace, *Trends Biochem. Sci.*, 2006, **31**, 333.
2. A.V. Kazantsev and N.R. Pace, *Nat. Rev. Microbiol.*, 2006, **4**, 729.
3. S.C. Walker and D.R. Engelke, *Crit. Rev. Biochem. Mol. Biol.*, 2006, **41**, 77.
4. A. Torres-Larios, K.K. Swinger, T. Pan and A. Mondragon, *Curr. Opin. Struct. Biol.*, 2006, **16**, 327.
5. C. Guerrier-Takada, K. Gardiner, T. Marsh, N. Pace and S. Altman, *Cell*, 1983, **35**, 849.
6. J.A. Pannucci, E.S. Haas, T.A. Hall, J.K. Harris and J.W. Brown, *Proc. Natl. Acad. Sci. U.S.A.*, 1999, **96**, 7803.
7. E. Kikovska, S.G. Svard and L.A. Kirsebom, *Proc. Natl. Acad. Sci. U.S.A.*, 2007, **104**, 2062.
8. J.M. Goodenbour and T. Pan, *Nucleic Acids Res.*, 2006, **34**, 6137.
9. K.A. Peck-Miller and S. Altman, *J. Mol. Biol.*, 1991, **221**, 1.
10. A.L.M. Bothwell, R.L. Garber and S. Altman, *J. Biol. Chem.*, 1976, **251**, 7709.
11. C. Guerrier-Takada, A. van Belkum, C.W.A. Pleij and S. Altman, *Cell*, 1988, **53**, 267.
12. Y. Komine, M. Kitabatake, T. Yokogawa, K. Nishikawa and H. Inokuchi, *Proc. Natl. Acad. Sci. U.S.A.*, 1994, **91**, 9223.
13. P. Alifano, F. Rivellini, C. Piscitelli, C.M. Arraiano, C.B. Bruni and M.S. Carlomagno, *Gene Develop.*, 1994, **8**, 3021.
14. R.K. Hartmann, J. Heinrich, J. Schlegl and H. Schuster, *Proc. Natl. Acad. Sci. U.S.A.*, 1995, **92**, 5822.
15. S. Altman, D. Wesolowski, C. Guerrier-Takada and Y. Li, *Proc. Natl. Acad. Sci. U.S.A.*, 2005, **102**, 11284.
16. E. Hartmann and R.K. Hartmann, *Trends Genet.*, 2003, **19**, 561.

17. P. Schedl and P. Primakoff, *Proc. Natl. Acad. Sci. U.S.A.*, 1973, **70**, 2091.
18. D. Apirion and N. Watson, *FEBS Lett.*, 1980, **110**, 161.
19. D.S. Waugh and N.R. Pace, *J. Bacteriol.*, 1990, **172**, 6316.
20. J.R. Chamberlain, Y. Lee, W.S. Lane and D.R. Engelke, *Genes Dev.*, 1998, **12**, 1678.
21. M. Gossringer, R. Kretschmer-Kazemi Far and R.K. Hartmann, *J. Bacteriol.*, 2006, **188**, 6816.
22. E. Kovrigina, D. Wesolowski and S. Altman, *Proc. Natl. Acad. Sci. U.S.A.*, 2003, **100**, 1598.
23. B.C. Thomas, J. Chamberlain, D.R. Engelke and P. Gegenheimer, *RNA*, 2000, **6**, 554.
24. S. Altman and J.D. Smith, *Nat. New Biol.*, 1971, **233**, 35.
25. J.A. Beebe and C.A. Fierke, *Biochemistry*, 1994, **33**, 10294.
26. A.G. Cassano, V.E. Anderson and M.E. Harris, *Biochemistry*, 2004, **43**, 10547.
27. C. Guerrier-Takada, K. Haydock, L. Allen and S. Altman, *Biochemistry*, 1986, **25**, 1509.
28. D. Smith and N.R. Pace, *Biochemistry*, 1993, **32**, 5273.
29. D. Smith, A.B. Burgin, E.S. Haas and N.R. Pace, *J. Biol. Chem.*, 1992, **267**, 2429.
30. M. Brannvall, N.E. Mikkelsen and L.A. Kirsebom, *Nucleic Acids Res.*, 2001, **29**, 1426.
31. S. Cuzic and R.K. Hartmann, *Nucleic Acids Res.*, 2005, **33**, 2464.
32. E. Kikovska, N.E. Mikkelsen and L.A. Kirsebom, *Nucleic Acids Res.*, 2005, **33**, 6920.
33. X.W. Fang, T. Pan and T.R. Sosnick, *Nat. Struct. Biol.*, 1999, **6**, 1091.
34. O. Kent, S.G. Chaulk and A.M. MacMillan, *J. Mol. Biol.*, 2000, **304**, 699.
35. T. Pan, *Biochemistry*, 1995, **34**, 902.
36. J.A. Beebe, J.C. Kurz and C.A. Fierke, *Biochemistry*, 1996, **36**, 10493.
37. D.E. Draper, *RNA*, 2004, **10**, 335.
38. J.M. Warnecke, J.P. Furste, W. Hardt, V.A. Erdmann and R.K. Hartmann, *Proc. Natl. Acad. Sci. U.S.A.*, 1996, **93**, 8924.
39. J.M. Warnecke, R. Held, S. Busch and R.K. Hartmann, *J. Mol. Biol.*, 1999, **290**, 433.
40. T.A. Steitz and J.A. Steitz, *Proc. Natl. Acad. Sci. U.S.A.*, 1993, **90**, 6498.
41. W. Yang, J.Y. Lee and M. Nowotny, *Mol. Cell*, 2006, **22**, 5.
42. T. Persson, S. Cuzic and R.K. Hartmann, *J. Biol. Chem.*, 2003, **278**, 43394.
43. G.W. Fox and C.R. Woese, *Nature*, 1975, **256**, 505.
44. N.R. Pace, D.K. Smith, G.J. Olsen and B.D. James, *Gene*, 1989, **82**, 65.
45. C.R. Woese and N.R. Pace, in *Probing RNA Structure, Function and History by Comparative Analysis*, eds. R.F. Gesteland and J.F. Atkins, Cold Spring Harbor, NY1993.
46. J.W. Brown, J.M. Nolan, E.S. Haas, M.A.T. Rubio, F. Major and N.R. Pace, *Proc. Natl. Acad. Sci. U.S.A.*, 1996, **93**, 3001.
47. M.A. Tanner and T.R. Cech, *RNA*, 1995, **1**, 349.

48. S.M. Marquez, J.K. Harris, S.T. Kelley, J.W. Brown, S.C. Dawson, E.C. Roberts and N.R. Pace, *RNA*, 2005, **11**, 739.
49. P. Piccinelli, M.A. Rosenblad and T. Samuelsson, *Nucleic Acids Res.*, 2005, **33**, 4485.
50. J.K. Harris, E.S. Haas, D. Williams, D.N. Frank and J.W. Brown, *RNA*, 2001, **7**, 220.
51. J.-L. Chen and N.R. Pace, *RNA*, 1997, **3**, 557.
52. E.S. Haas, A.B. Banta, J.K. Harris, N.R. Pace and J.W. Brown, *Nucleic Acids Res.*, 1996, **24**, 4775.
53. M. Marszalkowski, J.H. Teune, G. Steger, R.K. Hartmann and D.K. Willkomm, *RNA*, 2006, **12**, 1915.
54. E.S. Haas and J.W. Brown, *Nucleic Acids Res.*, 1998, **26**, 4093.
55. D.S. Waugh, C.J. Green and N.R. Pace, *Science*, 1989, **244**, 1569.
56. R.W. Siegel, A.B. Banta, E.S. Haas, J.W. Brown and N.R. Pace, *RNA*, 1996, **2**, 452.
57. S.M. Marquez, J.L. Chen, D. Evans and N.R. Pace, *Mol. Cell*, 2006, **24**, 445.
58. B.K. Oh, D.N. Frank and N.R. Pace, *Biochemistry*, 1998, **37**, 7277.
59. M.E. Harris, J.M. Nolan, A. Malhotra, J.W. Brown, S.C. Harvey and N.R. Pace, *EMBO J.*, 1994, **13**, 3953.
60. E. Westhof and S. Altman, *Proc. Natl. Acad. Sci. U.S.A.*, 1994, **91**, 5133.
61. M.E. Harris, A.V. Kazantsev, J.-L. Chen and N.R. Pace, *RNA*, 1997, **3**, 561.
62. B.C. Thomas, A.V. Kazantsev, J.L. Chen and N.R. Pace, *Methods Enzymol.*, 2000, **318**, 136.
63. J.L. Chen, J.M. Nolan, M.E. Harris and N.R. Pace, *EMBO J.*, 1998, **17**, 1515.
64. C. Massire, L. Jaeger and E. Westhof, *J. Mol. Biol.*, 1998, **279**, 773.
65. A. Loria and T. Pan, *RNA*, 1996, **2**, 551.
66. A.S. Krasilnikov, Y. Xiao, T. Pan and A. Mondragon, *Science*, 2004, **306**, 104.
67. A.S. Krasilnikov, X. Yang, T. Pan and A. Mondragon, *Nature*, 2003, **421**, 760.
68. A. Torres-Larios, K.K. Swinger, A.S. Krasilnikov, T. Pan and A. Mondragon, *Nature*, 2005, **437**, 584.
69. A.V. Kazantsev, A.A. Krivenko, D.J. Harrington, S.R. Holbrook, P.D. Adams and N.R. Pace, *Proc. Natl. Acad. Sci. U.S.A.*, 2005, **102**, 13392.
70. X.W. Fang, X.J. Yang, K. Littrell, S. Niranjanakumari, P. Thiyagarajan, C.A. Fierke, T.R. Sosnick and T. Pan, *RNA*, 2001, **7**, 233.
71. A.H. Buck, A.B. Dalby, A.W. Poole, A.V. Kazantsev and N.R. Pace, *EMBO J.*, 2005, **24**, 3360.
72. E.S. Haas, D.W. Armbruster, B.M. Vucson, C.J. Daniels and J.W. Brown, *Nucl. Acids Res.*, 1996, **24**, 1252.
73. P.B. Moore, *Acc. Chem. Res.*, 1995, **28**, 251.
74. N.B. Leontis and E. Westhof, *Curr. Opin. Struct. Biol.*, 2003, **13**, 300.

75. S.R. Holbrook, *Curr. Opin. Struct. Biol.*, 2005, **15**, 302.
76. J.H. Cate, A.R. Gooding, E. Podell, K. Zhou, B.L. Golden, C.E. Kundrot, T.R. Cech and J.A. Doudna, *Science*, 1996, **273**, 1678.
77. I. Tinoco, Jr., *Curr. Biol.*, 1996, **6**, 1374.
78. M. Tamura and S.R. Holbrook, *J. Mol. Biol.*, 2002, **320**, 455.
79. P. Nissen, J.A. Ippolito, N. Ban, P.B. Moore and T.A. Steitz, *Proc. Natl. Acad. Sci. U.S.A.*, 2001, **98**, 4899.
80. J.H. Cate, A.R. Gooding, E. Podell, K. Zhou, B.L. Golden, A.A. Szewczak, C.E. Kundrot, T.R. Cech and J.A. Doudna, *Science*, 1996, **273**, 1696.
81. P.S. Klosterman, D.K. Hendrix, M. Tamura, S.R. Holbrook and S.E. Brenner, *Nucleic Acids Res.*, 2004, **32**, 2342.
82. T.E. LaGrandeur, A. Hüttenhofer, H.F. Noller and N.R. Pace, *EMBO J.*, 1994, **13**, 3945.
83. L. Odell, V. Huang, M. Jakacka and T. Pan, *Nucleic Acids Res.*, 1998, **26**, 3717.
84. A.B. Burgin and N.R. Pace, *EMBO J.*, 1990, **9**, 4111.
85. M.E. Harris and N.R. Pace, *RNA*, 1995, **1**, 210.
86. D.N. Frank and N.R. Pace, *Proc. Natl. Acad. Sci. U.S.A.*, 1997, **94**, 14355.
87. E.L. Christian, N.M. Kaye and M.E. Harris, *RNA*, 2000, **6**, 511.
88. M. Schmitz and I. Tinoco, Jr., *RNA*, 2000, **6**, 1212.
89. E.L. Christian, N.M. Kaye and M.E. Harris, *EMBO J.*, 2002, **21**, 2253.
90. S.M. Crary, J.C. Kurz and C.A. Fierke, *RNA*, 2002, **8**, 933.
91. M.M. Getz, A.J. Andrews, C.A. Fierke and H.M. Al-Hashimi, *RNA*, 2007, **13**, 251.
92. A.H. Buck, A.V. Kazantsev, A.B. Dalby and N.R. Pace, *Nat. Struct. Mol. Biol.*, 2005, **12**, 958.
93. T. Stams, S. Niranjanakumari, C.A. Fierke and D.W. Christianson, *Science*, 1998, **280**, 752.
94. A.V. Kazantsev, A.A. Krivenko, D.J. Harrington, R.J. Carter, S.R. Holbrook, P.D. Adams and N.R. Pace, *Proc. Natl. Acad. Sci. U.S.A.*, 2003, **100**, 7497.
95. C. Spitzfaden, N. Nicholson, J.J. Jones, S. Guth, R. Lehr, C.D. Prescott, L.A. Hegg and D.S. Eggleston, *J. Mol. Biol.*, 2000, **295**, 105.
96. B. Wegscheid, C. Condon and R.K. Hartmann, *EMBO Rep.*, 2006, **7**, 411.
97. J.W. Brown and N.R. Pace, *Nucleic Acids Res.*, 1992, **20**, 1451.
98. V. Gopalan, A.D. Baxevanis, D. Landsman and S. Altman, *J. Mol. Biol.*, 1997, **267**, 818.
99. V. Gopalan, H. Kuhne, R. Biswas, H. Li, G.W. Brudvig and S. Altman, *Biochemistry*, 1999, **38**, 1705.
100. R. Biswas, D.W. Ledman, R.O. Fox, S. Altman and V. Gopalan, *J. Mol. Biol.*, 2000, **296**, 19.
101. H.Y. Tsai, B. Masquida, R. Biswas, E. Westhof and V. Gopalan, *J. Mol. Biol.*, 2003, **325**, 661.
102. S. Niranjanakumari, T. Stams, S.M. Crary, D.W. Christianson and C.A. Fierke, *Proc. Natl. Acad. Sci. U.S.A.*, 1998, **95**, 15212.

103. E. Westhof, D. Wesolowski and S. Altman, *J. Mol. Biol.*, 1996, **258**, 600.
104. A. Loria, S. Niranjanakumari, C.A. Fierke and T. Pan, *Biochemistry*, 1998, **37**, 15466.
105. S.M. Sharkady and J.M. Nolan, *Nucleic Acids Res.*, 2001, **29**, 3848.
106. C. Rox, R. Feltens, T. Pfeiffer and R.K. Hartmann, *J. Mol. Biol.*, 2002, **315**, 551.
107. S. Niranjanakumari, J.J. Day-Storms, M. Ahmed, J. Hsieh, N.H. Zahler, R.A. Venters and C.A. Fierke, *RNA*, 2007.
108. C. Reich, G.J. Olsen, B. Pace and N.R. Pace, *Science*, 1988, **239**, 178.
109. L. Sun, F.E. Campbell, N.H. Zahler and M.E. Harris, *EMBO J.*, 2006, **25**, 3998.
110. W.H. McClain, C. Guerrier-Takada and S. Altman, *Science*, 1987, **238**, 527.
111. S.G. Svard and L.A. Kirsebom, *J. Mol. Biol.*, 1992, **227**, 1019.
112. Y. Yuan and S. Altman, *EMBO J.*, 1995, **14**, 159.
113. T. Pan, A. Loria and K. Zhong, *Proc. Natl. Acad. Sci. U.S.A.*, 1995, **92**, 12510.
114. T. Pan, *Biochemistry*, 1995, **34**, 8458.
115. A. Loria and T. Pan, *Biochemistry*, 1997, **36**, 6317.
116. L.A. Kirsebom and S.G. Svard, *EMBO J.*, 1994, **13**, 4870.
117. B.-K. Oh and N.R. Pace, *Nucleic Acids Res.*, 1994, **22**, 4087.
118. W.D. Hardt, J.M. Warnecke, V.A. Erdmann and R.K. Hartmann, *EMBO J.*, 1995, **14**, 2935.
119. O. Pellegrini, J. Nezzar, A. Marchfelder, H. Putzer and C. Condon, *EMBO J.*, 2003, **22**, 4534.
120. K. Nakanishi and O. Nureki, *Mol. Cells*, 2005, **19**, 157.
121. A. Vogel, O. Schilling, B. Spath and A. Marchfelder, *Biol. Chem.*, 2005, **386**, 1253.
122. S.G. Svard, U. Kagardt and L.A. Kirsebom, *RNA*, 1996, **2**, 463.
123. S. Busch, L.A. Kirsebom, H. Notbohm and R.K. Hartmann, *J. Mol. Biol.*, 2000, **299**, 941.
124. M. Brannvall, B.M. Pettersson and L.A. Kirsebom, *J. Mol. Biol.*, 2003, **325**, 697.
125. B. Wegscheid and R.K. Hartmann, *RNA*, 2006, **12**, 2135.
126. M. Brannvall and L.A. Kirsebom, *J. Mol. Biol.*, 2005, **351**, 251.
127. M. Brannvall, E. Kikovska and L.A. Kirsebom, *Nucleic Acids Res.*, 2004, **32**, 5418.
128. N.H. Zahler, E.L. Christian and M.E. Harris, *RNA*, 2003, **9**, 734.
129. N.H. Zahler, L. Sun, E.L. Christian and M.E. Harris, *J. Mol. Biol.*, 2005, **345**, 969.
130. M. Brannvall, J.G. Mattsson, S.G. Svard and L.A. Kirsebom, *J. Mol. Biol.*, 1998, **283**, 771.
131. A. Hansen, T. Pfeiffer, T. Zuleeg, S. Limmer, J. Ciesiolka, R. Feltens and R.K. Hartmann, *Mol. Microbiol.*, 2001, **41**, 131.
132. D. Rueda, J. Hsieh, J.J. Day-Storms, C.A. Fierke and N.G. Walter, *Biochemistry*, 2005, **44**, 16130.

133. E.L. Christian, K.M. Smith, N. Perera and M.E. Harris, *RNA*, 2006, **12**, 1463.
134. N. Jarrous, *RNA*, 2002, **8**, 1.
135. M.A. Rosenblad, M.D. Lopez, P. Piccinelli and T. Samuelsson, *Nucleic Acids Res.*, 2006, **34**, 5145.
136. J.Y. Lee, C.E. Rohlman, L.A. Molony and D.R. Engelke, *Mol. Cell Biol.*, 1991, **11**, 721.
137. T.A. Hall and J.W. Brown, *RNA*, 2002, **8**, 296.
138. N. Jarrous, J.S. Wolenski, D. Wesolowski, C. Lee and S. Altman, *J. Cell Biol.*, 1999, **146**, 559.
139. M.R. Jacobson, L.G. Cao, K. Taneja, R.H. Singer, Y.L. Wang and T. Pederson, *J. Cell Sci.*, 1997, **110**, 829.
140. W.A. Ziehler, J. Morris, F.H. Scott, C. Millikin and D.R. Engelke, *RNA*, 2001, **7**, 565.
141. T.J. Welting, B.J. Kikkert, W.J. van Venrooij and G.J. Pruijn, *RNA*, 2006, **12**, 1373.
142. H.L. True and D.W. Celander, *J. Biol. Chem.*, 1996, **271**, 16559.
143. T. Jiang, C. Guerrier-Takada and S. Altman, *RNA*, 2001, **7**, 937.
144. F. Houser-Scott, S. Xiao, C.E. Millikin, J.M. Zengel, L. Lindahl and D.R. Engelke, *Proc. Natl. Acad. Sci. U.S.A.*, 2002, **99**, 2684.
145. T.A. Hall and J.W. Brown, *Archaea*, 2004, **1**, 247.
146. M. Kifusa, H. Fukuhara, T. Hayashi and M. Kimura, *Biosci. Biotechnol. Biochem.*, 2005, **69**, 1209.
147. S. Kawano, T. Nakashima, Y. Kakuta, I. Tanaka and M. Kimura, *J. Mol. Biol.*, 2006, **357**, 583.
148. H. Mann, Y. Ben-Asouli, A. Schein, S. Moussa and N. Jarrous, *Mol. Cell*, 2003, **12**, 925.
149. C. Srisawat, F. Houser-Scott, E. Bertrand, S. Xiao, R.H. Singer and D.R. Engelke, *RNA*, 2002, **8**, 1348.
150. H.Y. Tsai, D.K. Pulukkunat, W.K. Woznick and V. Gopalan, *Proc. Natl. Acad. Sci. U.S.A.*, 2006, **103**, 16147.
151. Y. Kouzuma, M. Mizoguchi, H. Takagi, H. Fukuhara, M. Tsukamoto, T. Numata and M. Kimura, *Biochem. Biophys. Res. Commun.*, 2003, **306**, 666.
152. A. Terada, T. Honda, H. Fukuhara, K. Hada and M. Kimura, *J. Biochem. (Tokyo)*, 2006, **140**, 293.
153. H. Fukuhara, M. Kifusa, M. Watanabe, A. Terada, T. Honda, T. Numata, Y. Kakuta and M. Kimura, *Biochem. Biophys. Res. Commun.*, 2006, **343**, 956.
154. D.T. Nieuwlandt, E.S. Haas and C.J. Daniels, *J. Biol. Chem.*, 1991, **266**, 5689.
155. E. Sharin, A. Schein, H. Mann, Y. Ben-Asouli and N. Jarrous, *Nucleic Acids Res.*, 2005, **33**, 5120.
156. W.P. Boomershine, C.A. McElroy, H. Y. Tsai, R.C. Wilson, V. Gopalan and M.P. Foster, *Proc. Natl. Acad. Sci. U.S.A.*, 2003, **100**, 15398.
157. T. Numata, I. Ishimatsu, Y. Kakuta, I. Tanaka and M. Kimura, *RNA*, 2004, **10**, 1423.

158. D.J. Sidote and D.W. Hoffman, *Biochemistry*, 2003, **42**, 13541.
159. D.J. Sidote, J. Heideker and D.W. Hoffman, *Biochemistry*, 2004, **43**, 14128.
160. Y. Kakuta, I. Ishimatsu, T. Numata, K. Kimura, M. Yao, I. Tanaka and M. Kimura, *Biochemistry*, 2005, **44**, 12086.
161. H. Takagi, M. Watanabe, Y. Kakuta, R. Kamachi, T. Numata, I. Tanaka and M. Kimura, *Biochem. Biophys. Res. Commun.*, 2004, **319**, 787.
162. R.C. Wilson, C.J. Bohlen, M.P. Foster and C.E. Bell, *Proc. Natl. Acad. Sci. U.S.A.*, 2006, **103**, 873.
163. B.C. Thomas, X. Li and P. Gegenheimer, *RNA*, 2000, **6**, 545.
164. T. Gill, T. Cai, J. Aulds, S. Wierzbicki and M.E. Schmitt, *Mol. Cell Biol.*, 2004, **24**, 945.

CHAPTER 10
Group I Introns: Biochemical and Crystallographic Characterization of the Active Site Structure

BARBARA L. GOLDEN

Department of Biochemistry, Purdue University, 175 South University Street, West Lafayette, IN 47907-2063, USA

10.1 Group I Intron Origins

Group I introns, like their protein counterparts, the inteins, are embedded within essential genes and are found in organisms that range from bacteriophage to plants and fungi. Some encode homing endonucleases within their primary structures and thereby retain the ability to "home", or transfer laterally from organism to organism. Although these mobile genetic elements are located in the coding region of essential genes, their self-splicing ability removes the foreign sequences from the gene once it is transcribed. This property has the potential to render the presence of a group I intron genetically neutral.

10.2 Group I Intron Self-splicing

Group I introns orchestrate two consecutive phosphotransesterification reactions that result in excision of the intron and ligation of the flanking RNA sequences (Figure 10.1). As observed with spliceosomal introns and group II introns, the first step in this process is recognition and cleavage of the 5′-splice site. In group I introns, the 5′-splice site is found within a short stem-loop (P1) formed by base pairing between the 5′-exon and the 5′-end of the ribozyme (Figure 10.2). The length of the helix is variable between introns, but the last nucleotide in the 5′-exon is always a uridine and it is base paired to a guanosine

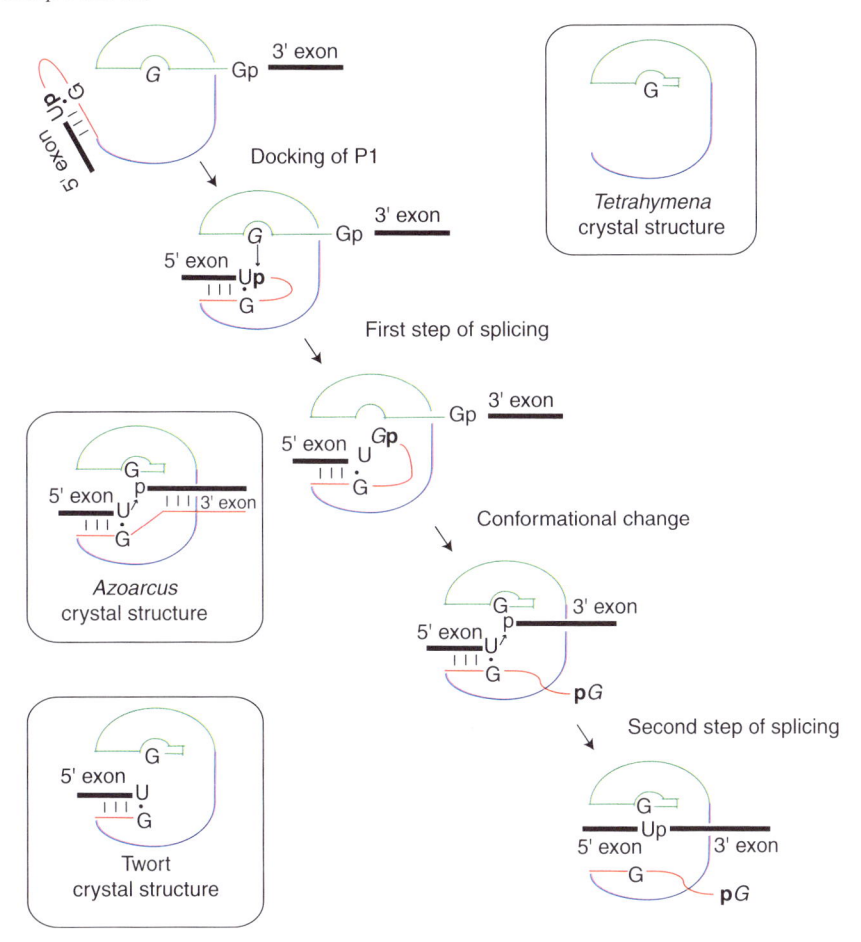

Figure 10.1 Group I intron self-splicing. Group I introns bind two substrates, a guanosine nucleoside and a short duplex, called P1, formed by base-pairing of the 5′-exon to the 5′-end of the intron. Before splicing, the duplex must dock into the catalytic core of the intron, thereby positioning the 5′-splice site adjacent to the guanosine nucleoside. The first phosphotransesterification reaction links the guanosine nucleoside to the 5′-end of the intron and releases the 5′-exon. This chemical reaction is followed by a conformational rearrangement in which the exogenous guanosine nucleoside exits the guanosine binding site and is replaced by a conserved guanosine residue at the 3′-end of the intron. The second phosphotransesterification reaction can then occur, linking the two exons and releasing the intron. In this cartoon, regions of the intron corresponding to the P1-P2, P3-P9, and P4-P6 domains are colored red, green, and blue respectively. The three available crystal structures represent three states that mimic parts of the self-splicing pathway. These are illustrated by diagrams of the RNAs contained within crystal structures of the *Tetrahymena*, *Azoarcus*, and Twort introns.

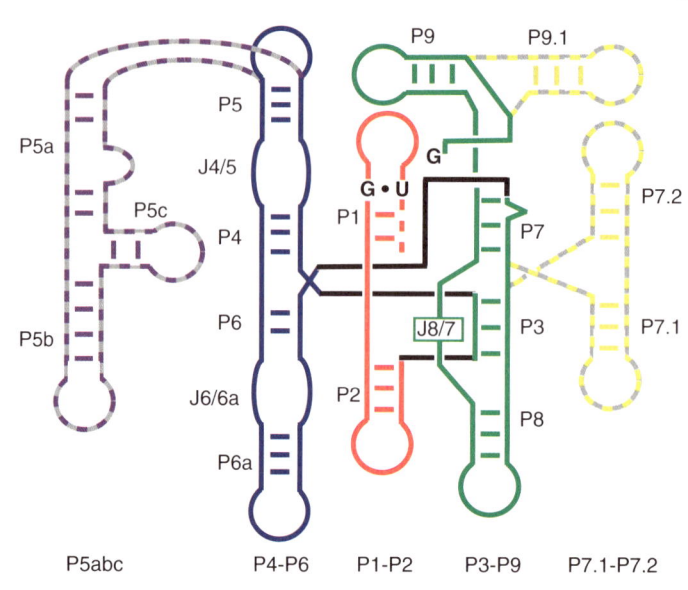

Figure 10.2 Secondary structure of group I introns. Group I introns fold into a series of short helices numbered P1-P9. Helices are linked by junctions that are named for the two helices that they connect. These elements are organized into the three domains called P1-P2 (red), P3-P9 (green) and P4-P6 (blue) that make up the core of all group I introns. The *Azoarcus* intron is composed of only these core structures. In addition to these conserved elements, many introns have insertions within the RNA primary sequence. The P5abc extension of the *Tetrahymena* intron is drawn in purple, and the P7.1–P7.2–P9.1 extension of the Twort intron is shown in yellow. These extra elements pack tightly against the core structure and serve as buttresses that aid in the kinetic folding pathway and provide thermodynamic stability to the intron. The P2.1, P9.1 and P9.2 insertions of the *Tetrahymena* intron are not shown.

within the intron. The scissile phosphate follows this uridine. In the first phosphotransesterification reaction, the group I intron binds a guanosine nucleoside (or nucleotide) within the guanosine binding site and activates its 3′-hydroxyl for nucleophilic attack at the scissile phosphate. In this first chemical reaction, the 3′-hydroxyl of the guanosine becomes linked to the 5′-end of the intron and the 3′-hydroxyl of the 5′-exon is freed.[1,2]

Prior to the second step of splicing, the complex must undergo a conformational switch. In this process, the guanosine substrate, now linked to the 5′-end of the intron, exits the guanosine binding site and the last nucleotide of the intron, a conserved guanosine termed ωG, is bound within the same guanosine binding site.[3,4] In the second chemical step of splicing, the 3′-hydroxyl of the newly freed 5′-exon attacks at the phosphate following ωG. This reaction splices the 5′-exon onto the 3′-exon and frees the intron. This second phosphotransesterification reaction is chemically equivalent to the reverse of the first step of splicing.

10.3 What has Changed in Group I Intron Knowledge in the Last Decade

It has been nearly 25 years since the catalytic activity of group I introns was recognized.[5] At the time of publication of the last version of this text,[6] a mechanistic framework for studying the group I introns had been developed[7,8] and was being used to understand the energetic contributions of individual atoms, primarily on the substrate RNAs, to catalysis.[9–15] The structure of group I introns had been extensively characterized using biochemical mapping techniques, including mutagenesis,[16–18] chemical footprinting[19] and covalent cross-linking.[20–22] These data were integrated into molecular models that, while not atomic resolution, were sufficiently detailed to guide experimental strategies.[23,24]

In the intervening years, the active site of group I introns has been examined by functional analyses and structural studies. Modified nucleotides have been harnessed to probe the roles of individual functional groups in the active site. In a series of intricate and elegant experiments, these have been used to tease out a network of molecular interactions that occur within the heart of the active site of the ribozyme. Such studies provide atomic resolution detail of how atoms within the ribozyme interact and can potentially quantitate the energetic contribution that a functional group contributes to catalysis. These elegant analyses can be difficult and tedious and have only been applied to a handful of atoms within the active site. In a complementary approach, over a decade of effort has been invested to analyze the three-dimensional structure of a group I intron using X-ray crystallography. These studies provide a picture of the entire ribozyme, but only at moderate resolution (3.8–3.1 Å). This chapter focuses on how these two complementary approaches have led to the current understanding of the mechanism of group I intron catalysis and the evolution of group I introns.

10.4 Structure of Group I Introns

Group I introns vary significantly in their size, catalytic properties and self-splicing abilities. Nevertheless, all retain a central conserved secondary structure composed of a series of base-paired helices numbered P1 through P9 (Figure 10.2).[25,26] Helices are connected by non-Watson–Crick base pairs or single-stranded junction (J) segments, named for the two helices that are linked. Helices are organized into three domains, named P1-P2, P4-P6, and P3-P9.[27–29] The conserved regions of the P1-P2 and P4-P6 domains consist of 2–3 helices that are nearly coaxially stacked. The P3-P9 domain is a more irregular structure that is interrupted by the P4-P6 domain. The P4-P6 domain serves as a scaffold upon which to build the active site. Two covalent linkages, a tertiary contact between P5 and P9 and a tertiary contact between P3 and J6/6a, allow the P3-9 domain to wrap around the P4-P6 domain, defining the gross architecture of the active site. P1-P2 sits on top of the P3-P9 domain. These secondary structural elements represent the catalytic core of the intron and are conserved in the vast majority of self-splicing group I introns.

In addition to the enzymatic core, many introns have insertions on their periphery that serve to enhance the stability of the folded structure and/or help the intron to fold into its native structure efficiently. Introns in the IC1 family, such as the *Tetrahymena* LSU intron, have evolved an independently folding subdomain called P5abc (Figure 10.2) that packs against the core elements of the P4-P6 domain and thereby stabilizes the active site conformation.[30] Likewise the P7.1, P7.2 and P9.1 helices of the IA2 family of introns, such as the Twort orf142-I2 intron, provide stability to the P3-P9 domain (Figure 10.2).[31]

10.5 Crystallography of Group I Introns

Over a decade of research by a multitude of laboratories has been invested studying the three-dimensional structure of group I introns. Since the last version of this book was published,[6] the crystal structures of an independently folding domain and three different group I introns have been published. The introns represent three distinct taxonomical families, have their origins within three distinct types of exons (rRNA, mRNA and tRNA), and are trapped in three distinct chemical states (Figure 10.1).

10.5.1 *Tetrahymena* LSU P4-P6 Domain

The *Tetrahymena* LSU intron is the most extensively characterized of the group I introns. The first glimpse of the active site of a group I intron came from the crystal structure of the P4-P6 domain from the *Tetrahymena* intron (pdb accession codes 1GID, 1HR2).[32,33] The P4-P6 domain is a rod-like structure composed of coaxially stacked Watson–Crick duplexes interrupted by junctions containing non-Watson–Crick base pairs. Like many introns, the *Tetrahymena* P4-P6 domain contains a short insertion, called P5abc (Figure 10.2). These nucleotides create a second set of co-axially stacked helices that pack side-by-side with the conserved helices P4, P5 and P6, thereby stabilizing the global tertiary structure of the P4-P6 domain, even in isolation from the remainder of the intron.[28] While globally the conformation of this domain is the same in isolation as it is within the intact intron, there are regions of the RNA in which the structure depends on its environment. Notably, the J6/6a internal loop and the 5'- and 3'-termini contact the P3-P9 domain and are in distinctly different conformations when the P3-P9 domain is absent. In contrast, the J4/5 internal loop, which is involved in 5'-splice site recognition,[22,34,35] is in the same conformation in the isolated domain and in the intact ribozyme. In the crystal, the J4/5 internal loop makes a crystal contact that mimics the interactions normally seen between J4/5 and the 5'-splice site.[32,35] 5'-Splice site recognition involves docking of the P1 hairpin into the ribozyme active site. To facilitate this process, it may be energetically favorable for the ribozyme to maintain the conformation of the J4/5 internal loop, even when P1 is not present.

10.5.2 *Tetrahymena* Intron Catalytic Core

To build upon the structure of the P4-P9 domain, an RNA that spanned both the P4-P6 and P3-P9 domains of the *Tetrahymena* intron was constructed (Figure 10.1).[36] This RNA spans the group I intron active site and possesses ribozyme activity. It can bind the P1-P2 domain in trans using tertiary contacts and perform a biologically relevant phosphotransesterification reaction. The crystal structure of this ribozyme was initially determined at 5.0 Å resolution, revealing the global conformation of the ribozyme active site (pdb accession code 1GRZ).[37]

To derive a clearer picture of this *Tetrahymena* ribozyme, Guo *et al.* used *in vitro* selection to identify catalytically active variants of the *Tetrahymena* intron that were more thermostable than the wild-type sequence.[38] One of these variants produced crystals with improved diffraction characteristics. This resulted in the structure of this ribozyme being solved at 3.8 Å resolution, which is sufficient to accurately position nucleotides in register, provide a clear view of the atomic structure, and to identify the position of catalytic metal ions (pdb accession code 1X8W).[39]

Was the time and effort that went into solving the structure at low resolution worth it? Many RNA crystals are found to diffract X-rays only to ~5 Å resolution. In these cases, is it worthwhile to pursue a crystal form that only diffracts to modest resolution? Comparison of the high resolution structure of the *Tetrahymena* ribozyme with the early low resolution structure provides some insight into these questions. At low resolution, several features were clear. The guanosine binding site was clearly distorted from A-form geometry, providing a snug binding pocket for the guanosine substrate. The atomic details of how the bases packed together to create this pocket were not apparent in the structure and had to await higher resolution. The path of the backbone was relatively accurate, although the register of the nucleotides was shifted in some regions of the molecule. Nevertheless, several ligands to the catalytic metal were correctly identified in this preliminary structure. If the biochemical and crystallographic data are detailed enough to accurately trace the path of the backbone, low resolution models appear to be sufficient to be used to guide further analyses.

10.5.3 Twort orf142-I2 Ribozyme

Bacteriophage such as T4 and Twort are a rich source of group I introns, and the T4 nrD intron, previously known as SunY,[40] was the subject of many extensive biochemical investigations. Phage introns are relatively small in size, but they have characteristic insertions that follow the conserved helices P7 and P9. The bacteriophage Twort orf142-I2 intron is a typical phage intron, containing the P7.1-7.2 subdomain and the P9.1 stem loop (Figure 10.2).[41] The structure of a ribozyme derived from the Twort orf142-I2 was also determined (pdb accession code 1Y0Q).[42] This ribozyme contains nucleotides 9–252 of the intron, including the terminal guanosine ωG, and is bound to a four nucleotide product-analog RNA to reconstitute the P1 helix (Figure 10.1). While the 3′-hydroxyls that serve as the attacking and leaving groups during the phosphotransesterification reaction are present in this RNA, the scissile phosphate is absent.

10.5.4 *Azoarcus* sp. BBH72 tRNA^{Ile} Intron

The *Azoarcus* sp. BBH72 tRNA^{Ile} intron is found embedded within the anticodon loop of a tRNA. Remarkably, although this intron is only 206 nucleotides in length and contains no peripheral extensions, it is extremely thermostable. To generate crystals, the intron was extensively engineered to trap a state that resembles the intron after the first step of splicing but prior to exon ligation. To accomplish this, additional nucleotides were added at the 5′-end of the ribozyme and the sequence was truncated within the P9 stem loop. An RNA that mimicked the 3′-end of the ribozyme, including ωG and a 3′-exon sequence, was added in trans. The resulting RNA allows formation of the helix P10, which is formed by base pairing of the 3′-exon with the loop region of P1 (Figure 10.1). This molecule contains a 5′-exon with a free 3′-hydroxyl that can react at the intron-3′-exon boundary. The structure was solved with both deoxyguanosine (pdb accession code 1U6B) and guanosine (pdb accession code 1ZZN) nucleotides in the ωG position.[43,44]

10.6 Structural Basis for Group I Intron Self-splicing

The three group I intron structures were obtained using three widely diverged primary sequences, spanning three different group I intron families, in three different catalytic states and under vastly different crystallization conditions. Nevertheless, the catalytic core structures are remarkably similar in all three molecules (Figure 10.3).[45,46] The Watson–Crick helices within each domain are often stacked upon each other. Both duplex and single-stranded junction regions mediate domain–domain interactions. Double stranded junctions, such as J4/5 and J6/6a, are composed of stacked bases that hydrogen bond to form non-Watson–Crick base pairs. These junctions are all involved in mediating minor groove–minor groove contacts between two domains. J4/5 and J6/6a bind to P1 and P3 respectively.

Single-stranded junctions, including the J6/7, J3/4, and J8/7 nucleotides are not simple tethers to connect helices. J3/4 and J6/7 were predicted to form base triples at the P4-P6 interface. In the crystal structures, we see that these facilitate contacts between the P4-P6 domain and the P3-P9 domain. Likewise, the J8/7 and J2/3 junctions allow P1-P2 to pack against the P3-P9 domain (Figure 10.4). In each case, the single stranded junctions line one side of a domain, providing a complementary surface for interdomain packing.

The three structures provide significant insight into how group I introns bind their substrates, position catalytic metal ions, and catalyze the phosphotransesterification reaction. Since the bulk of biochemical studies on group I

Group I Introns 185

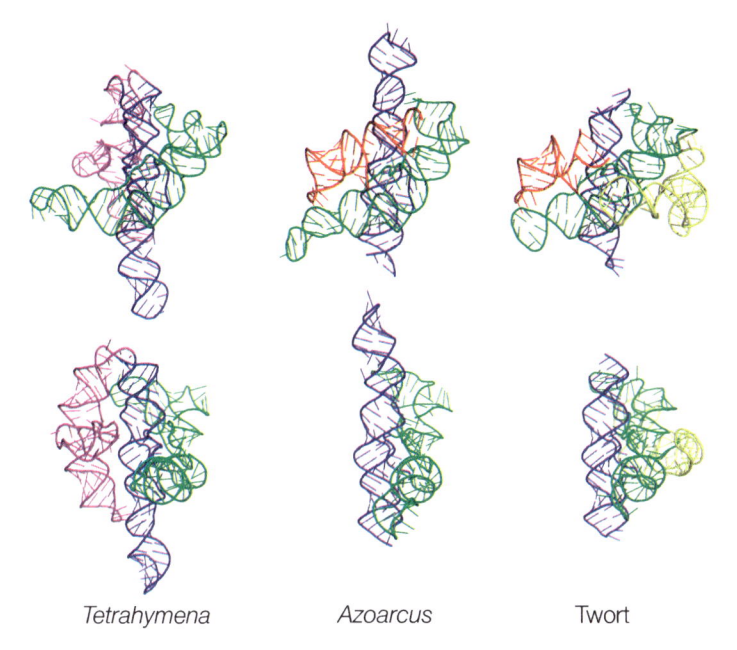

Figure 10.3 Crystal structures of the *Tetrahymena*, *Azoarcus*, and Twort introns. The intron structures were oriented to emphasize the similarity of all three molecules (pdb accession codes for these models are 1X8W, 1U6B, and 1Y0Q, respectively). The top row shows the introns from the "side" looking through the P3-P9 domain towards the P4-P6 domain. The bottom row shows the intron from the "front". The P1-P2 domain has been deleted from the *Azoarcus* and Twort structures for clarity. This view illustrates the side-by-side packing of the P3-P9 domain with the P4-P6 domain. Introns are colored as in Figure 10.2.

introns have focused on the *Tetrahymena* LSU, the nucleotide positions will correspond to the numbering system from that intron.

10.6.1 Recognition of the 5′-Splice Site

Group I introns recognize the 5′-splice site in the context of a G·U wobble base pair. Three distinct elements are involved in determining the accuracy of 5′-splice site selection. First, P2 serves as a base upon which the P1 helix stacks. In the *Azoarcus* and Twort introns, P2 anchors P1 by docking into the minor groove of P8. The P2 region of the *Tetrahymena* ribozyme contains two helices, P2 and P2.1, and is predicted to make a different tertiary structure.[24] Mutations in the junctions surrounding the *Tetrahymena* P2 and P2.1 helices compromise the accuracy of 5′-splice site selection,[47] suggesting that the P2 region serves the same role in this intron. Second, the nucleotides of J2/3 and J8/7 decorate the minor groove of the P1-P2 domain (Figure 10.4).[48,49] Finally, the G·U wobble pair that

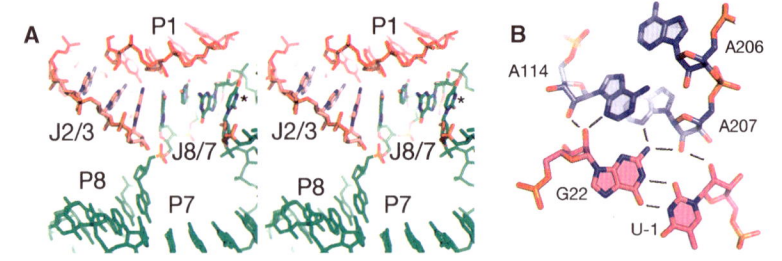

Figure 10.4 (*A*) "Single-stranded" nucleotides line domain interfaces. Nucleotides in single-stranded junctions J2/3, J8/7, and nucleotide U160 stack to create an interface between the P1-P2 and P3-P9 domains in a manner that is reminiscent of the hydrophobic core of a protein. This illustration was prepared using the Twort ribozyme. (*B*) Recognition of the 5'-splice site by the P4-P6 domain. Conserved tandem A·A base pairs in the J4/5 junction make a complementary surface for recognition of the G·U wobble that defines the 5'-splice site. Residue numbers are taken from the *Tetrahymena* ribozyme to match the majority of the literature; the illustration was made using the *Azoarcus* intron.

defines the 5'-splice site is docked into the P4-P6 domain, as predicted by previous biochemical and crystallographic experiments.[22,32,34,35] Tandem A·A base pairs in the J4/5 junction each extrude one adenosine into the minor groove of the P4-P6 domain. These exposed nucleotides make A-minor type interactions[50] with the G·U wobble pair. The N3 and 2'-hydroxyl of A114 hydrogen bond to the 2'-hydroxyl, and the N3 and 2'-hydroxyl of A207 bind the exocyclic amine of the wobble guanosine (Figure 10.4).[35]

10.6.2 Does the Ribozyme Undergo Conformational Changes upon P1 Docking?

The *Tetrahymena* ribozyme structure spans only the P3-P9 and P4-P6 domains of the intron. The substrate domain, P1-P2, is missing. The structure of this ribozyme is likely to represent the structure of the intact ribozyme prior to docking of the P1 helix into the active site.[51,52]

Docking of P1 into the catalytic core involves recognition of the G·U wobble by the J4/5 region of the ribozyme and formation of a network of hydrogen bonds between the minor groove of P1 and the J8/7 region of the helix.[22,34,35,48,49] The *Tetrahymena* ribozyme structure provides an opportunity to understand how much of the ribozyme core is preorganized for binding P1, and what conformational rearrangements are required to accommodate P1 docking.

Comparison of the *Tetrahymena* ribozyme structure to the Twort and *Azoarcus* structures reveals that the catalytic core of the *Tetrahymena* ribozyme retains its conformation despite deletion of ~160 nucleotides, including the P1 helix (Figure 10.3).[45,46] Thus, no gross conformational changes occur upon P1 docking.

Group I Introns 187

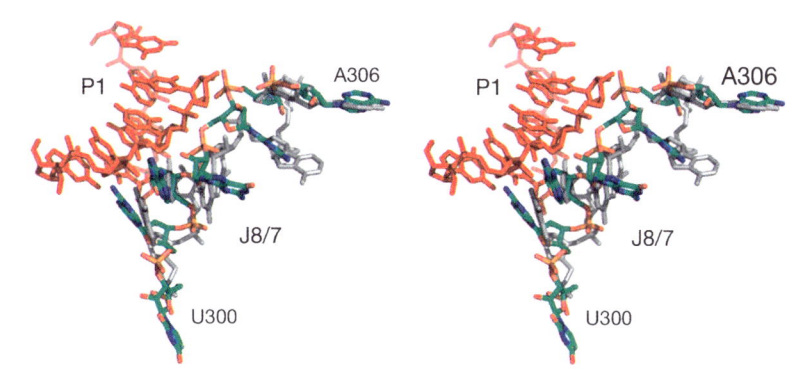

Figure 10.5 Conformational changes upon P1 docking. The P1 helix (red) and J8/7 region (grey) of the Twort ribozyme are superposed on the J8/7 junction from the *Tetrahymena* intron (green). In the absence of P1, there are only minor changes in the structure of J8/7 and little rearrangement is necessary to accommodate docking of the 5′-splice site. This figure was generated using the Twort intron structure, but comparison of the *Tetrahymena* and *Azoarcus* structures leads to similar conclusions.

As expected from the earlier structural studies of the P4-P6 domain, the J4/5 junction is largely preorganized to recognize the G · U wobble at the 5′-splice site, and no rearrangement of the P4-P6 domain is required to facilitate P1-docking. In contrast, the nucleotides of J8/7 that interact with the P1 helix are not in the same conformation as those seen in the docked structures (Figure 10.5).

In the *Tetrahymena* intron, the nucleotides of J8/7 that are involved in docking P1 are anchored to the ribozyme core by base-triples on either end.[49,53,54] U300 is located within the major groove of the P3 helix, and U305 is interacting with the P4-P6 domain. Thus, even in the absence of P1, there is significant conformational restriction on the intervening nucleotides. As a result, the difference in the path of the RNA backbone between the docked and undocked structures is quite modest. Interestingly, in the undocked ribozyme, bases within J8/7 that will make contacts with the minor groove of P1 are pointed out into solution and poised for recognition (Figure 10.5).

10.6.3 A Binding Pocket for Guanosine

The guanosine binding site is located within the P7 helix, a highly conserved 6 base pair duplex with a single bulged base. Michel *et al.* first identified the P7 helix as the guanosine binding site and identified a conserved G-C base pair within this helix as the major recognition determinant for the guanine base, the G264-C311 base pair in the *Tetrahymena* intron.[3] They predicted that this G-C base pair would form a major groove base triple with the guanosine nucleotide substrate. As A-form duplexes have no measurable affinity for binding guanosine in their major groove, it was clear that this base triple interaction is not sufficient to provide the affinity and specificity of nucleotide binding that is

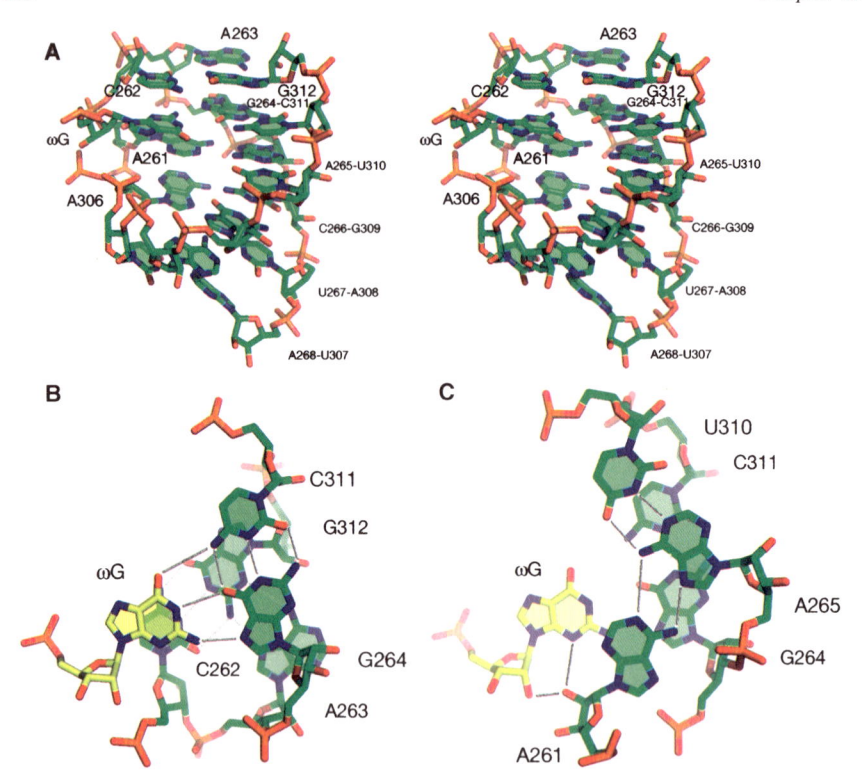

Figure 10.6 Guanosine binding site of the *Tetrahymena* intron. (*A*) Stereoview of the guanosine binding site. Guanosine substrate docks in a cleft composed of stacked base triples. (*B*) Stacking of the ωG · G264-C311 base triple upon the C262 · A263-G312 base triple. (*C*) Stacking of the A261 · A265-U310 base triple upon the ωG · G264-C311 triple. The A261 · A265-U310 triple positions the 2′-hydroxyl of A261 in hydrogen bonding distance of the 2′-hydroxyl and N3 of ωG. In (*B*) and (*C*), ωG is highlighted in yellow. This figure was made using the *Tetrahymena* structure to allow the nucleotide labels to match the majority of the available biochemical literature. Similar structures are observed in the *Azoarcus* and Twort introns.

observed in group I introns. Clearly, additional factors must contribute to this binding pocket. As the single nucleotide bulge is a conserved feature of all group I introns, it must be an important facet of the guanosine binding site.

As predicted, the base triple between ωG (or guanosine substrate in the first step of splicing) is observed in all three crystal structures (Figure 10.6). The single base-bulge (A263 in *Tetrahymena*) precedes the G264-C311 base pair. Surprisingly, A263 is neither extruded from the helix nor stacked in between base pairs as would be predicted from the behavior of bulged bases in other crystal structures. Rather, it is coplanar with the nucleotide that precedes it and stacked upon G264. The interaction between the C262-G312 base pair and the bulged A263 forms a second base triple that stacks extensively upon the ωG · G264-C311 base triple. This interaction forms the "top" of the guanosine binding site.

Similar structures are formed in all three structures even though the identity of the corresponding bases is different.

Surprisingly, a second nucleotide occupies the major groove of P7. In the secondary structure, A261 is located within a three nucleotide junction linking the P6 helix to the P7 helix. A261 spans the major groove of P7, linking the P3-P9 domain to the P4-P6 domain. A261 makes an amino-N7 base pair with the A265-U310 forming a third base triple in the guanosine binding site (Figure 10.6C). In contrast to the triple interaction at the top of P7, there is very little stacking between A261 and ωG. The role of this A261 is not simply structural. The 2′-hydroxyl of A261 makes hydrogen bonds to the 2′-hydroxyl and the N3 of ωG. The importance of this interaction is underscored by biochemical analysis that reveals severe catalytic consequences of removing the 2′-hydroxyl at either the substrate guanosine or A261.[55,56] Intertwining of A261 with ωG juxtaposes the negatively-charged phosphate group of A262 with the 2′-hydroxyl of ωG. This interaction provides a structural basis for the observation that protonated 2′-aminoguanosine, where the 2′-hydroxyl of guanosine is replaced by a positively charged NH_3^+ group, binds 200-fold more tightly than the natural substrate.[57] A306 and A308 are side-by side, allowing A306 to make a fourth base triple within the P7 major groove. This geometry contributes to the twisted structure of the P7 helix and surrounds the ribose group of ωG with a cluster of three phosphate groups from nucleotides A262, A306 and U307.

10.6.4 Packed Stacks

Like many other nucleobase-binding RNAs, group I introns make use of stacks of base triples to facilitate recognition of the purine base. This strategy appears to be a powerful method for generating purine binding sites. It has been observed in artificially evolved RNA's such as the FMN, theophylline, and malachite green binding aptamers,[58–60] and naturally occurring RNAs such as the guanine riboswitch[61] use this strategy to achieve tight binding pockets for their substrates.

Among the RNA active sites for which structural information is available, the group I introns are unique. The guanosine binding pocket must bind its substrate snugly and specifically such that it can attract the nucleoside substrate from within the cellular environment. Yet, after the first step of splicing, the active site must disgorge this nucleotide, allowing ωG to bind in its place. Furthermore, once bound, the active site must activate the nucleoside for catalysis and accommodate the conformational changes that accompany the self-splicing reaction.

A clue that the group I intron binding site is different from the nucleobase-binding RNAs comes from comparison of the guanosine-bound Twort, *Tetrahymena* and *Azoarcus* structures with the deoxyguanosine-bound *Azoarcus* structure. In contrast to guanosine, deoxyguanosine is a potent competitive inhibitor of group I introns and no reaction with this nucleoside has ever been observed.[55,62] This is a surprising result. Unlike an RNA polymerase there is no apparent evolutionary or functional imperative to exclude deoxynucleosides from the self-splicing reaction. Furthermore, deletion of a single hydroxyl from the reaction is not expected to have this profound effect on reactivity. As an example, deletion of the 2′-hydroxyl at the 5′-splice site reduces the reactivity only 3-fold.[63] These data suggest that deletion of the 2′-hydroxyl of guanosine substrate induces a more significant change in the active site than simple loss of an atom.

In the crystal structures, a small but important structural change is observed when the 2′-hydroxyl guanosine is removed from the active site. Deletion of this 2′-hydroxyl disrupts a network of interactions between the substrate and nucleotide 261 (Figure 10.6C). The consequences of this include rotation of the ωG ribose, thereby moving the 3′-hydroxyl of the substrate from its optimal catalytic position. In addition, binding of guanosine results in opening of the active site (Figure 10.7). The backbone atoms of P7 move subtly but distinctly further from the guanosine binding pocket and the phosphate of A262 swings away from the bound deoxyribose. With the active site propped open, a monovalent cation can bind a site normally occupied by Mg^{2+}, rendering the group I intron utterly inactive. This observation is supported biochemically. Although guanosine and deoxyguanosine bind with similar rate constants, significant differences in the manner in which these nucleotides bind in the active site have been observed.[64,65]

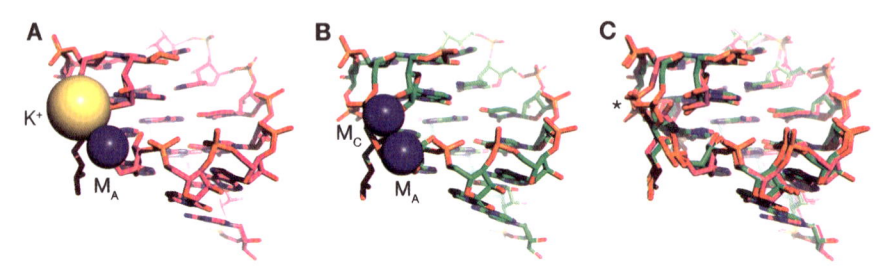

Figure 10.7 Metal ion binding sites and the active site conformation depend on the identity of the nucleotide substrate. (*A*) *Azoarcus* intron with deoxyguanosine bound. When the 2′-hydroxyl of the guanosine substrate is missing, the metal ion C (M_C) binding pocket is expanded, allowing monovalent ions, such as K^+, to occupy the magnesium ion binding site. (*B*) *Azoarcus* intron with guanosine bound. When the bona fide substrate is bound in the active site, two metal ions, A (M_A) and C (M_C), are observed in the crystal structure. (*C*) Comparison of the active site structure with deoxyguanosine (pink) and guanosine (green) bound. Two subtle changes are observed when the two conformations are superposed. When deoxyguanosine is bound, the phosphate corresponding to C262 of the *Tetrahymena* intron (marked with *) is displaced away from the ribose. This phosphate is known to be a ligand to the catalytic metal ion named M_C. In addition, the active site is modestly ajar, with the bulk of the backbone atoms surrounding ωG substrate displaced away from the substrate. Similar comparisons between the deoxyguanosine-bound *Azoarcus* intron and the guanosine-bound Twort and *Tetrahymena* introns reveal the same conformational changes.

The plasticity of the active site as revealed by structures with different nucleosides bound may be a characteristic that is crucial for catalysis, facilitating conformational changes that occur as the reactants approach the transition state and during the various steps of splicing.

10.7 Biochemical Characterization of the Structure

Solution characterization of the structure is an essential complement to the crystal structures. Although the crystal structures can provide a detailed view of the atoms in the active site, biochemical analysis is crucial to establish the catalytic relevance of the snapshot of the molecule trapped within the crystal.

10.7.1 Metal Ion Binding and Specificity Switches

Group I introns are obligate metalloenzymes that, like many of their protein counterparts, position catalytic magnesium ions within their active site. These ions play multiple roles in catalysis, including organizing the conformation of the active site, activating the nucleophile, and stabilizing the leaving group.

The interaction between active site metal ions and substrates has been probed extensively using specificity switches. In these experiments, a functional group on the substrate is changed to a sulfur or a nitrogen to weaken its ability to serve as an inner-sphere ligand to Mg^{2+}. Such modifications dramatically interfere with the ribozyme reaction by disrupting crucial interactions between the substrates and metal co-factors. "Soft" metal ions such as Mn^{2+} or Cd^{2+} are better able to coordinate to sulfur or nitrogen. Thus, in the presence of these ions, the metal–substrate bond with modified substrate is restored and the reaction is "rescued". Such experiments provided a detailed inventory of the four atoms in the substrates that are directly coordinated to Mg^{2+}: the 3'-hydroxyl of the uridine at the 5'-splice site, the 2'- and 3'-hydroxyls of the guanosine substrate and one of the non-bridging oxygens of the scissile phosphate.[66-68] The next step was to determine how many metal ions were participating in these four interactions with the substrates.

The active site magnesium ion cluster was further characterized by quantitative analysis of thiophilic metal ion binding using a series of doubly-modified substrates. If two modified ligands coordinate the same metal ion, only a single metal ion is required for rescue. If two modified ligands coordinate two distinct metal ions, two metal ions are required for rescue. By monitoring the degree of rescue as a function of catalytic metal ion concentration, apparent metal ion binding constants and the number of thiophilic metal ions required for rescue can be analyzed. In this way, a constellation of three metal ions, metal A, metal B, and metal C surrounding the group I intron substrates was biochemically mapped (Figure 10.8).[69,70] These are predicted to interact with the 3'-hydroxyl of the uridine at the 5'-splice site, the 3'-hydroxyl of the guanosine substrate, and the 2'-hydroxyl of the guanosine substrate respectively.

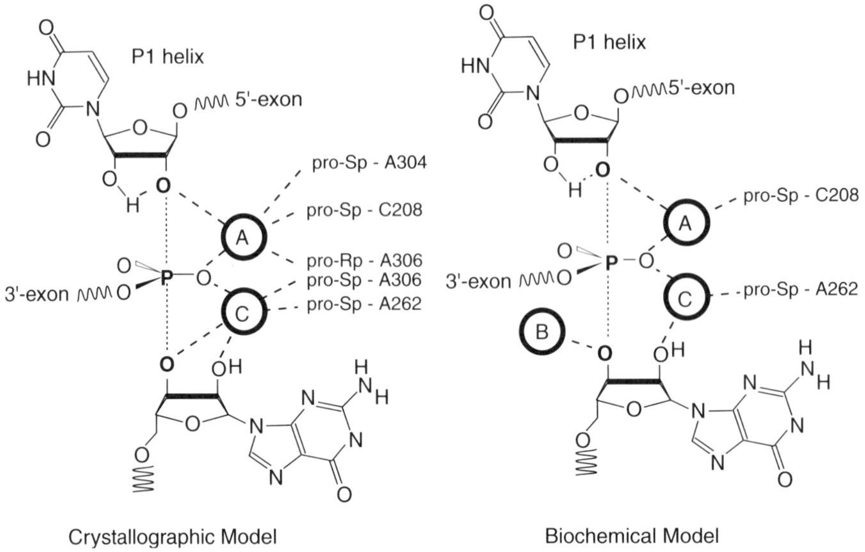

Figure 10.8 Interaction maps for metal ions and their ligands. Right: A biochemical model for metal–ligand interactions in the *Tetrahymena* intron.[70,73,74] Left: A crystallographic model for metal–ligand interactions in the *Tetrahymena* intron derived from the crystal structure of the *Azoarcus* intron.[44]

10.7.2 Identification of Ligands to the Catalytic Metal Ions

Once the cluster of catalytic metal ions in the active site was described, it was of great interest to determine the mechanism by which the ribozyme would bind and position these ions. Early experiments used T7 RNA polymerase and α-nucleoside thiotriphosphates to generate libraries in which single sulfur substitutions at *pro*-R$_p$ oxygens were incorporated into each RNA. The *pro*-S$_p$ oxygens cannot be investigated in this manner because T7 polymerase will not accommodate this stereochemistry. These libraries were used to identify the sulfur substitutions that interfered with ribozyme activity.[71,72] The oxygens at these positions are presumably involved either in essential hydrogen bonds or in contacts to metal ions. To help differentiate between the two possibilities, the reaction was repeated in the presence of manganese ion. At some positions, the presence of the thiophilic metal ion suppressed the deleterious effect of sulfur modifications.[72] The sulfur atoms so identified, and therefore the corresponding oxygen atoms in the unmodified ribozyme, are likely to be involved in coordinating divalent cations. Unfortunately, few ligands to the catalytic metal ions were observed in these experiments. This is likely to arise from the geometry of RNA helices. In an A-form helix, the *pro*-S$_p$ oxygen points out towards the solvent and is relatively accessible. In contrast, the *pro*-R$_p$ oxygen points into the deep major groove of the helix; this location precludes the generation of ion binding sites using pro-R_p oxygens and nucleotides that are distant in the primary sequence.

Ultimately, it was necessary to generate semisynthetic ribozymes in which single enantiomerically-pure phosphorothioate modifications were introduced into the ribozyme backbone. The locations of these modifications were guided by models[49] and locations where phosphorothioates were known to interfere with reactivity.[71] These variants were then screened for reactivity in the presence and absence of catalytic metal ions. These experiments identified the pro-S_p oxygens at C208 and A262 of the *Tetrahymena* intron as ligands to catalytic metal ions.[73,74] These ligands were then correlated with two distinct metal ions, metal ions A and C, respectively, using modified substrates and thermodynamic signatures (Figure 10.8).

10.7.3 Correlation with Metal Ion Binding Sites within the Crystal Structures

How do the biochemical data compare with the various magnesium ion binding sites observed in the various crystal structures (Figure 10.8)? The location of the biochemically observed metal ion A is straightforward to observe in crystal structures of the Twort and *Azoarcus* introns. As predicted biochemically, one of its ligands is C208 pro-S_p oxygen,[73] allowing the biochemically observed metal ion to be correlated with ions observed in the crystal structure. Metal ion A is buried in a snug pocket and appears to be integral to the structure of the ribozyme, bringing together ligands from the P1-P2 domain, the P4-P6 domain and the P3-P9 domain. It is the only catalytic metal ion observable in the Twort structure.

The second catalytic Mg^{2+} binding site observed in the crystal structures is located near the guanosine substrate. Coordination of this ion to the pro-S_p oxygen of A262 links the crystallographically observable metal ion to metal ion C.[74] In contrast to metal ion A, the ligands to metal ion C are localized within the P3-P9 domain. All non-water ligands to metal ion C are either atoms from the substrates or atoms from nucleotides juxtaposed in the primary sequence to the guanosine binding site. Perhaps as a consequence, this metal ion binding site is malleable. When deoxyguanosine inhibitor is bound in the guanosine binding site, structural rearrangements at this site occur that allow larger, monovalent ions to bind.[43] In the *Tetrahymena* structure, metal ion C was first observed to interact with both the 2'- and 3-hydroxyls of the guanosine substrate,[39] a feature later confirmed in the *Azoarcus* intron.[44] In the structure of the Twort ribozyme, this metal ion is not observable – likely caused by the high ionic strength of the crystallization buffer.

The third metal ion, metal ion B, has not been identified in the group I intron crystal structures. Furthermore, the 2'- and 3'-hydroxyls of guanosine substrate are both observed to interact with metal ion C. Do these structural data rule out the presence of a metal ion B in the ribozyme reaction? This is not crystal clear. First, the crystal structures give us information about what is presumably a ground state, while the biochemistry may tell us about a different low energy state or the transition state. We know from the structural data that deletion of a single atom of the substrate leads to significant conformational change and a change in the metal ion binding properties. In the phosphotransesterification reaction there are significant conformational changes that have to occur upon approach to the transition state. Ordering of solvent molecules or loosely bound ions could occur only on the approach to the transition state. Second, inability to observe an ion or water in the crystal does not rule out its presence. Metal ion B may be bound in multiple conformations or may even be completely disordered in the available crystal structures, and therefore not observable. Finally, the RNA within the *Azoarcus* intron crystals, in which both metal ion A and metal ion C are visible, was designed to mimic the step prior to the attack of the 5′-exon upon the 3′-splice site (Figure 10.1). In a typical splicing reaction, this intermediate is expected to be rapidly converted into ligated exon and released intron products. Under crystallization conditions, this reactivity is inhibited in an ill-understood manner. It is possible that minor, but catalytically significant, rearrangements of the active site and binding of additional ions or solvent are necessary for maximal activity. This is an area where our picture of group I intron catalysis is still blurred.

10.7.4 Nucleotide Analog Interference Techniques

Nucleotide analog interference mapping (NAIM) provides a high-throughput approach to characterizing the structure of RNA active sites. α-Phosphorothioate-tagged modified nucleotides are incorporated randomly to make a library of RNAs, and the library is screened to separate those ribozymes that retain activity from inactive RNAs. Sites that are tolerant of the modification are identified using the phosphorothioate tag, which is sensitive to cleavage by iodine. By comparison of the active pool of ribozymes with an unselected pool of ribozymes, modifications that interfere with activity can be identified.[75–77] This technique allows rapid identification of atoms involved in crucial interactions within the RNA active site.

Using a derivative of this technique, the interaction partners for these atoms were identified. In nucleotide analog interference suppression (NAIS), the NAIM experiment is performed, but it is in the background of an RNA where the interfering modification is always present. In this pool, a critical interaction is broken, and the population is less active. Now when the NAIM screen is performed, one partner in an interaction has been disrupted. Therefore, in most cases, the other partner in the interaction does not cause further destabilization when it is modified. The partner atom is therefore identified by a modification that interferes with maximal activity in the wild-type sequence, but does not further reduce activity in the chemically modified ribozyme.

NAIS studies have been used to characterize atomic level details of how the P4-P6 domain interacts with the 5′-splice site, how the J8/7 junction docks in the minor groove of P1, and how U300 binds in the major groove of P3.[35,49,54]

Group I Introns 195

These data were used to generate a three-dimensional model of the heart of the group I intron active site that had features not consistent with *Tetrahymena* crystal structures.[37,39] In particular, U300 is predicted both to make a base triple in the major groove of P3 and to interact *via* its 2′-hydroxyl with the 2′-hydroxyl of G26. However, docking of a P1 helix into the structures of the *Tetrahymena* ribozyme suggests that the distance between U300 and G26 is too long for direct interaction. As the recent structures, both in the presence and absence of P1, are very similar, it is not likely that NAIS data can be reconciled by invoking a significant conformational change upon P1 docking as previously suggested.[54]

Integration of the structural and biochemical data can be accomplished by superposing the Twort or *Azoarcus* intron structures upon the *Tetrahymena* intron structure. In the superposition, the 2′-hydroxyl of U300 is in a position to interact with not the P1 helix, but the last nucleotide J2/3 junction (Figure 10.9). Indeed, this interaction was identified in the NAIS analysis of the *Azoarcus* ribozyme.[78] Interestingly, in both the Twort and *Azoarcus* structures this nucleotide in J2/3 also interacts with the P1-P2 domain at the nucleotide that corresponds to G26 (Figure 10.9). This analysis suggests that the interaction between U300 and P1 uncovered by NAIS is indirect, mediated through the J2/3 junction. Additionally, mutations in the J2/3 region of the *Tetrahymena* intron reduce the accuracy of 5′-splice site selection, leading Downs *et al.* to predict

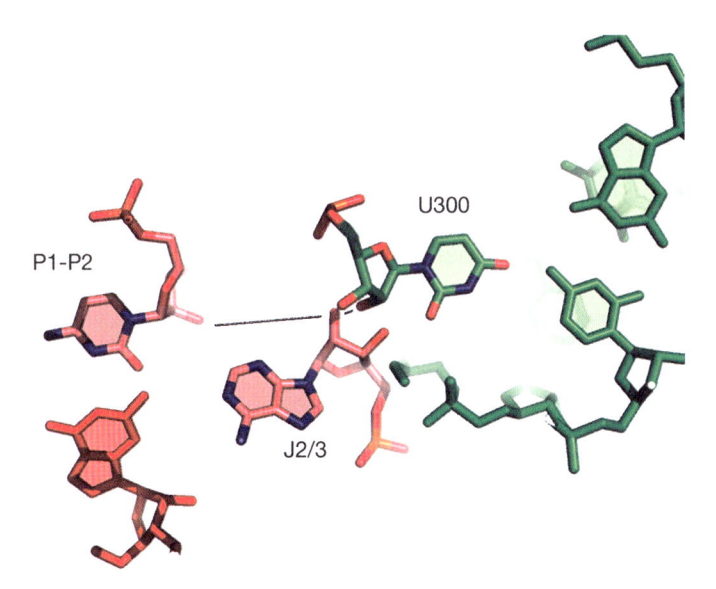

Figure 10.9 Potential interaction partners for U300. As predicted from NAIS analysis,[49,54] U300 is involved in a base triple with the A97-U277 base pair in P3 (green). Superposition of the *Tetrahymena* intron structure with the Twort (shown here) or *Azoarcus* structure suggests that the 2′-hydroxyl of U300 cannot interact directly with the P1-P2 domain as previously described.[49,54] Instead, the interaction is indirect, mediated by the J2/3 junction.

that, as seen in the superposition, the J2/3 region would be involved in a tertiary interaction linking the P1 helix to the P3-P9 domain.[47]

NAIM and NAIS are very powerful methodologies and have provided significant atomic-resolution detail about active site interactions. These interaction maps must be interpreted with a caveat. The contacts that are identified may be indirect, mediated through a third nucleotide or, perhaps, even a metal ion or solvent molecule.

10.8 What Makes a Catalytic Site?

Is the unusual geometry of the P7 helix intrinsic to the local sequence, or is it imposed by interactions with the rest of the introns catalytic core? Studies by Kitamura et al. suggest that at least part of the guanosine binding properties of this region are not dependent on the presence of long-range tertiary interactions within the intron.[79] They used NMR spectroscopy to study a 22-nucleotide RNA that mimicked four base pairs and the bulged nucleotide of helix P7, plus nucleotides corresponding to P9.0 and ωG. Remarkably, although this molecule is isolated from the remainder of the intron and is missing key residues involved in guanosine binding, including A261 and A306, it displays several features characteristic of the native guanosine binding site (Figure 10.10). First, although this molecule has three consecutive G-C or C-G base pairs with, presumably, equal ability to form a major groove base triple with guanosine, binding of ωG at the G264-C311 base pair is detectable and specific. Second, as observed in the crystal structures, there is a large helical twist between the G264-C311 base pair and the C262-G312 base pair. In the crystal structures the angle between the two base pairs is $\sim 60°$, twice that expected for an A-form helix. In the NMR structure, this angle is reported to be $\sim 40°$. Although the model for P7 calculated in the NMR structure (pdb accession code 1K2G) is less twisted than that observed in the crystal structures, stacking is observed

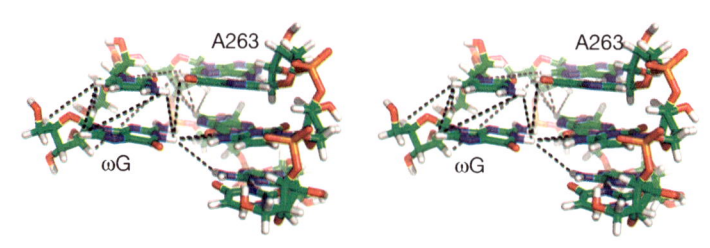

Figure 10.10 Crystallographic and spectroscopic characterization of the guanosine binding site. An RNA spanning much of the P7 helix and ωG was used as a model system to explore the structure of the group I intron guanosine binding site by NMR.[79] The NOE distance constraints are shown as dotted lines plotted on top of the crystallographic model of the *Tetrahymena* ribozyme guanosine-binding sites.[39] Only one weak NOE distance constraint is violated in the crystallographic model. This suggests that the model RNA used in the NMR studies retains many of the features of the guanosine binding site embedded within the intron.

between C262 and ωG, similar to what is observed in the intact introns. This suggests that the solution structure of the model RNA possesses geometry similar to that of the guanosine binding site within the intron. Finally, while the NMR constraints are insufficient to define the structure of A263, the pattern of NOEs observed between A263 and G264 is consistent with stacking interactions, and NOEs between A263 and C262 are in agreement with the side-by-side conformation observed in the crystal structures (Figure 10.10). Only one of the observed NOEs is violated in the conformation observed in the crystal structure, and this NOE was very weak. Overall, the 22-nucleotide model RNA used in this study has many of the features of the guanosine binding site in the context of the entire intron. These data suggest that many of the structural characteristics of the P7 helix are determined locally and are not dependent on long-range interactions within the intron.

10.9 Back to the Origins

The self-splicing reaction of the group I introns is complex, requiring the binding of multiple substrates and coordination of two chemical reactions. It seems unlikely that such a molecule would have evolved in a single step. It is tempting to speculate about the path by which this RNA machine could have evolved. Examination of the NMR structure of the guanosine binding site and the crystallographic structures of the introns suggest that a guanosine binding site capable of catalyzing a phosphotransesterification reaction could have evolved within a small RNA. The NMR structure demonstrates that the unusual conformation of P7 is largely determined locally, and in the crystal structures the additional features that contribute to the catalytic properties and structural integrity of the guanosine binding pocket are nearby in the primary sequence. A261 and A306 stack below ωG to fill the major groove of the P7 helix; A262 and A306 provide ligands to metal C. This suggests that a guanosine binding site capable of activating guanosine for nucleophilic attack could have evolved very early.

Once the guanosine binding pocket had evolved, a mechanism of 5'-splice site selection was needed. Examination of the crystal structures reveals that, by far, the majority of interaction partners to atoms within the P1 helix and the P1-P2 domain come from nucleotides local to the P3-P9 domain. The J2/3, J6/7, and J8/7 provide the majority of hydrogen bonds to the P1-P2 domain and the majority of ligands to the catalytic metal ions. The P4-P6 domain contributes hydrogen bonds to a single nucleotide in P1[35,80] and a single ligand to a catalytic metal ion.[73] This suggests that the intron could have evolved catalytic activity in the absence of the P4-P6 domain. This hypothesis was tested by Inoue et al. who showed that catalytic activity is retained even when the P4-P6 domain is deleted.[81]

The group I intron may well have evolved in three steps. In the first step, a guanosine binding pocket appeared. In the second, additional elements of the P3-P9 domain involved in 5'-splice site recognition came into being. In the final step, the P4-P6 domain was acquired, providing additional structural stability and a scaffold upon which to fold the P3-P9 domain. Obviously this path cannot be tested in the absence of time travel. It is clear though that evolution has created a robust RNA catalyst that has passed the test of many eons of evolution.

References

1. T.R. Cech, A.J. Zaug and P.J. Grabowski, *Cell*, 1981, **27**, 487.
2. A.J. Zaug and T.R. Cech, *Nucleic Acids Res.*, 1982, **10**, 2823.
3. F. Michel, M. Hanna, R. Green, D.P. Bartel and J.W. Szostak, *Nature*, 1989, **342**, 391.
4. M.D. Been and T.R. Cech, *Cell*, 1986, **47**, 207.
5. K. Kruger, P.J. Grabowski, A.J. Zaug, J. Sands, D.E. Gottschling and T.R. Cech, *Cell*, 1982, **31**, 147.
6. *Catalytic RNA*, ed. F. Eckstein and D. Lilley, Springer-Verlag, Berlin, 1996.
7. D. Herschlag and T.R. Cech, *Biochemistry*, 1990, **29**, 10159.
8. K. Karbstein, K.S. Carroll and D. Herschlag, *Biochemistry*, 2002, **41**, 11171.
9. P.C. Bevilacqua and D.H. Turner, *Biochemistry*, 1991, **30**, 10632.
10. A.M. Pyle and T.R. Cech, *Nature*, 1991, **350**, 628.
11. D. Herschlag, F. Eckstein and T.R. Cech, *Biochemistry*, 1993, **32**, 8299.
12. S.A. Strobel and T.R. Cech, *Biochemistry*, 1993, **32**, 13593.
13. A.M. Pyle, S. Moran, S.A. Strobel, T. Chapman, D.H. Turner and T.R. Cech, *Biochemistry*, 1994, **33**, 13856.
14. D.S. Knitt, G.J. Narlikar and D. Herschlag, *Biochemistry*, 1994, **33**, 13864.
15. S.A. Strobel and T.R. Cech, *Science*, 1995, **267**, 675.
16. R.B. Waring, J.A. Ray, S.W. Edwards, C. Scazzocchio and R.W. Davies, *Cell*, 1985, **40**, 371.
17. J.M. Burke, K.D. Irvine, K.J. Kaneko, B.J. Kerker, A.B. Oettgen, W.M. Tierney, C.L. Williamson, A.J. Zaug and T.R. Cech, *Cell*, 1986, **45**, 167.
18. S. Couture, A.D. Ellington, A.S. Gerber, J.M. Cherry, J.A. Doudna, R. Green, M. Hanna, U. Pace, J. Rajagopal and J.W. Szostak, *J. Mol. Biol.*, 1990, **215**, 345.
19. D.W. Celander and T.R. Cech, *Science*, 1991, **251**, 401.
20. W.D. Downs and T.R. Cech, *Biochemistry*, 1990, **29**, 5605.
21. J.F. Wang and T.R. Cech, *Science*, 1992, **256**, 526.
22. J.F. Wang, W.D. Downs and T.R. Cech, *Science*, 1993, **260**, 504.
23. F. Michel and E. Westhof, *J. Mol. Biol.*, 1990, **216**, 585.
24. V. Lehnert, L. Jaeger, F. Michel and E. Westhof, *Chem. Biol.*, 1996, **3**, 993.
25. F. Michel, A. Jacquier and B. Dujon, *Biochimie*, 1982, **64**, 867.
26. R.W. Davies, R.B. Waring, J.A. Ray, T.A. Brown and C. Scazzocchio, *Nature*, 1982, **300**, 719.

27. S.H. Kim and T.R. Cech, *Proc. Natl. Acad. Sci. U.S.A.*, 1987, **84**, 8788.
28. F.L. Murphy and T.R. Cech, *Biochemistry*, 1993, **32**, 5291.
29. T.R. Cech, S.H. Damberger and R.R. Gutell, *Nat. Struct. Biol.*, 1994, **1**, 273.
30. G. van der Horst, A. Christian and T. Inoue, *Proc. Natl. Acad. Sci. U.S.A.*, 1991, **88**, 184.
31. J.A. Doudna and J.W. Szostak, *Mol. Cell Biol.*, 1989, **9**, 5480.
32. J.H. Cate, A.R. Gooding, E. Podell, K.H. Zhou, B.L. Golden, C.E. Kundrot, T.R. Cech and J.A. Doudna, *Science*, 1996, **273**, 1678.
33. K. Juneau, E. Podell, D.J. Harrington and T.R. Cech, *Structure*, 2001, **9**, 221.
34. K.P. Williams, D.N. Fujimoto and T. Inoue, *J. Biochem. (Tokyo)*, 1994, **115**, 126.
35. S.A. Strobel, L. Ortoleva-Donnelly, S.P. Ryder, J.H. Cate and E. Moncoeur, *Nat. Struct. Biol.*, 1998, **5**, 60.
36. B.L. Golden, E.R. Podell, A.R. Gooding and T.R. Cech, *J. Mol. Biol.*, 1997, **270**, 711.
37. B.L. Golden, A.R. Gooding, E.R. Podell and T.R. Cech, *Science*, 1998, **282**, 259.
38. F. Guo and T.R. Cech, *Nat. Struct. Biol.*, 2002, **9**, 855.
39. F. Guo, A. Gooding and T.R. Cech, *Mol. Cell*, 2004, **16**, 351.
40. P. Young, M. Ohman, M.Q. Xu, D.A. Shub and B.M. Sjoberg, *J. Biol. Chem.*, 1994, **269**, 20229.
41. M. Landthaler and D.A. Shub, *Proc. Natl. Acad. Sci. U.S.A.*, 1999, **96**, 7005.
42. B.L. Golden, H. Kim and E. Chase, *Nat. Struct. Mol. Biol.*, 2005, **12**, 82.
43. P.L. Adams, M.R. Stahley, A.B. Kosek, J. Wang and S.A. Strobel, *Nature*, 2004, **430**, 45.
44. M.R. Stahley and S.A. Strobel, *Science*, 2005, **309**, 1587.
45. Q. Vicens and T.R. Cech, *Trends Biochem. Sci.*, 2006, **31**, 41.
46. M.R. Stahley and S.A. Strobel, *Curr. Opin. Struct. Biol.*, 2006, **16**, 319.
47. W.D. Downs and T.R. Cech, *Genes Dev.*, 1994, **8**, 1198.
48. A.M. Pyle, F.L. Murphy and T.R. Cech, *Nature*, 1992, **358**, 123.
49. A.A. Szewczak, L. Ortoleva-Donnelly, S.P. Ryder, E. Moncoeur and S.A. Strobel, *Nat. Struct. Biol.*, 1998, **5**, 1037.
50. P. Nissen, J.A. Ippolito, N. Ban, P.B. Moore and T.A. Steitz, *Proc. Natl. Acad. Sci. U.S.A.*, 2001, **98**, 4899.
51. P.C. Bevilacqua, R. Kierzek, K.A. Johnson and D.H. Turner, *Science*, 1992, **258**, 1355.
52. D. Herschlag, *Biochemistry*, 1992, **31**, 1386.
53. M.A. Tanner and T.R. Cech, *Science*, 1997, **275**, 847.
54. A.A. Szewczak, L. Ortoleva-Donnelly, M.V. Zivarts, A.K. Oyelere, A.V. Kazantsev and S.A. Strobel, *Proc. Natl. Acad. Sci. U.S.A.*, 1999, **96**, 11183.
55. B.L. Bass and T. Cech, *Biochemistry*, 1986, **25**, 4473.
56. L. Ortoleva-Donnelly, A.A. Szewczak, R.R. Gutell and S.A. Strobel, *RNA*, 1998, **4**, 498.

57. S.O. Shan, G.J. Narlikar and D. Herschlag, *Biochemistry*, 1999, **38**, 10976.
58. P. Fan, A.K. Suri, R. Fiala, D. Live and D.J. Patel, *J. Mol. Biol.*, 1996, **258**, 480.
59. G.R. Zimmermann, R.D. Jenison, C.L. Wick, J.P. Simorre and A. Pardi, *Nat. Struct. Biol.*, 1997, **4**, 644.
60. C. Baugh, D. Grate and C. Wilson, *J. Mol. Biol.*, 2000, **301**, 117.
61. R.T. Batey, S.D. Gilbert and R.K. Montange, *Nature*, 2004, **432**, 411.
62. S. Moran, R. Kierzek and D.H. Turner, *Biochemistry*, 1993, **32**, 5247.
63. D. Herschlag, F. Eckstein and T.R. Cech, *Biochemistry*, 1993, **32**, 8312.
64. L.A. Profenno, R. Kierzek, S.M. Testa and D.H. Turner, *Biochemistry*, 1997, **36**, 12477.
65. P.C. Bevilacqua, K.A. Johnson and D.H. Turner, *Proc. Natl. Acad. Sci. U.S.A.*, 1993, **90**, 8357.
66. J.A. Piccirilli, J.S. Vyle, M.H. Caruthers and T.R. Cech, *Nature*, 1993, **361**, 85.
67. A.S. Sjogren, E. Pettersson, B.M. Sjoberg and R. Stromberg, *Nucleic Acids Res.*, 1997, **25**, 648.
68. L.B. Weinstein, B.C. Jones, R. Cosstick and T.R. Cech, *Nature*, 1997, **388**, 805.
69. S. Shan, A. Yoshida, S. Sun, J.A. Piccirilli and D. Herschlag, *Proc. Natl. Acad. Sci. U.S.A.*, 1999, **96**, 12299.
70. S. Shan, A.V. Kravchuk, J.A. Piccirilli and D. Herschlag, *Biochemistry*, 2001, **40**, 5161.
71. E.L. Christian and M. Yarus, *J. Mol. Biol.*, 1992, **228**, 743.
72. E.L. Christian and M. Yarus, *Biochemistry*, 1993, **32**, 4475.
73. A.A. Szewczak, A.B. Kosek, J.A. Piccirilli and S.A. Strobel, *Biochemistry*, 2002, **41**, 2516.
74. J.L. Hougland, A.V. Kravchuk, D. Herschlag and J.A. Piccirilli, *PLoS Biol*, 2005, **3**, e277.
75. W.D. Hardt, V.A. Erdmann and R.K. Hartmann, *RNA*, 1996, **2**, 1189.
76. S.A. Strobel and K. Shetty, *Proc. Natl. Acad. Sci. U.S.A.*, 1997, **94**, 2903.
77. C.S. Vortler, O. Fedorova, T. Persson, U. Kutzke and F. Eckstein, *RNA*, 1998, **4**, 1444.
78. J.K. Strauss-Soukup and S.A. Strobel, *J. Mol. Biol.*, 2000, **302**, 339.
79. A. Kitamura, Y. Muto, S. Watanabe, I. Kim, T. Ito, Y. Nishiya, K. Sakamoto, T. Ohtsuki, G. Kawai, K. Watanabe, K. Hosono, H. Takaku, E. Katoh, T. Yamazaki, T. Inoue and S. Yokoyama, *RNA*, 2002, **8**, 440.
80. S.A. Strobel and L. Ortoleva-Donnelly, *Chem. Biol.*, 1999, **6**, 153.
81. Y. Ikawa, H. Shiraishi and T. Inoue, *Nat. Struct. Biol.*, 2000, **7**, 1032.

CHAPTER 11
Group II Introns: Catalysts for Splicing, Genomic Change and Evolution

ANNA MARIE PYLE

Department of Molecular Biophysics and Biochemistry, Yale University; Howard Hughes Medical Institute, New Haven, CT

11.1 Introduction: The Place of Group II Introns Among the Family of Ribozymes

Group II introns are among the largest ribozymes in nature. At 400-1000 nucleotides in size, these multi-domain RNAs are second only to the ribosome in terms of sheer mass. Perhaps as a result of their structural complexity, group II introns are the swiss army knives of the ribozyme universe: They catalyze a diversity of reactions with tremendous efficiency and site selectivity.[1,2] Like other large ribozymes such as RNase P and group I introns, group II introns are metalloenzymes that catalyze hydrolysis and transesterification reactions with release of a 3′-hydroxyl leaving group. And yet group II introns are unusual in that they react efficiently with both RNA and DNA; indeed they are the only ribozyme for which DNA is a natural substrate.[1] Group II introns are extremely widespread among the various kindoms of life, displaying a rich phylogenetic and evolutionary heritage while continuing to play an important role in the metabolism of modern organisms.[3-5]

11.2 The Basic Reactions of Group II Introns

Group II introns are well-known for their ability to catalyze self-splicing from precursor RNA transcripts.[6] It is most common for the two steps of splicing to initiate with a "branching" reaction, in which a 2′-hydroxyl near the intron terminus attacks the 5′-splice site (Figure 11.1A). This releases the 5′-exon and

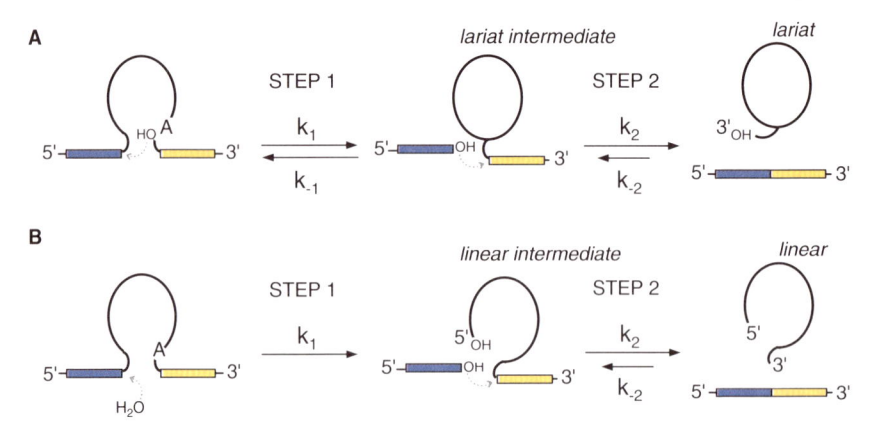

Figure 11.1 Two pathways for group II intron self-splicing. (*A*) Branching pathway and reverse-splicing. The 2′-hydroxyl group of a bulged adenosine in D6 is the nucleophile in splicing step 1, while the 3′-hydroxyl group of the free 5′-exon is the nucleophile in step 2. Both steps are highly reversible (*i.e.*, $k_1 \sim k_{-1}$). Intron mobility involves the reversal of this pathway (*i.e.*, free lariat inserts into RNA or DNA). (*B*) Hydrolysis pathway. Water is the nucleophile during the first step of splicing. The second step proceeds normally (and remains highly reversible).

results in a partially circularized form of the intron, known as a "lariat". In the second step of splicing, the 5′-exon terminus attacks the 3′-splice site, thereby joining the exons and releasing a free lariat intron.[1] It is notable that each of these two steps is highly reversible and reverse-splicing by group II introns is quite prevalent.[7,8] Indeed, it is the reactivity of liberated group II introns and their tendency to reverse-splice that explains their profound importance as mobile genetic elements that can invade both RNA and DNA (*vida infra*).[3]

Many group II introns can catalyze self-splicing through an alternative pathway in which water serves as the nucleophile during the first step (Figure 11.1B).[9–14] This results in release of a linear intron after the second step of splicing. The "hydrolytic" splicing pathway occurs both *in vitro* and *in vivo*, even for group II introns that tend to splice predominantly via branching.[10,11,15] It is not known if this alternative reaction provides some form of metabolic flexibility to organisms containing group II introns, if it is an ancestral form of reactivity related to RNase P, or if it is simply a widespread biochemical accident, but it is now clear that certain classes of group II introns splice exclusively through the hydrolytic pathway.[12–14] Linear introns readily undergo the first step of reverse splicing,[16] suggesting that the hydrolytic pathway does not preclude mobility.

The literature contains many examples of other reactions catalyzed by group II introns which, whether biologically relevant or not, underscore the adaptability of the intron active-site and the potential for involvement in numerous pathways for RNA processing and decay.[2] For example, group II introns have been shown to cyclize, forming stable circles that cannot be debranched.[17–19]

Certain constructs will attack a triphosphate, in a reaction reminiscent of polymerase.[16,20,21]

But perhaps the most important reaction catalyzed by group II introns, aside from RNA splicing, is the invasion of duplex DNA (Figure 11.2).[22] Indeed, this reaction is likely to have played a major role in the dispersal of ancestral introns and in the evolution of group II introns themselves.[23] A remarkable collaboration between RNA and protein enzymes, the invasion reaction is catalyzed by a complex between lariat RNA and two molecules of an intron-encoded enzyme called a "maturase" (*vida infra*).[8,23–25] While the intron

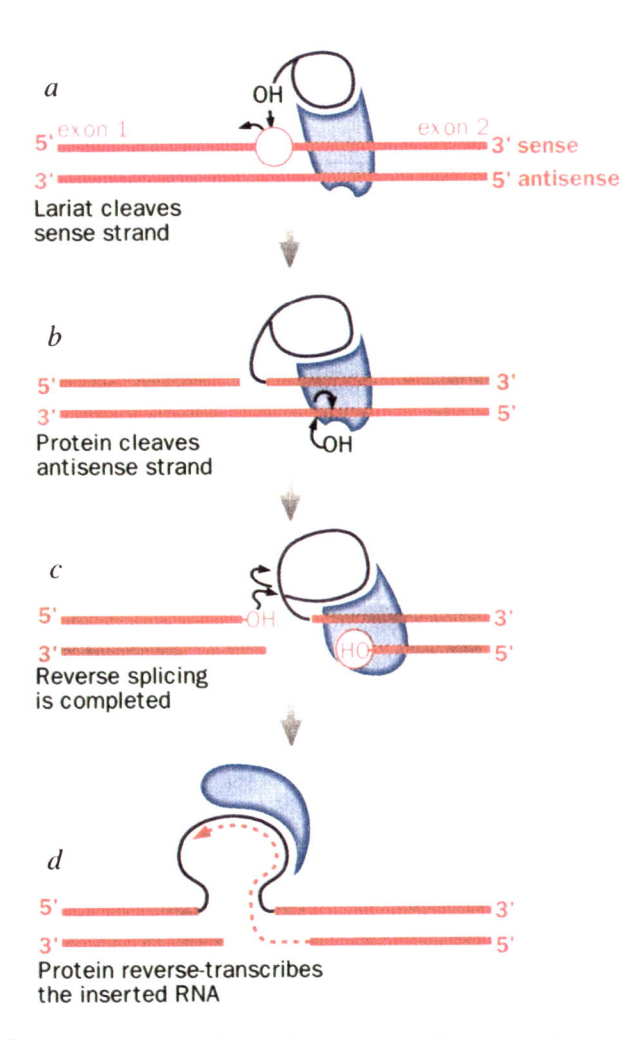

Figure 11.2 Catalysis of DNA invasion by group II introns. Red lines are duplex DNA, blue shapes are maturase protein dimer. Reproduced with permission from ref. 142.

RNA reverse-splices into one strand of DNA, the maturase protein cuts the second strand and, using the DNA end as a primer, makes a stable DNA copy of the intron through reverse-transcription.[26,27]

11.3 The Biological Significance of Group II Introns

11.3.1 Evolutionary Significance

Based on similarities in reaction mechanism, domain substructure or encoded sequence, much of our metabolic equipment is hypothesized to derive from ancestral group II introns.[28] For example, the eukaryotic spliceosome and our modern introns may have derived from group II introns (*vida infra*).[29] This is supported by similarities in reaction mechanism, splice site sequence and by analogous structures and functions for catalytic domains in both the spliceosome and group II introns (*i.e.*, U6 and D5, respectively). Similarities in the sequence of the encoded maturase protein and that of other cellular proteins, such as telomerase and some reverse-transcriptases, suggest that all of these molecules once shared a common ancestor.[28,30] If these familial connections are actually true, then approximately one third of our genetic material has originated from group II introns. While their precise historical impact may never be completely unravelled, there is little doubt that group II introns have played a major role in the evolution of bacterial and eukaryotic genomes.

11.3.2 Significance and Prevalence in Modern Genomes

Group II introns are common in all three branches of life.[4,31,32] Archaebacterial and eubacterial genomes often contain multiple group II introns and although these are usually in noncoding regions,[5,31,33–35] there is at least two cases in which the intron resides in a housekeeping gene.[36] For example, an intron from the soil bacterium *Azotobacter vinlandii* is unusually thermostable and requires high temperatures for self-splicing.[37] The fact that it lies within the gene for a heat shock protein suggests that intron thermostability can directly regulate metabolic response to the environment, and more broadly, that group II introns may be adapted to serve the needs of a host organism.[38] Nonetheless, it is more typical that group II introns in bacteria function strictly as retroelements (a form of genomic parasite).[5,33] Eukaryotic organisms such as plants, fungi, yeasts and protists contain numerous group II introns within their mitochondrial and chloroplast genomes.[23,39,40] These introns are typically found within genes that encode proteins essential for respiration and also within tRNAs and rRNAs.

11.3.3 The Potential Utility of Group II Introns

Group II introns react with extreme accuracy, they can be targeted to almost any sequence, they can cleave genes or introduce new genes by reacting either with RNA or DNA.[1] It is therefore not suprising that they are being extensively developed as biotechnological tools for manipulating gene sequences[41,42] and as molecular scalpels in new forms of gene therapy.[43,44] There has been particular focus on the development of group II intron/maturase complexes that selectively modify duplex DNA, which has led to powerful new reagents for the routine transformation of bacterial genomes.[41]

11.4 Domains and Parts: The Anatomy of a Group II Intron

Group II introns do not contain an abundance of highly conserved sequence (except in domains 5 (D5) and the ORF of domain 4 (D4), but they do adopt a highly conserved secondary structure in which six helical domains are arranged in specific positions around a large central junction.[39] Each of these domains plays a particular role in structure, catalysis or mobility by group II introns. The nucleotides within inter-helical junctions are also important, and many of these substructures are critical elements of the intron active-site[1,2] (Figure 11.3).

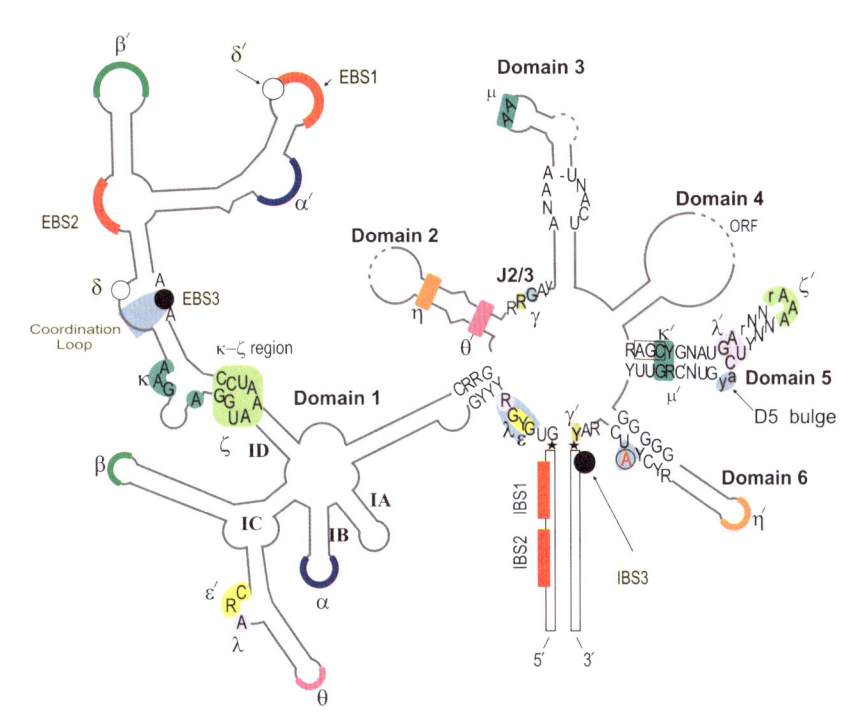

Figure 11.3 Consensus secondary structure of the group IIB introns. Exons are represented as gray bars. Greek letters indicate the position of known tertiary interaction partners, which are color coded. Short tertiary interactions (such as EBS-IBS3) are indicated with circles. Sequence conservation: Upper-case letters indicate >90% conservation (an exception are the two adenosines shown in the D3a loop, as these are important in ai5γ, but their generality is not known).

11.4.1 Domain 1

D1 is the largest domain of the intron and it contains nucleotides that are essential for the recognition of both exons and the branch-site nucleophile.[45–49] Group II intron folding hinges on the initial formation of D1,[50,51] which forms a compact scaffold that provides recognition surfaces and docking sites for most of the other intronic elements (*vida infra*).[51,52]

11.4.2 Domain 2

D2 is highly variable in composition, its first stem mediates a tertiary interaction (θ–θ′) that is important for docking and orientation of catalytically important functional groups in adjacent J2/3 and domain 3 (D3).[53,54]

11.4.3 Domain 3

D3 contains nucleotides that are important for active-site formation and catalysis. The first stem of D3 contains a loop region that is conserved throughout all families of group II introns,[32,39] and which is deeply buried within the folded core.[55] Group IIB introns (*vida infra*) contain additional D3 substructures (such as the D3a stem-loop),[32] that are important for core structure assembly.[56] Importantly, the functional domain in this region of the intron actually includes the first stem of D2, the J2/3 linker and D3,[54] which fold together into a single structural unit that can recapitulate catalysis in-trans.

11.4.4 Domain 4

D4 does not contain nucleotides that facilitate intron folding or catalysis, but it plays a major role in group II intron mobility by encoding a large open reading frame (ORF) for expression of the cognate maturase protein.[3,23] Each maturase protein binds and collaborates with its parent intron, and the two components have been shown to closely co-evolve.[32] Thus, D4 is an information-carrying region of the intron that can bring new genes into sites that are invaded by group II introns.

11.4.5 Domain 5

D5 is the most conserved and catalytically important region of a group II intron.[2] This small hairpin is almost invariant among the diverse family of group II introns and the importance of this domain was noted in the very earliest investigations of intron phylogeny and catalytic mechanism.[39] The extended D5 hairpin has a "back", which forms extensive contacts with a set of receptor structures in the central stem of D1 (the κ–κ′ and ζ–ζ′ interactions[57–60]), and a "front" that participates in active-site assembly and presents critical active-site functional groups that mediate chemical catalysis together with atoms in D1, J2/3 and D3 (*vida infra*).[59,60] Both sides of D5 bind Mg^{2+} ions that are critical for proper docking and for catalysis, respectively.[61–63] The importance of metal ion binding in the D5 bulge is apparent from both crystallographic and NMR studies of free D5 molecules,[63,64] and was evident from Tb^{3+} probing experiments.[62] The various high resolution structures of D5 indicate that there are several conformations of the undocked molecule and that the D5 bulge region is a site of dynamic flexibility.[63–67] The D5 bulge, which is particularly important for chemical catalysis, is likely to adopt its proper conformation only upon assembly of the intact active-site. Any discussion of D5 would be incomplete without mentioning the functional coupling between this domain and the adjacent D6, which contains the branch-site nucleophile. D5 is essential for the proper orientation of D6,[68] to which it is linked by a short tether (J5/6) that serves as a molecular caliper for determining the site of branch-site choice during the first step of splicing.[69]

11.4.6 Domain 6

D6 is essential for the branching reaction, and it forms a substructure that uniquely specifies a single 2′-hydroxyl group as the nucleophile during transesterification.[15] A functional branch site involves much more than a bulged adenosine within a duplex: the nucleophilic adenosine is usually flanked by noncannonical base pairs that promote conformational flexibility of the adenosine, such as base-flipping,[70] that enables the nucleophilic 2′-OH to adopt precisely the correct orientation for nucleophilic attack.[69,71,72] The branch-site itself is imbedded in a conserved polypyrimidine sequence, which helps to specify the correct nucleophile. Branch-site choice is incredibly precise and it is tightly controlled by the distance between D5 and D6[69] and by the steric "fit" of the nucleophilic adenosine into a specific docking site that is located within D1 (the coordination loop).[72,73] The importance of conformational dynamics at the branch-site, and evidence for divalent cation binding in this region,[62] have been underscored by recent NMR studies on isolated D6 molecules.[74] These studies also indicate that D6 adopts the overall secondary structure that has long been inferred from phylogenetic analysis.[69–70,74]

11.4.7 Other Domains and Insertions

While the domain architecture of group II introns appears to be tightly constrained, and the domains are always found in exactly the same order,[32] deviations from this basic organization are sometimes observed. For example, a group II intron was recently identified that contains a seventh domain (D7) at the extreme downstream terminus of the intron.[75] Although the D7 hairpin has been inserted within a relatively inconvenient spot (between D6 and the 3′-splice site), the intron has adapted, and it splices accurately. Large insertions are often tolerated within domains 1,[37] 2 and 3, provided that they do not prevent proper intron folding.[39]

11.4.8 Alternative Structural Organization and Split Introns

Even more striking is the fact that plant introns are often transcribed as two, or even three separate pieces, which either interact through tertiary interactions or anneal through pairings that recapitulate positions of the D2 and/or D4 stems.[40,76-78] The group III class of introns is likely to be derived from separated group II intron domains.[79] Indeed, intron domains *in vitro* have long been known to be physically separable[80] They can be transcribed as separate molecules, which readily assemble into active complexes that accurately recapitulate hydrolysis, branching, or even the second step of splicing.[7,54,68,80-83] These approaches have been critical for applying enzymology to the study of group II introns, thereby revealing the thermodynamics of inter-domain interactions, mechanisms of transition-state stabilization and the functional groups that mediate chemical catalysis.[2]

11.5 A Big, Complicated Family: The Diversity of Group II Introns

Because group II introns are so abundant in nature, it has recently been possible to conduct rigorous phylogenetic analyses, and to classify group II introns into major families and subfamilies[32,39,84,85] As a result, we no longer combine all group II intron sequences and structures together as we attempt to deduce the architectural features and chemical strategies of these enzymes. There are important distinctions between the major intron families, and their differences reveal much about their individual behaviors, biological functions and evolutionary histories.

While the precise familial relationships among group II introns have been described in detail elsewhere, it is important to note that three major classes have been distinguished thus far: Group IIA, B and C.[32] The group IIA and B families are both large introns (600-1000 nts, exclusive of ORF) with important differences in subdomain architecture.[32] For example, the group IIB introns contain an obvious coordination loop for organization of the splice sites and branch-site (Figure 11.3), while the analogous region in group IIA introns is undefined. Group IIB introns tend to self-splice more efficiently *in vitro*, while Group IIA introns are more maturase-dependent. The group IIC class is the most distinctive in that the introns are small ($\sim$450 nts exclusive of ORF), they react *in vitro* through hydrolysis, and they are only found in bacteria (Figure 11.4).[13,32,84,86] This class is considerably less sequence-specific than IIA and IIB introns, as EBS1 is short and EBS2 is absent. Phylogenetic analysis indicates that group IIC introns are the most ancient form,[31] and that they diverged into group IIB and group IIA introns prior to the evolution of eukaryotes.

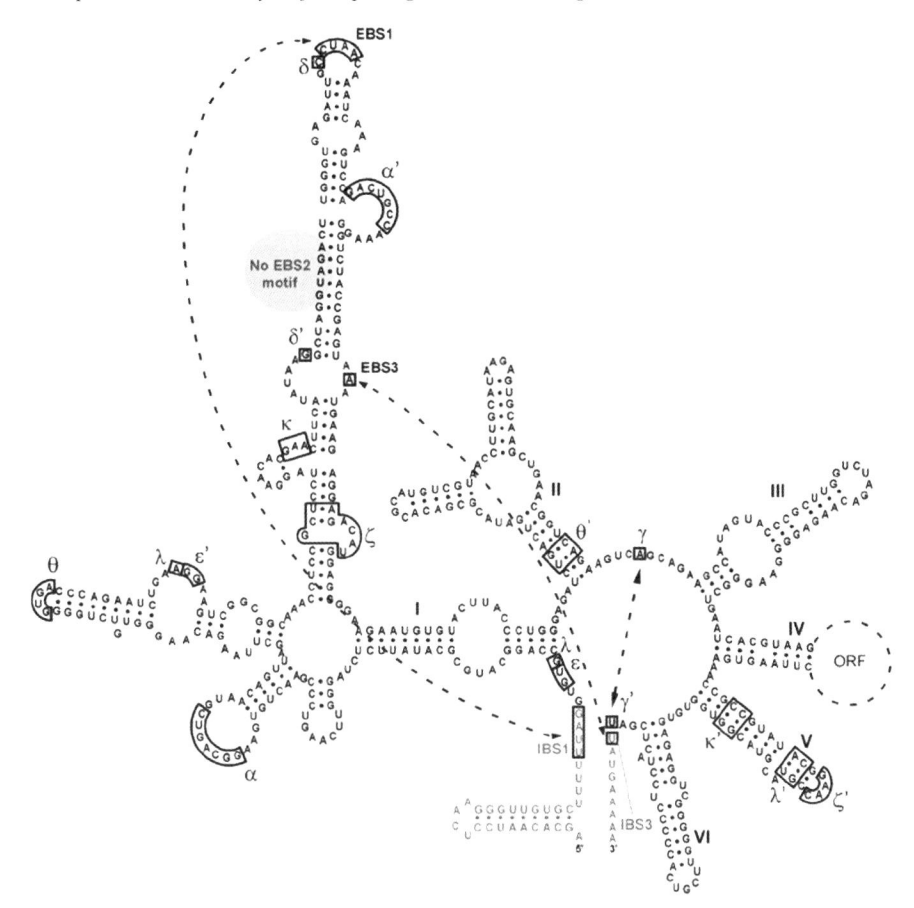

Figure 11.4 Secondary structure of a representative group IIC intron. The B.h.I1 intron Bacillus halodurans is located downstream from a translation terminator structure (gray hairpin). Domains are indicated by Roman numerals and tertiary interactions by Greek letters. Dotted lines indicate the positions of long-range pairings. DIV contains an open reading frame (ORF) that encodes a maturase protein.

11.6 Group II Intron Tertiary Structure

While the crystal structure of a group II intron is not yet available, a considerable accumulation of biochemically-determined distance constraints has facilitated the modeling of group IIA and group IIB structures.[47,55,87,88] The largest set of distance restraints has been collected for the ai5γ group IIB1 intron,[88] from the mitochondrial genome of *S. cerevisiae*. This intron has been subject to decades of genetic, phylogenetic, chemogenetic and biochemical study that have yielded a wealth of information on functional, long-range interactions.[1] Collectively, this information has facilitated the modeling of an almost intact intron and provided a particularly detailed view of proximity

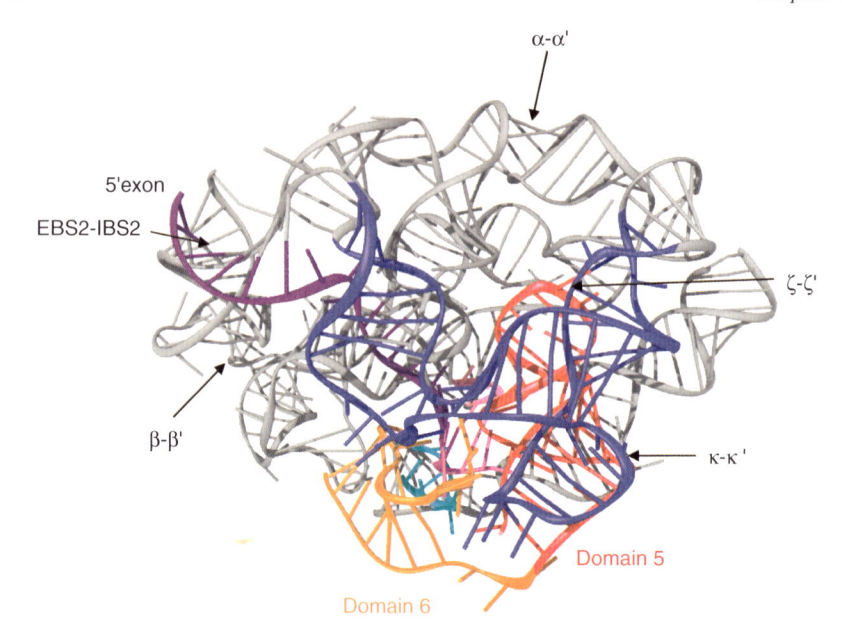

Figure 11.5 Three-dimensional model of the ai5γ group II intron active site. (Adapted from ref. 88.) Long-range tertiary interactions are indicated with Greek letters. Note that D5 is shown in red, D1 is shown in blue and gray, the junction between Domains 2 and 3 is green, D6 is yellow, and the 5′-exon is purple.

among active-site elements (Figure 11.5).[88] The model shows that the D1 scaffold provides a preassembled, external groove for docking of the exons and that D5 is almost entirely buried within the structure. Most importantly, constraints on the model are consistent with only a single active-site for both steps of group II intron self-splicing.[88] Structural analysis of the intron at various stages of reaction indicate that that all reactants and major catalytic groups remain in approximately the same position from the beginning to the end of a complete cycle of splicing.[48,88]

A striking feature of group II intron structure is the importance of critical tertiary interactions that are mediated by 2′-hydroxyl groups and other backbone functionalities that do not display phylogenetic signatures in sequence databases.[57,60,89,90] Indeed, genetic and phylogenetic studies of intron tertiary structure were limited by the paucity of tertiary contacts based on Watson-Crick pairings and other known forms of base-base interaction. As a result, modeling of group II intron structure lagged behind that of other large RNAs, and progress awaited the development of chemogenetic techniques such as Nucleotide Analog Interference Suppression (NAIS) and site-directed crosslinking with thioketone nucleotide analogs (6-thioguanosine and 4-thiouracil).[48,91] These approaches made it possible to probe intron tertiary structure atom-by-atom. They confirmed the complexity of group II intron tertiary architecture and showed that 2′-hydroxyl contacts play a major

Group II Introns: Catalysts for Splicing, Genomic Change and Evolution 211

role in its organization. For example, one of the first demonstrations of a minor groove triplex (what is now called the A-minor motif) resulted from a NAIS study on interactions between D5 and D1 of intron ai5γ. This work showed that the λ–λ' interaction, which orients catalytic D5 functionalities within the intron core, is mediated by a specific minor groove triplex that is stabilized by 2'-hydroxyl interactions.[89] Additional studies have shown that backbone contacts involving ε–ε', θ–θ' and κ–κ' are essential for active-site assembly.[54,56,57,88] These studies, together with crystallographic and NMR studies on individual D5 and D6 molecules, suggest that group II introns will continue to be a rich source of information on tertiary structural motifs in RNA.

11.7 Group II Intron Folding Mechanisms

Group II introns assemble into their native structures through a variety of pathways that depend, to some extent, on the role of proteins in stabilization of the intron structure.[52] Most of the available information on kinetic folding pathways comes from studies on the isolated ai5γ intron, and from the D135 and D1356 ribozymes that were derived from it. These studies have revealed a novel paradigm for RNA folding in which the group II intron ribozyme folds directly, slowly and accurately to the native state (Figure 11.6), without becoming trapped in a series of misfolded intermediate states.[52] Given the large size of the D135 ribozyme, and the fact that smaller RNA molecules have difficulty in folding properly, these findings were surprising.

11.7.1 A Slow, Direct Path to the Native State

Biophysical analyses of D135 folding revealed a pathway that appears to be two-state: that is, all parts of the intron appear to fold simultaneously. Most remarkably, there are no subdomains of the intron that fold or collapse

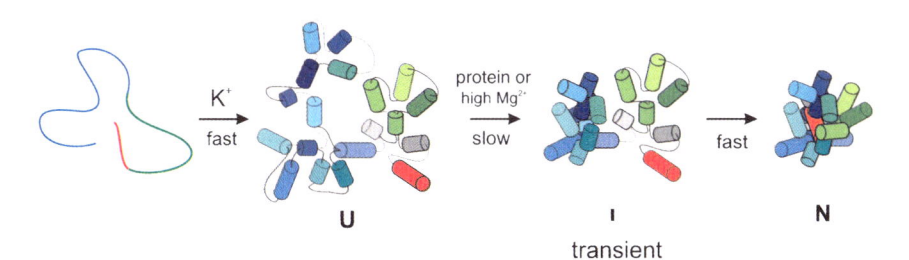

Figure 11.6 Folding pathway for group IIB intron ai5γ. After formation of the secondary structure (barrels indicate duplexes), the first step in tertiary folding involves the collapse of D1 to a compact scaffold structure (formation of the intermediate, I state). This step requires high Mg^{2+} ion concentration or the Mss116 protein. The second step of tertiary folding is the formation of the native (N) state, which involves docking of D3 (green) and D5 (red).

quickly, and the whole process is relatively slow ($\sim 1 \text{ min}^{-1}$).[92,93] This behavior is highly unusual, as most RNAs collapse rapidly and they contain small regions that fold more quickly than the rest of the molecule.[94] However, the deceptively simple folding kinetics of D135 masks a more complex mechanism, in which the ribozyme folds through a series of steps that involve sequential buildup of an active core structure. Two-state behavior is observed because the very first step in the process is slow, and all subsequent events are rapid.[52] The rate-limiting initial step in D135 folding is the collapse of intron D1,[51] which forms a compact scaffolding structure into which all of the other domains dock.[51]

11.7.2 A Folding Control Element in the Center of D1

In an effort to understand why D1 collapses slowly, and why the process requires so much Mg^{2+} ion, chemogenetic studies were designed to identify the substructure that controls folding within intron D1. Nucleotide analog interference mapping (NAIM), coupled with an electrophoretic selection for separating folded from unfolded molecules revealed that a small substructure in the center of D1 (κ–ζ) controls the entire folding process and is essential for molecular collapse.[95] Behaving like a Mg^{2+}-dependent riboswitch, the κ–ζ element must adopt exactly the correct conformation before any of the long-range tertiary interaction partners can bind each other correctly. Once κ–ζ forms and the D1 tertiary contacts are established, D1 becomes highly compact and the other domains dock rapidly into well-defined binding sites on the well-folded D1 scaffold.[95,96] (Figure 11.6). This remarkable folding pathway represents an effective, high fidelity mechanism for keeping a large RNA out of trouble (*i.e.*, misfolding) and shepherding the molecule into a single native conformation. Indeed, the behavior of intron ai5γ may be paradigmatic for other large, multidomain RNA molecules that have necessarily evolved faithful, direct folding pathways.[52]

11.7.3 Proteins and Group II Intron Folding

Many group II introns have recruited or co-evolved with proteins that facilitate their folding and promote stability.[23,32] Under physiological conditions, even the ai5γ intron has difficulty folding due to a high kinetic barrier for D1 collapse and unstable docking of the catalytic domains.[96] It is therefore not surprising that ai5γ recruits a yeast nuclear protein to facilitate intron stability and reactivity (Mss116p).[97–99] The LtrB group II intron from *Lactococcus lactis* requires interactions with its encoded maturase protein (LtrA) to catalyze efficient splicing and mobility under physiological ionic conditions.[23,24] Plant introns, such as atpF from maize chloroplast, form a folded core structure that is similar to that of the yeast ai5γ intron,[100] and yet folding requires formation of a large ribonucleoprotein complex that contains multiple proteins.[101–103] This suggests that some aspects of group II intron folding are context-dependent, with variation that depends on intron subtype and the cellular environment of the host organism.

11.8 Setting the Stage for Catalysis: Proximity of the Splice Sites and Branch-site

Once the intron has folded, the "reactants" for splicing must be brought together. From a chemical perspective, there are three reactive groups that participate in splicing by a group II intron: The phosphodiester linkages at both the 5′ and 3′ splice sites (*i.e.*, the substrates) and the adenosine 2′-hydroxyl group at the branch-site (*i.e.*, the nucleophile in the first step). Recent work has elucidated the spatial positioning of these three reactants relative to each other, and relative to active-site functional groups that stimulate catalysis.

11.8.1 Recognition of Exons and Ribozyme Substrates

The 5′-exon and the adjacent 5′-splice site are specifically recognized and held in place by two sets of base-pairings between the exon binding sequences of the intron (EBS1 and EBS2) and complementary sites on the exon (IBS1 and IBS2).[45,46,49] These sequences need only co-vary and preserve base-pairing; there is no absolute sequence requirement for 5′-exon or oligonucleotide substrate recognition.[39] As a result, group II introns can be designed to target, or splice from, any sequence of interest.[44,104] The total number of EBS-IBS base pairs can be quite large (up to 15 bp), which ordinarily would result in binding so strong that mispaired substrates would be cleaved by ribozyme constructs of the intron (or by lariat introns that are reacting with RNA or DNA in-trans). However, group II intron ribozymes are exceptionally sequence-specific because they lack tertiary contacts with the ribose backbone (unlike group I introns)[105] and because EBS-IBS pairing is accompanied by an energetic penalty that reduces the overall binding affinity between ribozyme and substrate.[106] The actual 5′-splice site linkage is chosen motif-selectively, as the intron core specifically targets the phosphodiester junction between single and double-stranded RNA at the EBS1 terminus.[107] A final element that is important for positioning the scissile phosphodiester is the δ–δ′ interaction, which joins EBS1 with a loop in the D1 main stem (Figure 11.3).[47] Adjacent to the δ–δ′ interaction is a pairing called EBS-IBS3, that links D1 with the first nucleotide of the 3′-exon.[47] In addition, the 3′-splice site is determined by its distance from the D6 branch-site and by the small γ–γ′ pairing that joins J2/3 with the intron terminus (Figure 11.3).[45,108]

11.8.2 Branch-site Recognition and the Coordination Loop

During splicing through branching, the 2′-hydroxyl group of the nucleophilic adenosine must be brought into close proximity with the 5′-splice site. Despite the central importance of the branch-site, the intronic substructures that are required for docking and activating it remained obscure until very recently. The genetic, phylogenetic and chemogenetic approaches that had revealed the intron binding sites for the two splice sites and many other tertiary interactions had failed to identify intronic functional groups that mobilize the branch-site, which is a poorly conserved region of the intron that does not bind with high affinity to the core.[7] To identify the branch-site receptor within the ai5γ intron, proximity-sensitive photocrosslinking approaches were employed, as these do not depend on the formation of strong hydrogen bonds. The branch-site adenosine and flanking uridines were replaced by photoreactive 6-thioguanosine and 4-thiouridine, respectively, and photoreaction products were mapped by primer extension.[48]

Strikingly, the branch-site nucleotides all crosslinked with a defined polarity to a set of corresponding nucleotides in the D1 main stem.[48] This receptor structure is an internal loop that had never been previously implicated in branch-site function and which is not conserved in sequence (Figure 11.7). However, this same loop also contains nucleotides that participate in the δ–δ′ and EBS-IBS3 pairings that contribute to specification of both splice sites.[47]

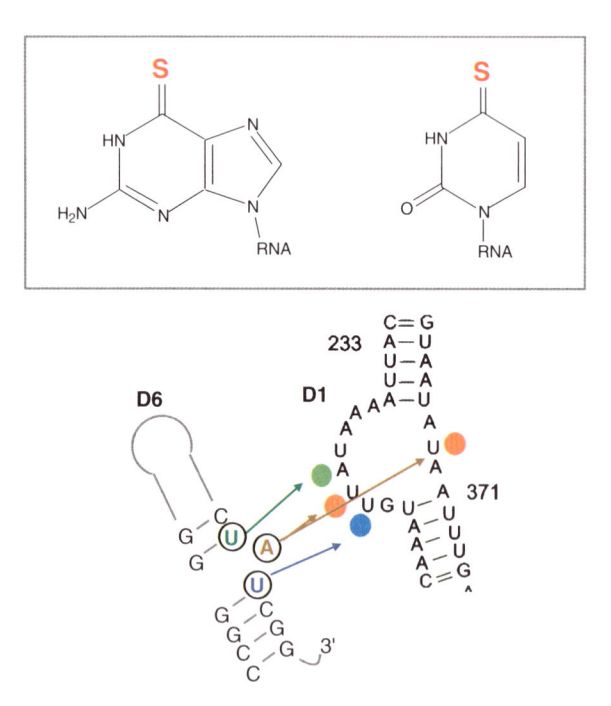

Figure 11.7 Identification of the branch-site receptor through thionucleobase photocrosslinking. The nucleoside analogs 6-thioguanosine (left) and 4-thiouridine (right) were incorporated at the circled positions in D6, and they crosslink to the indicated positions within the coordination loop of D1.

Given that a central D1 loop coordinates the position of the branch-site and both splice sites, it has been designated the coordination loop[48] (Figure 11.3). Because these results were based on crosslinking, and not on genetics or NAIS, it was of interest to determine if the crosslinked species were functional and whether they could participate in splicing. Indeed, branch-sites that are crosslinked to the coordination loop in the precatalytic state (*i.e.*, before either step of splicing) are readily chased through both the first and the second step of splicing.[48] These results indicate that the branch-site and the phosphodiester linkages at both splice sites maintain the same position through the entire cycle of splicing. Indeed, 6-thiouridine at the 3′-splice site crosslinks with the same nucleotides as the branch site, and the photoproducts are catalytically active. This indicates that the branch-site is held in close proximity with residues that are essential for both the first and the second step of splicing. The branch-site crosslinking studies therefore not only identified the position of the branch-site receptor, but they provided strong evidence that a single group II intron active site catalyzes both steps of splicing.[48,88]

11.9 A Single Active-site for Group II Intron Catalysis

If there is a single active-site region for group II intron self-splicing, not only should the reactants be held in close proximity, but major catalytic groups within the intron core would also be expected to retain similar positions through both steps of splicing. Such a finding would be consistent with early functional group analyses of D5, which showed that the same D5 atoms were important for both steps of splicing.

To explore this hypothesis, the short-range photocrosslinkable nucleotides 6-thioguanosine and 4-thiouridine were synthetically incorporated at catalytically important regions adjacent to the 5′-splice site and at D5 positions that are known to be important for chemical catalysis by group II introns. Upon irradiation and crosslink mapping, it was found that all of these residues crosslink to similar nucleotides, at regions that are essential for both the first and second steps of splicing (*vida infra*). As in the previous case, the crosslinked species are active. Therefore, not only are all the reactive groups proximal through the entire cycle of splicing, but catalytically important active-site groups on D5, D1 and J2/3 are proximal as well. Modeling studies based on these and all other existing restraints resulted in only a single intron conformation (Figure 11.5) with a well-defined active-site that contains the functional groups essential for catalysis (Figure 11.8).[88]

These data suggest that a single active-site catalyzes both steps of splicing and that a major restructuring of the core does not occur between the two transesterifications.[88] This finding is consistent with the fact that the first step of group II intron splicing is always rate-limiting and that the second step is extremely rapid.[45] However, it is interesting in light of the fact that the first and second steps of ai5γ self-splicing proceed with similar stereoisomeric preference

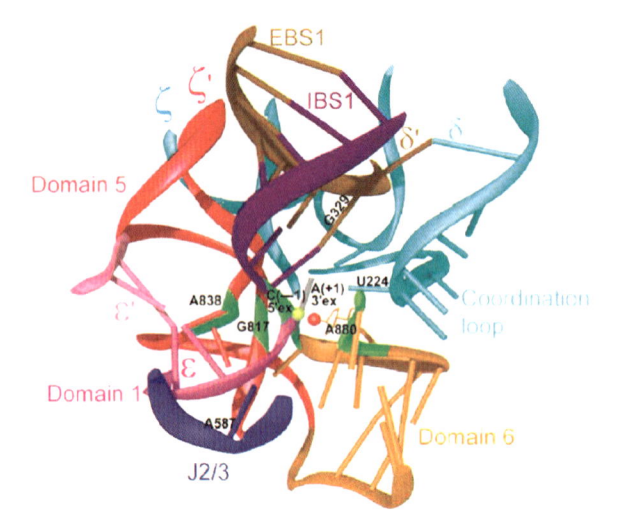

Figure 11.8 Close-up of a IIB intron active-site, from the biochemical constraints and ai5γ intron model shown in Figure 11.5. Proximal active-site functional groups and reactive groups are shown in green (G817 of the D5 AGC triad; A838 of the D5 bulge, both splice site and the branch-site). The scissile phosphate is a yellow ball and the branch 3′-OH is a red ball. Color scheme: IBS1, purple; ε and the 5′ terminus of D1 are magenta; ε′ is pink, the region spanning the coordination loop and ζ is cyan; EBS1 is brown, J2/3 is blue, D5 is red, and the lower stem of D6 is orange.

(they do not appear to be simple reversals of one another, as in the case of group I introns),[109] which would suggest that transition-state for the two steps is not completely identical (*vida infra*).

11.10 The Group II Intron Active-site: What are the Players?

For almost two decades, our notions about the group II intron active-site have centered on a very small part of the intron: D5. Furthermore, most studies of D5 catalytic function and most phylogenetic parallels with the spliceosome have centered on a very small sector of D5: a conserved AGC sequence at the base of stem 1.[59,110–112] However, recent work has demonstrated that D5 is not the heart of a group II intron. Important active-site functional groups are also contributed by motifs in D1, the J2/3 linker, and nucleotides in D3. In fact, the most important part of D5 is probably not the AGC sequence, but the bulged elbow region that separates the two helices of D5.[60,62,65,66,88] These findings indicate that we should think more carefully about the composition of a group II intron active-site and take a fresh look at the players involved in its structure and catalytic mechanism.

11.10.1 Active-site Players in D1 and Surrounding Linker Regions

It is particularly useful to consider the catalytically important motifs in the upstream regions of the intron. Some of the most important functional groups in group II intron catalysis are located immediately downstream from the 5′-splice site (nts 3-6).[57,88,89,108] The short ε–ε′ interaction contains highly conserved nucleotides which,[108] when mutated, cause defects in the chemical step. Indeed, almost every nucleobase and backbone atom of ε–ε′ is important for the first step of group II intron splicing.[57] These findings make sense in light of recent work showing that ε–ε′ lies at the center of a large tertiary interaction network that includes the θ–θ′ pairing (from the base of D2) and nucleotides in D3,[54] all of which are important for organizing the core.

Immediately next to ε–ε′ are two nucleotides that are among the most conserved of the group II intron (G5 and A115 in ai5γ numbering).[39] These residues form a pair of long-range triple interactions with nucleotides in stem 2 of D5 (the λ–λ′ interaction).[89] Mutation of these nucleotides disrupts both chemical catalysis and the ability of D5 to properly dock with D1,[89] suggesting that λ–λ′ is important for active-site organization.

The D1 main stem contains scaffolding that holds major active-site residues and reactive groups in place. These regions are therefore essential for active-site organization, although they do not contribute directly to chemical catalysis. The κ–ζ region (Figure 11.3) binds the "back" side of D5 through two sets of inverted tetraloop-receptor motifs.[57,58] It is highly significant that the coordination loop lies immediately next to the κ–ζ receptor, as this brings the splice sites and branch-site into proximity with catalytic groups on D5.[48]

11.10.2 Domain 3 and the J2/3 Linker

D3 contains many highly conserved nucleotides and, like D5, it can be transcribed as a separate molecule that activates group II intron ribozyme reactions in-trans.[83] D3 and D5 have similar affinities for the intron core[83] and similar levels of hydroxyl radical protection.[55] All things being equal, there is little reason to consider D3 (together with the adjacent J2/3 linker) as any less important than D5 in catalytic function of the intron. Recent crosslinking and modification interference studies have underscored the importance of J2/3 and D3 in group II intron catalysis[48,54,56,88] (Figure 11.8).

The highly conserved AGA sequence in the J2/3 linker region was originally shown to play a role in the second step of splicing.[108,113] However, short-range crosslinking studies have shown that J2/3 is in close proximity with nucleotides in D5, the branch-site and at the 5′-splice site, during both steps of splicing.[48,87,88] Indeed, mutation of J2/3 residues inhibits chemical catalysis in ribozyme systems that mimic the first step of splicing (A. DeLencastre and A.M.P., unpublished results). Based on their location in the tertiary structure, conservation, and effects on catalysis, it is likely that the AGA nucleotides in J2/3 play a direct role in reaction chemistry. Intriguingly, the spliceosomal

U6 ACAGA box is also essential for catalysis,[112,114] and it is therefore possible that J2/3 is the analogous substructure in group II introns.

The first stem-loop structure at the base of D3 is conserved across all three intron families of group II intron.[32] Within intron ai5γ, it forms a series of tandem sheared A-A pairs which, when mutated, severely disrupt intron catalysis in-cis and by trans-D3 ribozyme constructs.[56] The precise role of this structure in active-site formation and/or catalysis is unknown and awaits NAIS and/or crystallographic analysis.

Group IIB introns contain additional hairpin-loops in D3, the most functionally important of which is stem D3a. Forward and reverse NAIS studies on D3a resulted in the first known tertiary interaction between D3 and another region of the intron (μ–μ').[56] This contact between the D3a loop and the minor groove of D5 stem 1 anchors D3 within the active-site, potentially bringing J2/3 and other D3 functional groups into position for participation in catalysis.

11.10.3 Domain 5: Structural and Catalytic Regions

D5 contains most of the highly conserved nucleotides in a group II intron, although most of these residues do not contribute directly to catalysis. Rather, they serve to hold D5 in precisely the correct orientation for participation in reaction chemistry. Nucleotides of the ζ–ζ', κ–κ', λ–λ' and μ–μ' interactions are essential for docking D5 properly within D1 and D3[56–58,89] (Figure 11.3).

The most important region of D5 for chemical catalysis is likely to be a small section of a widened major groove that extends from the highly conserved AGC sequence through the nucleotides of the D5 bulge.[59,60,66] Considerable attention has been focused on the AGC "catalytic triad" of D5, presumably because its importance can be discerned from the almost invariant base sequence and because an analogous sequence is observed in the spliceosome[59,110–112] (Chapter 13). Extensive studies conducted both *in vitro* and *in vivo* have shown that the GC nucleotides are most essential and that they influence reaction chemistry.[59,110,111] In particular, major groove functional groups specifically influence chemical catalysis, while minor groove substituents stabilize D5 binding.[59]

Of equal, and potentially greater importance to the mechanism of D5 catalysis are substituents in the D5 bulge,[57,60,65,66,88,89,115] as almost every functional group in this region has a strong and direct influence on the chemical step. Short-range photocrosslinking and NAIS studies indicate that the D5 bulge is in close proximity with catalytic regions of D1, D3 and the splice sites,[88] while evidence for proximity of the AGC is lacking (despite direct analysis by these same methods, A.M.P., Olga Fedorova and Alex DeLencastre, unpublished results). Atoms within the sugar-phosphate backbone of the bulge are particularly important[57,60,90,115] and structural work has shown that this region is a binding site for divalent metal ions.[62,63] Indeed, the bulge and AGC nucleotides are separated by only half a helical turn and they are likely to work together, creating a surface that binds catalytic metal ions and providing nucleotide functional groups that participate directly in the catalytic mechanism of the intron.

11.11 The Chemical Mechanism of Group II Intron Catalysis

Group II introns catalyze in-line attack of phosphodiester linkages,[109,116,117] through an S_N2 reaction that involves a nucleophilic attack by an alcohol or water molecule and the release of a 3′-hydroxyl leaving group. This reaction is exactly analogous to that of group I introns (Chapter 10), RNase P (Chapter 9), and the Group I-like ribozyme (GIR1, Chapter 12), whereby the nucleophile is an exogenous functional group that must be bound and stabilized by the active-site. It contrasts with the mechanism of the nucleolytic ribozymes, such as the hammerhead, in which the nucleophile is the 2′-hydroxyl group adjacent to the scissile linkage[118] (Chapter 3-6).

Through the development and study of ribozyme constructs for which the chemical step is rate-limiting,[81,82,104,105,119] it has been possible to deduce some of the features that promote transition-state stabilization by group II introns (Figure 11.9). Some of these strategies are shared by the other large ribozymes, while others are distinctly different.

Stabilizing the build-up of charge on the 3′-hydroxyl leaving group is particularly important for transition-state stabilization, and this is achieved

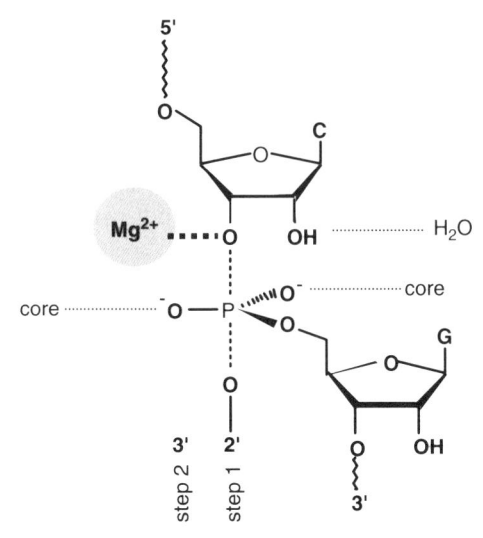

Figure 11.9 Transition-state stabilization by a group II intron core. The leaving-group 3′-OH is stabilized through interaction with Mg^{2+}. Both the pro-R_p and pro-S_p phosphoryl oxygens play a role in the transition-state, potentially through additional uncharacterized interactions. The solvent plays an important role, potentially through the adjacent 2′-hydroxyl group during RNA splicing.

by direct metal ion coordination with the 3′-oxygen. Elegant metal ion specificity switch experiments have demonstrated the importance of this interaction during both steps of splicing and they have provided strong evidence that group II introns are obligate metalloenzymes.[120,121]

During the first step of splicing, both of the nonbridging phosphoryl oxygens play a role in transition-state stabilization, as both of them are sensitive to substitution with phosphorothioates.[109,117] Sulfur substitution of the pro-R_p phosphoryl oxygen results in the largest effect, as it effectively eliminates ribozyme activity. Unlike group I introns, neither effect can be rescued by the addition of soft metal ions (e.g. Mn^{2+} or Cd^{2+}) indicating that any metal ion coordination to these atoms would necessarily be indirect an/or through a water molecule.[109,117]

During the second step of splicing, the pro-R_p phosphoryl oxygen is highly sensitive to sulfur substitution,[109,117] resulting in effects that are similar to those observed during spliceosomal catalysis. Given that both steps involve S_N2 reactions with inversion of configuration at phosphorus, the phosphorothioate effects suggest that the two reactions may not be exact reversals of one another.[122] An active-site functional group would need to change position by ~2Å in order to interact with the pro-R_p oxygen in both cases. Alternatively, the same intron functional group may interact with the pro-S_p oxygen in the first step and the pro-R_p oxygen in the second step, and the first step may differ in the presence of an additional transition-state contact with the pro-R_p oxygen. Neither of these possibilities requires massive restructuring of the intron active-site between the steps of splicing, which is contraindicated by the available structural data on position of active-site groups (see Sections 9 and 10).

The feature that most distinguishes group II introns from other ribozymes is that the 2′-hydroxyl group at the scissile linkage does not play a major role in transition-state stabilization.[105,123] In group I introns, RNase P, and even the ribosome, substitution of ribose at the scissile linkage with a deoxyribose causes a 10^6-fold reduction in the rate constant of the chemical step.[123] In these cases, the adjacent 2′-hydroxyl group participates in a network of interactions that mediate transition-state stabilization. However, analogous oligonucleotide cleavage studies by group II intron ribozymes show that the adjacent 2′-hydroxyl group stimulates the chemical step by only ~10-fold (although this effect is somewhat larger for a construct designed to mimic the second step).[105,123] These findings are consistent with the fact that DNA is the natural substrate for group II introns, and the central mechanism of intron mobility involves reverse-splicing into DNA.[1] The group II intron active-site has therefore evolved to tolerate DNA, and it does not mobilize the adjacent 2′-hydroxyl group for strong stabilization of the transition-state.

Given these findings, the chemical mechanism for group II intron catalysis remains largely mysterious. Known strategies for transition-state stabilization cannot account for the large reduction in activation energy that is provided by the enzyme. Interesting results have been reported from analysis of quantitative structure activity relationships (QSAR), which indicates that solvent interactions are extremely important during chemical catalysis by group II intron ribozymes.[123] However, a clear understanding of the reaction mechanisms awaits high resolution structural analysis of the intron core in the presence of well-designed transition-state analogs.

11.12 Proteins and Group II Intron Function

Although the role of proteins in splicing and mobility of diverse group II introns has been well-reviewed elsewhere,[1,3,23,40] it is notable that all group II introns have protein partners in the cell. In some cases, these proteins are recruited from the host organism and in other cases, the proteins are encoded by open-reading frames within the introns themselves (maturases). Although protein dependence can take many forms depending on intron subtype and cellular context, partner proteins are generally important for RNA structural stabilization. Splicing and reverse-splicing reactions are catalyzed by RNA functional groups within the intron.

11.12.1 Maturases

The most well-studied group II intron RNP is the 1:2 complex that is formed between the group IIA intron and its encoded maturase protein from the bacterium *L. lactis*.[1,3,23] The maturase is essential for self-splicing of the intron under physiological conditions and for reverse-splicing into duplex DNA. In addition, the maturase protein itself is an enzyme, catalyzing second-strand cleavage and reverse-transcription of the inserted intron RNA. Binding to intron domains 1 and 4,[124] the maturase is likely to significantly stabilize the intron core structure and thereby stimulate ribozyme catalysis of splicing and reverse-splicing reactions.[87] Most group II introns encode maturase proteins in D4, and their coevolution with these proteins is likely to have significantly impacted their function as ribozymes.[32]

11.12.2 CRM-domain Plant Proteins

Plant organelles encode many group II introns and these are highly dependent on a family of plant-specific splicing factors that facilitate core stabilization and potentially other functions.[40,101,103] The atpF intron from maize chloroplast interacts with a dimeric form of the host-encoded CRS1 protein.[100,102] Like other group II intron splicing factors in plants (such as CAF1 and CAF2[103]), CRS1 contains the CRM domain, which is believed to be an ancient RNA-binding motif that may be of general importance for the stabilization and activation of group II introns in diverse contexts.[101,102,125]

11.12.3 ATPase Proteins

Although the ai5γ intron from yeast mitochondria does not encode a maturase and it readily self-splices *in vitro* (with high Mg^{2+} ion and high temperatures), it recruits a yeast nuclear protein for stabilization and stimulation of splicing in normal contexts. Genetic studies have shown that the yeast Mss116 protein is required for ai5γ splicing *in vivo*[99] and biochemical studies have demonstrated that Mss116 is necessary for self-splicing *in vitro* under near-physiological conditions (*i.e.*, low Mg^{2+} ion concentration) at temperatures of yeast growth (30 °C).[97,98,126] Intriguingly, the Mss116 protein is an RNA-binding ATPase enzyme (a DEAD-box protein) that belongs to a larger family of proteins that can catalyze both the unwinding and/or the annealing of RNA duplexes. Given that ATP hydrolysis is necessary for Mss116-stimulated group II intron splicing, it is of great interest to determine whether the protein facilitates ai5γ folding, either by unwinding kinetically trapped structures,[98] or by stabilizing the weak interactions that are required for D1 intermediate and native-state formation.[97]

11.13 Group II Introns and Their Many Hypothetical Relatives

Ever since it was discovered that group II introns splice through branching, there has been a tremendous amount of text devoted to an ancestral relationship between group II introns and the eukaryotic spliceosome. Indeed, it is commonly said that their chemical mechanisms are "identical" and that group II introns are evolutionary predecessors of the more complex, protein-dependent spliceosome. However, it is now clear that branching can be catalyzed by diverse classes of ribozyme, irrespective of their ancestry.[127] The group I-like ribozyme (GIR) is almost certainly derived from group I introns,[128,129] and yet it catalyzes a branching reaction that caps adjacent transcripts with a small, stabilizing lariat structure.[128] Branching ribozymes[130] and even deoxyribozymes have been[131–133] created artificially through *in vitro* selection. Therefore branching is likely to be a common activity that is catalyzed by many RNA molecules that are under selective pressure to yield stable, nuclease-resistant products.[127]

That said, there is ample reason to suspect that group II introns share a common ancestor with the spliceosome (Chapter 13). The two systems share similar splice sites, similar reaction stereospecificity, a similar branch-site, and numerous catalytic substructures in common.[2] The U6 snRNA adopts a structure that is very similar to D5,[134–139] with similar placement of a conserved AGC sequence and a catalytic bulge region that binds catalytic metal ions.[140] Having modeled the group II intron core,[88] it is clear that the AGA of J2/3 is in a similar three-dimensional position to that of the U6 AGA sequence,[112,114] and both are of equivalent importance for catalysis. The U1 snRNA binds 5′-exon in a manner that is similar to EBS1,[141] and other short pairings are shared by both systems. A clear understanding of the evolutionary relationship between group II introns and spliceosomes awaits further growth in our structural and phylogenetic knowledge about these systems.

It is notable that group II introns do not require branching for the catalysis of RNA splicing and that some forms of group II intron (including the most ancient) splice predominantly through the hydrolytic pathway.[12-14] While this behavior is likely to have arisen spontaneously, it raises the possibility that group II introns arose from RNase P. Given the lively debate about spliceosomal comparisons, it might be just as valuable to consider chemical and structural parallels between the ancient RNase P class of ribozymes and group II introns.

11.14 Group II Introns: RNA Processing Enzymes, Transposons, or Tiny Living Things?

Having discussed the structure of group II introns, their abundance and the remarkable diversity of their chemical reactivities, it is worth considering our basic classification of these remarkable molecular machines. Are they simply ribozymes, or are they transposons, parasites or floating toolkits for evolutionary change? In many cases, group II introns have functioned as aggressive genomic parasites that have spread throughout Nature with an agenda that is all their own. And yet there are organisms that have clearly tamed them, and redirected group II intron activities to enhance or embellish their own metabolic processes. Whether as friend or foe, group II introns have clearly brought great evolutionary dynamism to terrestrial genomes.

Broadly speaking and with considerable help from encoded proteins, group II introns can replicate themselves, spread to new ecological niches, speciate and catalyze diverse reactions that are important for metabolism. These properties would suggest that group II introns are more than mechanical devices with catalytic potential: perhaps they deserve more respect. If one squints very hard and takes great liberties in interpretation, one can classify group II introns as tiny living things. This perspective makes it all the more important that we understand the great potential and influence of group II introns in Nature and in the laboratory.

References

1. A.M. Pyle and A.M. Lambowitz, in *The RNA World*, ed. R. Gesteland, T.R. Cech and J.F. Atkins, Cold Spring Harbor Press, Cold Spring Harbor, 2006, pp. 469–506.
2. K. Lehmann and U. Schmidt, *Crit. Rev. Biochem. Mol. Biol.*, 2003, **38**, 249–303.
3. A.M. Lambowitz and S. Zimmerly, *Annu. Rev. Genet.*, 2004, **38**, 1–35.
4. S. Zimmerly, G. Hausner and X. Wu, *Nucleic Acids Res.*, 2001, **29**, 1238–1250.
5. L. Dai and S. Zimmerly, *RNA*, 2003, **9**, 14–19.
6. C.L. Peebles, P.S. Perlman, K.L. Mecklenburg, M.L. Petrillo and J.H. Tabor *et al.*, *Cell*, 1986, **44**, 213–223.
7. K. Chin and A.M. Pyle, *RNA*, 1995, **1**, 391–406.

8. Y. Aizawa, Q. Xiang, A.M. Lambowitz and A.M. Pyle, *Mol. Cell*, 2003, **11**, 795–805.
9. K.A. Jarrell, C.L. Peebles, R.C. Dietrich, S.L. Romiti and P.S. Perlman, *J. Biol. Chem.*, 1988, **263**, 3432–3439.
10. D.L. Daniels, W.J. Michels Jr. and A.M. Pyle, *J. Mol. Biol.*, 1996, **256**, 31–49.
11. M. Podar, V.T. Chu, A.M. Pyle and P.S. Perlman, *Nature*, 1998, **391**, 915–918.
12. J. Vogel and T. Borner, *EMBO J.*, 2002, **21**, 3794–3803.
13. N. Toor, A.R. Robart, J. Christianson and S. Zimmerly, *Nucleic Acids Res.*, 2006, **34**, 6461–6471.
14. J. Li-Pook-Than and L. Bonen, *Nucleic Acids Res.*, 2006, **34**, 2782–2790.
15. R. van der Veen, J.H.J.M. Kwakman and L.A. Grivell, *EMBO J.*, 1987, **6**, 3827–3831.
16. M. Mörl, I. Niemer and C. Schmelzer, *Cell*, 1992, **70**, 803–810.
17. K.A. Jarrell, *Proc. Natl. Acad. Sci. U.S.A.*, 1993, **90**, 8624–8627.
18. H.L. Murray, S. Mikheeva, V.W. Coljee, B.M. Turczyk and W.F. Donahue et al., *Mol. Cell*, 2001, **8**, 201–211.
19. M.D. Molina-Sanchez, F. Martinez-Abarca and N. Toro, *J. Biol. Chem.*, 2006, **281**, 28737–28744.
20. M.W. Mueller, M. Hetzer and R.J. Schweyen, *Science*, 1993, **261**, 1035–1038.
21. M. Hetzer, R.J. Schweyen and M.W. Mueller, *Nucleic Acids Res.*, 1997, **25**, 1825–1829.
22. J. Yang, S. Zimmerly, P.S. Perlman and A.M. Lambowitz, *Nature*, 1996, **381**, 332–335.
23. M. Belfort, V. Derbyshire, M. Parker, B. Cousineau and A. Lambowitz, in *Mobile DNA II*, ed. N. Craig, R. Craigie, M. Gellert and A. Lambowitz, ASM Press, Washington DC, 2002.
24. R. Saldanha, B. Chen, H. Wank, M. Matsuura, J. Edwards and A.M. Lambowitz, *Biochemistry*, 1999, **38**, 9069–9083.
25. R.P. Rambo and J.A. Doudna, *Biochemistry*, 2004, **43**, 6486–6497.
26. S. Zimmerly, H. Guo, P.S. Perlman and A.M. Lambowitz, *Cell*, 1995, **82**, 545–554.
27. S. Zimmerly, J.V. Moran, P.S. Perlman and A.M. Lambowitz, *J. Mol. Biol.*, 1999, **289**, 473–490.
28. J.D. Boeke, *Genome Res.*, 2003, **13**, 1975–1983.
29. F. Michel and J.-L. Ferat, *Annu. Rev. Biochem.*, 1995, **64**, 435–461.
30. Y. Xiong and T.H. Eickbush, *EMBO J.*, 1990, **9**, 3353–3362.
31. J.S. Rest and D.P. Mindell, *Mol. Biol. Evol.*, 2003, **20**, 1134–1142.
32. N. Toor, G. Hausner and S. Zimmerly, *RNA*, 2001, **7**, 1142–1152.
33. N. Toro, *Environ. Microbiol.*, 2003, **5**, 143–151.
34. F. Martinez-Abarca and N. Toro, *Mol. Microbiol.*, 2000, **38**, 917–926.
35. N.J. Tourasse, F.B. Stabell, L. Reiter and A.B. Kolsto, *J. Bacteriol.*, 2005, **187**, 5437–5451.
36. X.Q. Liu, J. Yang and Q. Meng, *J. Biol. Chem.*, 2003, **278**, 46826–46831.

37. C. Adamidi, O. Fedorova and A.M. Pyle, *Biochemistry*, 2003, **42**, 3409–3418.
38. K.P. Jenkins, L. Hong and R.B. Hallick, *RNA*, 1995, **1**, 624–633.
39. F. Michel, K. Umesono and H. Ozeki, *Gene*, 1989, **82**, 5–30.
40. L. Bonen and J. Vogel, *Trends Genet.*, 2001, **17**, 322–331.
41. J. Yao and A.M. Lambowitz, *Appl. Environ. Microbiol*, 2007, **73**, 2735–2743.
42. J. Zhong, M. Karberg and A.M. Lambowitz, *Nucleic Acids Res.*, 2003, **31**, 1656–1664.
43. H. Guo, M. Karberg, M. Long, J.P. Jones, 3rd, B. Sullenger and A.M. Lambowitz, *Science*, 2000, **289**, 452–457.
44. J.P. Jones, 3rd, M.N. Kierlin, R.G. Coon, J. Perutka, A.M. Lambowitz and B.A. Sullenger, *Mol. Ther.*, 2005, **11**, 687–694.
45. A. Jacquier and N. Jacquesson-Breuleux, *J. Mol. Biol.*, 1991, **219**, 415–428.
46. A. Jacquier and F. Michel, *Cell*, 1987, **50**, 17–29.
47. M. Costa, F. Michel and E. Westhof, *EMBO J.*, 2000, **19**, 5007–5018.
48. S. Hamill and A.M. Pyle, *Mol. Cell*, 2006, **23**, 831–840.
49. R. van der Veen, A.C. Arnberg and L.A. Grivell, *EMBO J.*, 1987, **6**, 1079–1084.
50. P.Z. Qin and A.M. Pyle, *Biochemistry*, 1997, **36**, 4718–4730.
51. L.J. Su, C. Waldsich and A.M. Pyle, *Nucleic Acids Res.*, 2005, **33**, 6674–6687.
52. A.M. Pyle, O. Fedorova and C. Waldsich, *Trends Biochem. Sci.*, 2007, **32**, 138–145.
53. M. Costa, E. Deme, A. Jacquier and F. Michel, *J. Mol. Biol.*, 1997, **267**, 520–536.
54. O. Fedorova, T. Mitros and A.M. Pyle, *J. Mol. Biol.*, 2003, **330**, 197–209.
55. J. Swisher, C. Duarte, L. Su and A. Pyle, *EMBO J.*, 2001, **20**, 2051–2061.
56. O. Fedorova and A.M. Pyle, *EMBO J.*, 2005, **24**, 3906–3916.
57. M. Boudvillain and A.M. Pyle, *EMBO J.*, 1998, **17**, 7091–7104.
58. M. Costa and F. Michel, *EMBO J.*, 1995, **14**, 1276–1285.
59. B.B. Konforti, D.L. Abramovitz, C.M. Duarte, A. Karpeisky, L. Beigelman and A.M. Pyle, *Mol. Cell*, 1998, **1**, 433–441.
60. D.L. Abramovitz, R.A. Friedman and A.M. Pyle, *Science*, 1996, **271**, 1410–1413.
61. P.M. Gordon and J.A. Piccirilli, *Nat. Struct. Biol.*, 2001, **8**, 893–898.
62. R. Sigel, A. Vaidya and A. Pyle, *Nat. Struct. Biol.*, 2000, **7**, 1111–1116.
63. R.K. Sigel, D.G. Sashital, D.L. Abramovitz, A.G. Palmer, S.E. Butcher and A.M. Pyle, *Nat. Struct. Mol. Biol.*, 2004, **11**, 187–192.
64. L. Zhang and J.A. Doudna, *Science*, 2002, **295**, 2084–2088.
65. N.V. Eldho and K.T. Dayie, *J. Mol. Biol.*, 2007, **365**, 930–944.
66. O.H. Gumbs, R.A. Padgett and K.T. Dayie, *RNA*, 2006, **12**, 1693–1707.
67. M. Seetharaman, N.V. Eldho, R.A. Padgett and K.T. Dayie, *RNA*, 2006, **12**, 235–247.

68. S.D. Dib-Hajj, S.C. Boulanger, S.K. Hebbar, C.L. Peebles, J.S. Franzen and P.S. Perlman, *Nucl. Acids Res.*, 1993, **21**, 1797–1804.
69. V.T. Chu, C. Adamidi, Q. Liu, P.S. Perlman and A. Pyle, *EMBO J.*, 2001, **20**, 6866–6876.
70. J.C. Schlatterer, S.H. Crayton and N.L. Greenbaum, *J. Am. Chem. Soc.*, 2006, **128**, 3866–3867.
71. V.-T. Chu, Q. Liu, M. Podar, P.S. Perlman and A.M. Pyle, *RNA*, 1998, **4**, 1186–1202.
72. Q. Liu, V.T. Chu and A.M. Pyle, *J. Mol. Biol.*, 1997, **267**, 163–171.
73. Q. Liu, J.B. Green, A. Khodadadi, P. Haeberli, L. Beigelman and A.M. Pyle, *J. Mol. Biol.*, 1997, **267**, 163–171.
74. M.C. Erat, O. Zerbe, T. Fox and R.K. Sigel, *ChemBioChem*, 2007, **8**, 306–314.
75. F.B. Stabell, N.J. Tourasse, S. Ravnum and A.B. Kolsto, *Nucleic Acids Res.*, 2007, **35**, 1612–1623.
76. Y. Chapdelaine and L. Bonen, *Cell*, 1991, **65**, 465–472.
77. L. Bonen, *FASEB J.*, 1993, **7**, 40–46.
78. L. Merendino, K. Perron, M. Rahire, I. Howald, J.D. Rochaix and M. Goldschmidt-Clermont, *Nucleic Acids Res.*, 2006, **34**, 262–274.
79. D.W. Copertino, S. Shigeoka and R.B. Hallick, *EMBO J.*, 1992, **11**, 5041–5050.
80. K. Jarrell, R. Dietrich and P. Perlman, *Mol. Cell Biol.*, 1988, **8**, 2361–2366.
81. A.M. Pyle and J.B. Green, *Biochemistry*, 1994, **33**, 2716–2725.
82. W.J. Michels, Jr. and A.M. Pyle, *Biochemistry*, 1995, **34**, 2965–2977.
83. M. Podar, S. Dib-Hajj and P.S. Perlman, *RNA*, 1995, **1**, 828–840.
84. F. Michel, M. Costa, A.J. Doucet and J.L. Ferat, *Biochimie*, 2007, **89**, 542–553.
85. N. Toro, M.D. Molina-Sanchez and M. Fernandez-Lopez, *Gene*, 2002, **299**, 245–250.
86. M. Granlund, F. Michel and M. Norgren, *J. Bacteriol.*, 2001, **183**, 2560–2569.
87. J.W. Noah and A.M. Lambowitz, *Biochemistry*, 2003, **42**, 12466–12480.
88. A. DeLencastre, S. Hamill and A.M. Pyle, *Nat. Struct. Mol. Biol.*, 2005, **12**, 626–627.
89. M. Boudvillain, A. Delencastre and A.M. Pyle, *Nature*, 2000, **406**, 315–318.
90. G. Chanfreau and A. Jacquier, *Science*, 1994, **266**, 1383–1387.
91. O. Fedorova, M. Boudvillain, J. Kawaoka and A.M. Pyle, *Handbook of RNA Biochemistry*, ed. R.K. Hartmann *et al.*, Wiley-VCH, Weinheim, 2005, pp. 259–293.
92. J.F. Swisher, L.J. Su, M. Brenowitz, V.E. Anderson and A.M. Pyle, *J. Mol. Biol.*, 2002, **315**, 297–310.
93. L.J. Su, M. Brenowitz and A.M. Pyle, *J. Mol. Biol.*, 2003, **334**, 639–652.
94. T.R. Sosnick and T. Pan, *Curr. Opin. Struct. Biol.*, 2003, **13**, 309–316.
95. C. Waldsich and A.M. Pyle, *Nat. Struct. Mol. Biol.*, 2007, **14**, 37–44.

96. O. Fedorova, C. Waldsich and A.M. Pyle, *J. Mol. Biol.*, 2007, **366**, 1099–1114.
97. A. Solem, N. Zingler and A.M. Pyle, *Mol. Cell*, 2006, **24**, 611–617.
98. S. Mohr, M. Matsuura, P.S. Perlman and A.M. Lambowitz, *Proc. Natl. Acad. Sci. U.S.A.*, 2006, **103**, 3569–3574.
99. H.R. Huang, C.E. Rowe, S. Mohr, Y. Jiang, A.M. Lambowitz and P.S. Perlman, *Proc. Natl. Acad. Sci. U.S.A.*, 2005, **102**, 163–168.
100. O. Ostersetzer, A.M. Cooke, K.P. Watkins and A. Barkan, *Plant Cell*, 2005, **17**, 241–255.
101. B.D. Jenkins, D.J. Kulhanek and A. Barkan, *Plant Cell*, 1997, **9**, 283–296.
102. B. Till, C. Schmitz-Linneweber, R. Williams-Carrier and A. Barkan, *RNA*, 2001, **7**, 1227–1238.
103. G.J. Ostheimer, R. Williams-Carrier, S. Belcher, E. Osborne, J. Gierke and A. Barkan, *EMBO J.*, 2003, **22**, 3919–3929.
104. Q. Xiang, P.Z. Qin, W.J. Michels, K. Freeland and A.M. Pyle, *Biochemistry*, 1998, **37**, 3839–3849.
105. E.A. Griffin, Jr., Z. Qin, W.J. Michels, Jr. and A.M. Pyle, *Chem. Biol.*, 1995, **2**, 761–770.
106. P.Z. Qin and A.M. Pyle, *J. Mol. Biol.*, 1999, **291**, 15–27.
107. L. Su, P. Qin, W. Michels and A. Pyle, *J. Mol. Biol.*, 2001, **306**, 665–668.
108. A. Jacquier and F. Michel, *J. Mol. Biol.*, 1990, **213**, 437–447.
109. M. Podar, P.S. Perlman and R.A. Padgett, *Mol. Cell Biol.*, 1995, **15**, 4466–4478.
110. S.C. Boulanger, S. Belcher, U. Schmidt, S.D. Dib-Hajj, T. Schmidt and P.S. Perlman, *Mol. Cell, Biol.*, 1995, **15**, 4479–4488.
111. C.L. Peebles, M. Zhang, P.S. Perlman and J.F. Franzen, *Proc. Natl. Acad. Sci. U.S.A.*, 1995, **92**, 4422–4426.
112. H. Madhani and C. Guthrie, *Annu. Rev. Genet.*, 1994, **28**, 1–26.
113. S. Mikheeva, H.L. Murray, H. Zhou, B.M. Turczyk and K.A. Jarrell, *RNA*, 2000, **6**, 1509–1515.
114. C.F. Lesser and C. Guthrie, *Science*, 1993, **262**, 1982–1988.
115. J.-L. Jestin, E. Deme and A. Jacquier, *EMBO J.*, 1997, **16**, 2945–2954.
116. M.W. Müller, P. Stocker, M. Hetzer and R.J. Schweyen, *J. Mol. Biol.*, 1991, **222**, 145–150.
117. R.A. Padgett, M. Podar, S.C. Boulanger and P.S. Perlman, *Science*, 1994, **266**, 1685–1688.
118. A.M. Pyle, *Science*, 1993, **261**, 709–714.
119. P. Gordon, E. Sontheimer and J. Piccirilli, *Biochemistry*, 2000, **39**, 12939–12952.
120. E.J. Sontheimer, P.M. Gordon and J.A. Piccirilli, *Genes Develop.*, 1999, **13**, 1729–1741.
121. P.M. Gordon, E.J. Sontheimer and J.A. Piccirilli, *RNA*, 2000, **6**, 199–205.
122. M. Podar, P. Perlman and R. Padgett, *RNA*, 1998, **4**, 890–900.
123. P.M. Gordon, R. Fong, S.K. Deb, N.S. Li and J.P. Schwans *et al.*, *Chem. Biol.*, 2004, **11**, 237–246.

124. M. Matsuura, J.W. Noah and A.M. Lambowitz, *EMBO J.*, 2001, **20**, 7259–7270.
125. B.D. Jenkins and A. Barkan, *EMBO J.*, 2001, **20**, 872–879.
126. C. Halls, S. Mohr, M. Del Campo, Q. Yang, E. Jankowsky and A.M. Lambowitz, *J. Mol. Biol.*, 2007, **365**, 835–855.
127. A.M. Pyle, *Science*, 2005, **309**, 1530–1531.
128. H. Nielsen, E. Westhof and S. Johansen, *Science*, 2005, **309**, 1584–1587.
129. C. Einvik, H. Nielsen, E. Westhof, F. Michel and S. Johansen, *RNA*, 1998, **4**, 530–541.
130. T. Tuschl, P.A. Sharp and D.P. Bartel, *RNA*, 2001, **7**, 29–43.
131. R.L. Coppins and S.K. Silverman, *Biochemistry*, 2005, **44**, 13439–13446.
132. R.L. Coppins and S.K. Silverman, *Nat. Struct. Mol. Biol.*, 2004, **11**, 270–274.
133. Y. Wang and S.K. Silverman, *J. Am. Chem. Soc.*, 2003, **125**, 6880–6881.
134. H. Blad, N.J. Reiter, F. Abildgaard, J.L. Markley and S.E. Butcher, *J. Mol. Biol.*, 2005, **353**, 540–555.
135. D.G. Sashital, G. Cornilescu, C.J. McManus, D.A. Brow and S.E. Butcher, *Nat. Struct. Mol. Biol.*, 2004, **11**, 1237–1242.
136. G.C. Shukla and R.A. Padgett, *Mol. Cell*, 2002, **9**, 1145–1150.
137. H.D. Madhani and C. Guthrie, *Cell*, 1992, **71**, 803–817.
138. J.S. Sun and J.L. Manley, *Genes Dev.*, 1995, **9**, 843–854.
139. Y.-T. Yu, P.A. Maroney, E. Darzynkiewicz and T.W. Nilsen, *RNA*, 1995, **1**, 46–54.
140. S.L. Yean, G. Wuenschell, J. Termini and R.J. Lin, *Nature*, 2000, **408**, 881–884.
141. M. Hetzer, G. Wurzer and R.J. Schweyen, *Nature*, 1997, **386**, 417–420.
142. A.M. Pyle, *Nature*, 1996, **381**, 280–281.

CHAPTER 12
The GIR1 Branching Ribozyme

HENRIK NIELSEN,[a,b] BERTRAND BECKERT,[c] BENOIT MASQUIDA[c] AND STEINAR D. JOHANSEN[b]

[a] Department of Cellular and Molecular Medicine, The Panum Institute, University of Copenhagen, Copenhagen, DK-2200N, Denmark;
[b] Department of Molecular Biotechnology, Institute of Medical Biology, University of Tromsø, Tromsø, Norway; [c] Institut de Biologie Moléculaire et Cellulaire, Centre National de le Recherche Scientifique, Université Pasteur, Strasbourg, France

12.1 Introduction

The GIR1 branching ribozyme is a *ca.* 179 nt ribozyme with structural resemblance to group I ribozymes.[1] It is found within a complex type of group I introns, termed the twin-ribozyme introns.[2] Rather than splicing, it catalyses a branching reaction in which the 2'OH of an internal residue is involved in a nucleophilic attack at a nearby phosphodiester bond.[3] As a result, the RNA is cleaved at an internal processing site (IPS), leaving a 3'OH and a downstream product with a tiny lariat at its 5' end (Figure 12.1). The lariat has the first and the third nucleotide joined by a 2',5' phosphodiester bond and is referred to as "the lariat cap" because it caps an intron-encoded mRNA. The biological function of the GIR1 ribozyme thus appears to be in expression of an intron-encoded protein.

The GIR1 ribozyme was originally discovered during functional characterization of the *Didymium* twin-ribozyme intron. Combined deletion and *in vitro* self-splicing analyses revealed two distinct ribozyme domains within the intron.[2] Subsequently it was discovered that both ribozyme domains could be folded as group I ribozymes (GIRs), and named GIR1 and GIR2 according to the order of completion during transcription of the intron.[4] GIR2 was shown to be a conventional group IE splicing ribozyme and GIR1 a cleavage ribozyme acting at an intron internal site.

A description of the biology of GIR1 requires the introduction of a certain amount of nomenclature. Group I introns in ribosomal RNA (rRNA) genes are

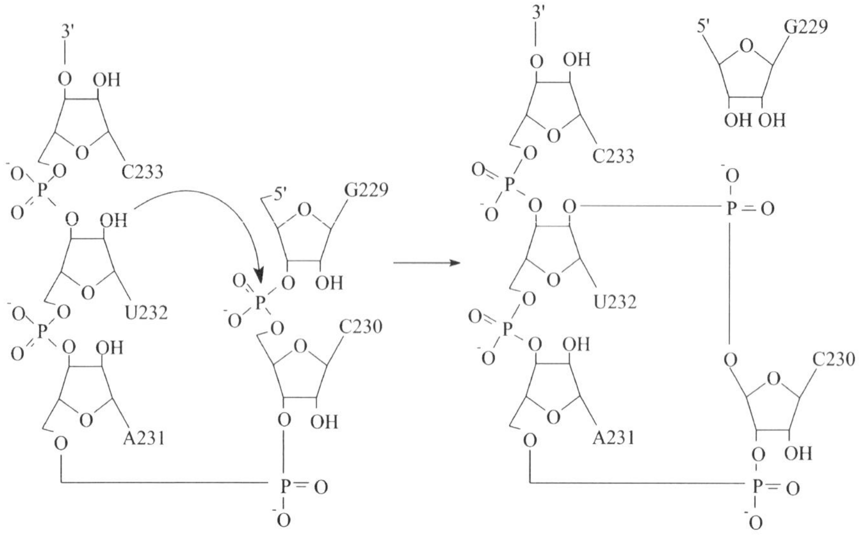

Figure 12.1 Branching reaction catalysed by GIR1. The 2′OH of the internal residue U232 makes a nucleophilic attack at the IPS. Bond lengths not drawn to scale. (The figure is reproduced from ref. 3 with permission.)

named according to Johansen and Haugen.[5] The name of host species and the insertion site within rRNA genes (*Escherichia coli* numbering) are reflected in the intron nomenclature. Dir.S956-1 is the twin-ribozyme intron inserted in the small rRNA gene of *Didymium iridis* at position 956. Since two different introns have been found in different isolates, this particular intron is numbered "-1". In a similar way, Nae.S516 is the twin-ribozyme intron from various species and isolates of *Naegleria*. Whenever a general feature of the branching ribozyme is described, it is referred to as "GIR1". When the description is specific for an individual ribozyme, or has only been investigated in one example, the species is indicated, *e.g.* "DiGIR1". Since several *Naegleria* ribozymes are known, "NaGIR1" refers to the *Naegleria* ribozyme in general and for example "NanGIR1" to the ribozyme from *Naegleria andersoni*. The classification of group I introns into subgroups follows the system by Michel and Westhof[6] based on differences in the structure of peripheral elements. DiGIR2 is a group IE,[7] and NaGIR2 a group IC ribozyme. Twin-ribozyme introns include a homing endonuclease gene (HEG). The transcript is referred to as the I-*Dir*I mRNA and the protein as the I-*Dir*I homing endonuclease following the rules for classification of these and similar enzymes.[8,9] Nucleotides in GIR1 are numbered according to their position within the twin-ribozyme intron. The minimal branching variant of DiGIR1 begins at pos. 73 and ends at pos. 251 in Dir.S956-1. The length variants are named according to how much sequence is included upstream, and downstream of the IPS. The minimal variant is thus named 157.22 to state that 157 nt upstream and 22 nt downstream of the IPS are included, respectively.

In this chapter, we describe the GIR1 ribozyme, focussing on GIR1 from the myxomycete *D. iridis* (DiGIR1). The emphasis is on describing GIR1 as a ribozyme for which a plausible hypothesis can be made as to how the structure was derived from group I introns and how this resulted in the gaining of a new reaction mechanism.

12.2 Distribution and Structural Organization of Twin-ribozyme Introns

The twin-ribozyme introns represent some of the most complex organized group I introns known and consist of a homing endonuclease gene (HEG) embedded in two functionally distinct catalytic RNA domains.[10] One of the catalytic RNAs is a conventional group I intron ribozyme (GIR2) responsible for intron splicing and reverse splicing, as well as intron RNA circularization. The other catalytic RNA domain is the group I-like ribozyme (GIR1) directly involved in homing endonuclease mRNA maturation.[3,4,11] Only two main natural variants of the twin-ribozyme group I introns are known,[1] and these include the Dir.S956-1 intron from the myxomycete *D. iridis* and the Nae.S516 from various species and isolates of *Naegleria* amoeboflagellates (see Table 12.1 below). Both introns are inserted into conserved regions of the nuclear small subunit rRNA (SSU rRNA) gene in their respective host organisms, and have a similar overall structural organization at the RNA level (Figure 12.2).

Several differences in distribution, inheritance, and structural organization are noted between the *Didymium* and *Naegleria* twin-ribozyme introns:

(i) The introns are located at different insertion sites in the SSU rRNA gene. Whereas the 1.4 kb *Didymium* intron is inserted after a U residue at position 956 (*E. coli* SSU rRNA numbering), the 1.2 kb *Naegleria* intron is located after residue 516. Both sites are frequently known to harbour group I introns in nuclear ribosomal DNA (rDNA) of eukaryotic microorganisms.

(ii) The pattern of intron distribution among host organisms is different. The Dir.S956-1 twin-ribozyme intron is unique to the Panama 2 isolate of *D. iridis*. In fact, very closely related *D. iridis* isolates either lack any group I intron at position 956 or harbour distantly related group I introns at this position.[12,13] The Nae.S516 intron, on the other hand, is restricted to the *Naegleria* genus, with a widespread but sporadic distribution that includes 21 of 70 strains analysed.[14]

(iii) The pattern of intron inheritance appears different. The *Didymium* intron is an example of an optional group I intron. The closest relatives known to the *Didymium* splicing ribozyme domain (DiGIR2) are found within different myxomycete genera, but not in a *Didymium* species, suggesting a recent gain by horizontal intron transfer. Interestingly, experimental support of Dir.S956-1 mobility has been obtained both at the RNA-level due to reverse splicing[15] and at the DNA-level due to

Table 12.1 GIR1 present in *Didymium* and *Naegleria* isolates.

Species/isolate	GIR1 size (nt)[a]	Acc no	Ref.
Didymium iridis Pan2-44	180	AJ938153	Johansen and Vogt 1994[2]
Naegleria andersoni A2	198	X78280	DeJonckheere 1994[46]
N. andersoni PPMFB-6	198	Z16417	DeJonckheere and Brown 1994[47]
N. carteri NG055	211	AM167878	Wikmark *et al.* 2006[14]
N. clarki RU30	197	AM167879	Wikmark *et al.* 2006[14]
N. clarki RU42	197	AM167880	Wikmark *et al.* 2006[14]
N. clarki Pd72Z/I	197	AF338417	Dykova *et al.* 2001[48]
N. clarki 4177/I	197	AF338418	Dykova *et al.* 2001[48]
N. clarki 4564/IV	197	AF338419	Dykova *et al.* 2001[48]
N. clarki 4709/I	197	AF338420	Dykova *et al.* 2001[48]
N. clarki Pd56Z/I	197	AF338422	Dykova *et al.* 2001[48]
N. clarki CB1S/I	197	DQ768725	Dykova *et al.* 2006[49]
N. gruberi CCAP1518-1D	208	X78278	DeJonckheere 1994[46]
N. italica AB-T-F3	209	U80249	Einvik *et al.* 1997[11]
N. jamiesoni T56E	197	U80250	Einvik *et al.* 1997[11]
N. martinezi NG872	219	AJ001399	DeJonckheere and Brown 1998[50]
N. philippinensis RJTM	209	AM167881	Wikmark *et al.* 2006[14]
N. pringsheimi 1D	208	AM167882	Wikmark *et al.* 2006[14]
Naegleria sp. CL/I	197	DQ768715	Dykova *et al.* 2006[49]
Naegleria sp. BCZ4/I	199	DQ768716	Dykova *et al.* 2006[49]
Naegleria sp. SUM3V/I	216	DQ768723	Dykova *et al.* 2006[49]
Naegleria sp. GP3/III	197	DQ768724	Dykova *et al.* 2006[49]
Naegleria sp. NG163	198	AM497929	S. Johansen and C. Einvik, unpublished
Naegleria sp. NG332	209	AM497930	S. Johansen and C. Einvik, unpublished
Naegleria sp. NG358	200	AM167883	Wikmark *et al.* 2006[14]
Naegleria sp. NG393	207	AM167884	Wikmark *et al.* 2006[14]
Naegleria sp. NG458	200	AM497931	S. Johansen and C. Einvik, unpublished
Naegleria sp. NG491	207	AM497932	S. Johansen and C. Einvik, unpublished
Naegleria sp. NG498	200	AM167885	Wikmark *et al.* 2006[14]
Naegleria sp. NG647	207	AM167886	Wikmark *et al.* 2006[14]

[a] The GIR1 sizes correspond to positions 1506–1685 of the *D. iridis* sequence (AJ938153), or positions 227–424 of the *N. andersoni* sequence (Z16417).

gene conversion initiated by the intron encoded I-*Dir*I homing endonuclease.[16] A very different inheritance pattern is noted among the *Naegleria* twin-ribozyme introns.[14] Here, the intron was apparently gained early in evolution of the *Naegleria* genus and co-evolved along with its host rDNA in a strict vertical inheritance pattern. Subsequently, the intron was lost by sporadic deletions in approximately 70% of the isolates.

The GIR1 Branching Ribozyme 233

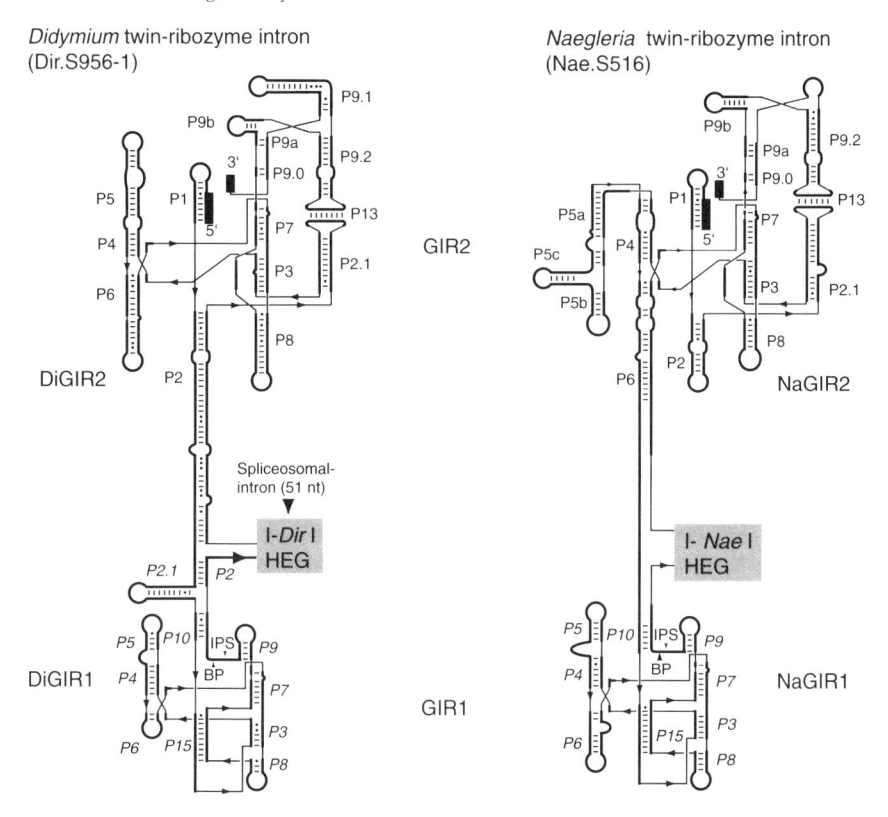

Figure 12.2 Structure diagrams showing the two known twin-ribozyme introns. GIR2 is the splicing ribozyme, GIR1 is the branching ribozyme, and HEG is a homing endonuclease gene. A spliceosomal intron is found in the I-*Dir*I HEG in *Didymium*. BP: branch point. IPS: internal processing site. Exons are indicated by thick lines.

(iv) The splicing ribozyme domains (GIR2) of Dir.S956-1 and Nae.S516 represent two of the most distantly related nuclear group I intron subgroups known. Whereas the *Didymium* GIR2 has a typical IE intron fold, the *Naegleria* GIR2 represents the common IC1 intron subgroup (Figure 12.2). Structural differences between these two intron subgroups are observed in the organization of the P4-P6 region (Figure 12.2), as well as in sequence motifs flanking the P7 guanosine binding site.

(v) The GIR1-HEG insertions are located at different helical segments within the *Didymium* and *Naegleria* GIR2 ribozymes (Figure 12.2), namely P2 in DiGIR2 and P6 in NaGIR2.

(vi) The HEGs encode different homing endonucleases. Both the *Didymium* and *Naegleria* homing endonucleases (I-*Dir*I and I-*Nae*I, respectively) are members of the His-Cys box family, but they are distantly related in sequence and possess different target DNA specificities.[12,14,17] Interestingly, the I-*Dir*I HEG, but not the I-*Nae*I HEG, is interrupted by a small spliceosomal intron (Figure 12.2) that is removed during mRNA maturation in *D. iridis*.[18]

(vii) Finally, the *Didymium* and *Naegleria* GIR1 ribozymes have notable structural differences. Despite the fact that DiGIR1 and NaGIR1 have very similar secondary structure folds of the ribozyme core domain (Figure 12.2), important sequence differences are noted at flanking regions, in the P4-P6 domain as well as at several peripheral loop regions. However, comparisons of 29 natural *Naegleria* GIR1 variants (Table 12.1) confirm a high degree of conservation with only a very few variable nucleotide positions. Minor variations in size were noted at only four regions (J5/4, L6, L8, and L9).

12.3 Biological Context

Didymium iridis is a myxomycete that preys on other microorganisms and dead organic matter on the forest floor. It has a complex life-cycle that includes haploid amoebae, flagellates, and cysts, diploid amoebae, and a syncytial plasmodium that can differentiate into sporangia with haploid spores. Our studies have so far focused on the haploid life forms that are easily grown in liquid culture. Incidentally, these are the only life-cycle forms that have been observed in *Naegleria*. Obviously, given the complexity of both the organism and the twin-ribozyme intron itself, much can be learnt from studying the biological context of the GIR1 ribozyme. In this section we describe how our characterization of the molecular biology of the processing of the twin-ribozyme in *D. iridis* haploid life forms provides clues to understanding the structure and mechanism of the GIR1 ribozyme.

12.3.1 Three Processing Pathways of a Twin-ribozyme Intron

From a combination of *in vitro* and *in vivo* studies we have mapped three different processing pathways of the Dir.S956-1 intron that applies to different cellular conditions (Figure 12.3). The first pathway involves intron excision and exon ligation catalysed by GIR2, and subsequent processing of free intron to form the I-*Dir*I mRNA.[18] The formation of the I-*Dir*I mRNA involves the GIR1 activity and is detailed in the next section. The order of the activity of the ribozymes is (1) GIR2 at the 5′ splice site (SS), (2) GIR2 at the 3′SS and (3) GIR1. This pathway benefits both the host and the intron, and is the dominant pathway during the growth phase of haploid amoebae and flagellates.

The second pathway results in the formation of full-length circular introns and un-ligated exons. This is a general pathway in nuclear group I introns.[19] The order of reactivity is (1) hydrolytic cleavage at the 3′SS catalysed by GIR2 and (2) transesterification at the 5′SS catalysed by GIR2. GIR1 is not active in this pathway and circle re-opening by GIR2 seems to be required for activation of the GIR1 activity. This pathway benefits the intron at the expense of the

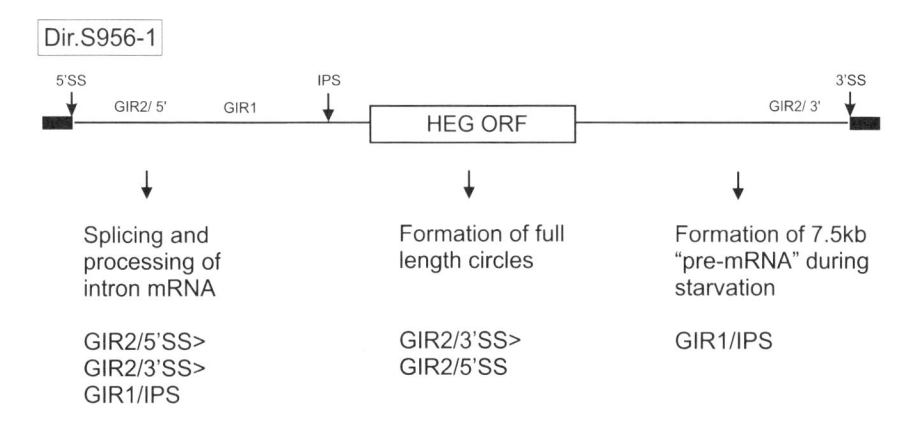

Figure 12.3 Ribozyme catalysed processing steps at the splice sites (SS) and the internal processing site (IPS) in three different processing pathways of the Dir.S956-1 intron. The order of the processing steps in each pathway is indicated. Black boxes: exons; Open box: The HEG open reading frame (ORF). Figure not drawn to scale.

host. The biological significance of the circularization pathway is not clear, but the circular introns are frequently suggested to be involved in the spreading of the intron. Interestingly, GIR2 adopts a particular conformation in the circles, and circularization is responsive to the cellular conditions (unpublished).

The third pathway is induced by starvation-induced encystment.[20] *D. iridis* undergoes frequent rounds of encystment and excystment in nature, depending on fluctuations in environmental conditions. During encystment, the rRNA precursor is processed into a 7.5 kb RNA product that accumulates within the cells to become the dominant intron containing molecular species. The processing is accomplished by the branching activity of GIR1 without prior GIR2 activity. The biological function is unknown but the current speculation is that the 7.5 kb RNA is stored as a precursor that will allow I-*Dir*I expression in the absence of rRNA expression. The pathway is at the expense of the host in the sense that functional rRNA is not being produced, but it could also be viewed as a mechanism to down-regulate rRNA expression during starvation.

From the observation of the three different pathways and the order of activity of the ribozyme activities, it follows that the two ribozymes are regulated with respect to each other. It is currently not known if this regulation involves protein factors, RNA–RNA interactions within the twin-ribozyme intron, or both. The regulation, furthermore, implies that both ribozymes can fold into an inactive, yet biologically significant conformation.

12.3.2 Processing of the I-*Dir*I mRNA

Several examples are known of nuclear protein-coding genes embedded in rDNA. These genes are transcribed as an integral part of an RNA polymerase I transcript. Normally protein-coding genes are transcribed by RNA polymerase II and their expression is facilitated by co- and post-transcriptional modifications that are specific for RNA polymerase II transcription. This raises the problem of how the protein-encoded genes within rDNA are brought on to the RNA polymerase II pathway. A recent survey of the best described examples[21] concluded that several different strategies are being used, and that those employed by the Dir.S956-1 intron in expression of the I-*Dir*I homing endonuclease rank among the most sophisticated.

The expression pathway is initiated by the splicing out of the intron by conventional group I intron splicing (Figure 12.4). Then, the 5' end of the mRNA is formed by cleavage catalysed by the immediate upstream GIR1 ribozyme. The cleavage by branching provides the mRNA with a lariat cap in place of the conventional m^{7}G cap. The 3' end is then formed by cleavage at site referred to as IPS3. This poorly characterized cleavage reaction does not occur in isolated RNA and is thus believed to depend on a host factor.[18] Following this, the mRNA is further processed by cleavage and polyadenylation at a *bona fide* polyA signal.[18] Next, a short (51 nt) spliceosomal intron is spliced out.[18] The spliceosomal intron harbours the conventional splice signals but the splicing mechanism has not been studied. Similar short spliceosomal introns have been found in other homing endonuclease genes encoded within group I introns.[13] It is possible that the acquisition of the splicing and polyadenylation signals by the intron serves to recruit protein factors that help guide the mRNA onto the polymerase II pathway.

The lariat capped, spliced, and polyadenylated form of the mRNA is the only molecular species derived from the intron that is found in the cytoplasm.[18] The mRNA becomes associated with ribosomes[18] but the protein product has not been directly demonstrated. However, its activity has been shown to be present by enzymatic assays in cellular extracts,[22] and by demonstration of endonuclease intron homing in genetic crosses.[16] The lariat cap has indirectly been shown to be required for expression of the protein in yeast.[22] It substitutes for the conventional cap in protection of the mRNA against 5' exonucleases and perhaps it even plays a role in recruitment of translation initiation factors.

12.3.3 Conformational Switching in GIR1

Group I introns are generally believed to fold directly into the active conformation *in vivo*, at least when the flanking exons are correctly folded. However, if GIR1 is folded in a similar way, this would result in cleavage of the ribosomal precursor and be detrimental to the cell. DiGIR1 and the I-*Dir*I HEG are inserted into P2 of DiGIR2. This positions DiGIR1 towards the 5' end of the 1436 nt Dir.S956-1 intron. If GIR1 cleaves *in vivo* at a rate that is comparable to the *Tetrahymena* intron self-splicing ($t_{1/2}$ of *ca.* 2 s;[23]) there is ample time to fold into an active conformation and cleave before transcription and folding of GIR2 is completed. GIR1 will thus most likely fold initially into an inactive conformation. In fact, some key components of an inactive GIR1

The GIR1 Branching Ribozyme

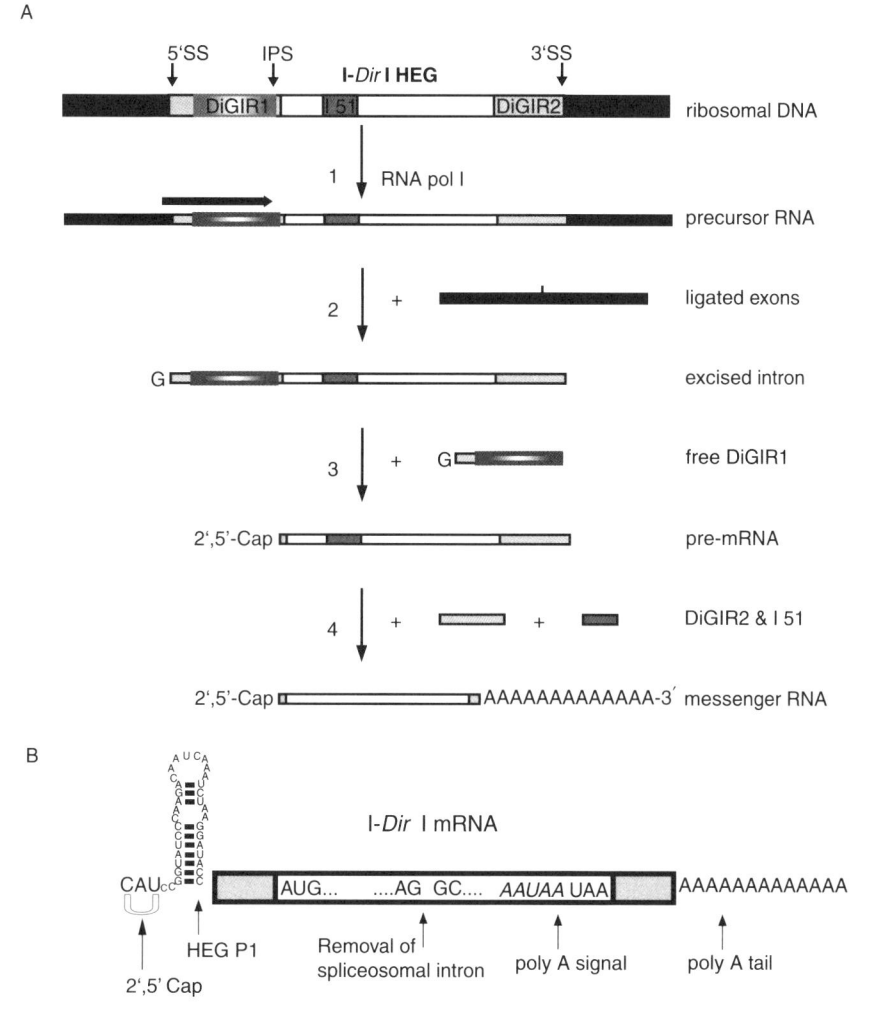

Figure 12.4 (*A*) Formation of the homing endonuclease I-*Dir*I mRNA by processing of the Dir.S956-1 intron. The homing endonuclease is coded by the sense strand of the intron and is transcribed by RNA pol I as part of the rRNA precursor (1). The intron is spliced out and the exons are ligated by the GIR2 ribozyme (2). The spliced out intron is cleaved by the GIR1 branching ribozyme, leaving the free ribozyme and a downstream pre-mRNA equipped with a lariat cap (3). Finally, in a series of steps dependent on host factors, the mRNA becomes polyadenylated, and a small spliceosomal intron is spliced out (4). (*B*) Detailed structure of the I-*Dir*I mRNA, including the lariat cap and HEG P1 hairpin found in the short 5′-UTR. (Figure 12.4B is reproduced from ref. 3 with permission.)

conformation have been identified, primarily from structure probing experiments. One is a hairpin formed by nucleotides 235–266 immediately downstream of the IPS. The formation of this hairpin (HEG P1) is co-transcriptionally favoured and precludes the formation of the catalytically

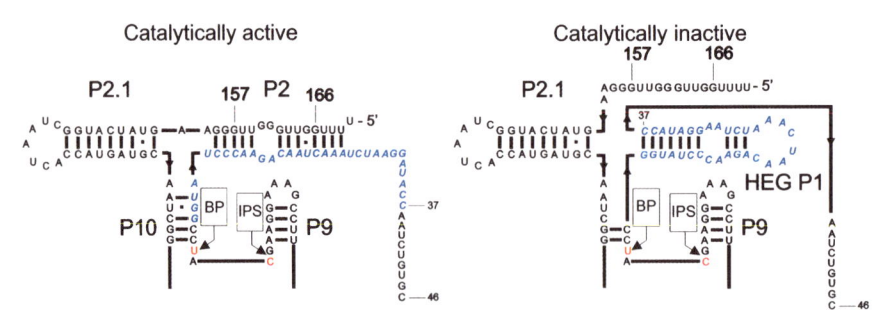

Figure 12.5 Alternative secondary structures of the part of DiGIR1 involved in conformational switching between catalytically active (including P2 and P10) and inactive (including HEG P1) conformations. The nucleotides involved in alternative pairings are in bold, blue italics. The nucleotides involved in formation of the 2′,5′ phosphodiester bond (C230 and U232) are in red. BP: branch point; IPS: internal processing site. The numbers indicate the distance in nt from the IPS.

active conformation (Figure 12.5). Incidentally, this hairpin is also found in the mature I-*Dir*I mRNA (Figure 12.4B). The conformational switching mechanism between P10-P2 of the catalytically active conformation and HEG P1 take place immediately following the branching reaction, where the formation of HEG P1 inhibits the reversal of the reaction and provides the I-*Dir*I mRNA with a pre-folded 5′-UTR.[24]

12.4 Biochemical Characterization

Characterization of a ribozyme often starts with the problem of delimitation of the functional unit. This is because many ribozymes are found as an integral part of a larger RNA molecule that is impractical to study. From looking at the structure diagrams of twin-ribozymes (Figure 12.2), isolation of a GIR1 functional unit appears to be straightforward. In the *Didymium* and *Naegleria* ribozymes, GIR1 is easily identified as a separate domain (Figure 12.2) presented on extended helices P2 and P6, respectively. This observation proved to be deceptive, mainly because of a misinterpretation of the activity of the ribozyme. The *Didymium* ribozyme was originally found to cleave at a single site (IPS1) by primer extension analysis of a construct carrying a deletion of most of the ORF and the 3′-part of GIR2.[2] Subsequently it has been shown that the full-length intron *in vitro* similarly is cleaved at a single site and with a relatively low efficiency.[24] Narrowing down by deletion studies to define a minimal version of GIR1 that could be studied in greater detail concluded that the 162.65 variant best reflected the functional unit.[10] This variant is cleaved at the same site as the full-length intron, but an additional stop site was mapped by primer extension analysis three nucleotides further downstream (IPS2). Incidentally, this is the only primer extension site observed in analysis of cellular RNA.[15,18,20] Subcloning of the *Naegleria* ribozyme (NanGIR1) led to a variant with 267 nt upstream and *ca.* 50 nt downstream of the IPS that showed the same cleavage pattern as the *Didymium* ribozyme.[11] This was further narrowed down to 178.19 by Jabri *et al.*[25] who found this variant to cleave at a high rate compared to several other length variant but with cleavage only at IPS1, as in the case of the full-length intron. At this stage, the cleavage at the processing site (IPS1) was demonstrated to leave a 3'OH and a 5' phosphate, as would be expected by hydrolytic cleavage.[25] The primer extension stop at IPS2 was not characterized, but it was inferred that this cleavage was hydrolytic as well.[10] In addition, it was concluded from a concatenation RT-PCR approach that the two cleavages occurred in an obligatory sequential manner.[10] The interpretation of the IPS2 primer extension stop proved to be incorrect (see below) and in this case it turned out that "less is more" in the sense that the full-length intron masks the fact that the reaction is actually highly reversible *in vitro*.

12.4.1 GIR1 Catalyzes Three Different Reactions

DiGIR1 catalyses three different reactions. The natural reaction is the branching reaction (1A in Figure 12.6) in which a transesterification at the IPS results in the cleavage of the RNA with a 3'OH and a downstream lariat cap made by joining of the first and the third nucleotide by a 2',5' phosphodiester bond.[3] These are the only products observed by analysis of cellular RNA.[18,20] *In vitro*, DiGIR1 catalyses the reverse reaction (1B), referred to as the ligation reaction. It is very efficient to the extent that the forward reaction is completely masked in reactions with full-length intron and length variants that include more than 166 nucleotides upstream of the IPS.[3,24] Finally, DiGIR1 catalyses hydrolytic cleavage at the IPS (2A) at a relatively low rate. This is the cleavage reaction observed with the full-length intron and several length variants.[24] The hydrolytic cleavage is irreversible and is considered an *in vitro* artefact resulting from a failure to present the branch nucleotide (BP) correctly for catalysis. NaGIR1 catalyses the same reactions as DiGIR1. Specifically, several length variants of NanGIR1, the smallest being 178.28, have been shown to catalyse branching (unpublished).

The three reactions can be experimentally separated. The branching reaction is isolated from the reverse reaction by cleavage in the presence of 2 M urea. This inhibits ligation completely and the contribution from hydrolytic cleavage

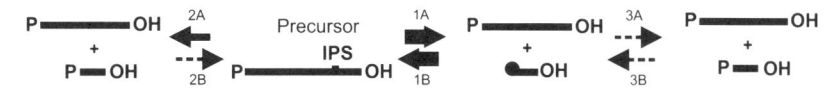

Figure 12.6 Reactions known to be catalysed by GIR1 as well as hypothetical reactions that have not been observed (dashed arrows). The main activity is the branching activity (1A) that is highly reversible *in vitro* (1B). A hydrolytic cleavage reaction (2A) is less pronounced and only observed *in vitro*.

at these conditions is negligible.[24] Ligation is studied under acidic conditions (e.g. pH 5.5) at which the forward reaction is inhibited.[24] Finally, the hydrolysis reaction is the only reaction observed at standard conditions with certain length variants. This is not a true isolation of the reaction because the major fraction of the molecules apparently is engaged in multiple rounds of branching and ligation reactions. The reaction rates vary considerably among the length variants. Analysed as described above, the 166.22 variant of DiGIR1 performs branching at a rate of $0.085\,\text{min}^{-1}$, ligation at a rate approaching $1\,\text{min}^{-1}$, and hydrolysis at $0.01\,\text{min}^{-1}$.[24] The branching rate is only one order of magnitude less than the cleavage rates of most optimized minimal cleavage ribozymes (e.g. $0.2-0.5\,\text{min}^{-1}$ for the hairpin,[26] $1.0\,\text{min}^{-1}$ for the VS,[27] and $0.5-2.0\,\text{min}^{-1}$ for the hammerhead ribozyme[28]).

12.4.2 Characterization of the Branching Reaction

The branching reaction is initiated by a nucleophilic attack involving the 2'OH of U232 at the phosphate of C230 (Figure 12.1). This was demonstrated in a trans-cleavage experiment using a ribozyme that was truncated at A222 in L9 (7 nt upstream of the IPS) and a substrate carrying the missing 7 nt followed by 22 nt downstream of the IPS.[3] Deoxy-substitutions were introduced in the substrate at positions corresponding to C230, A231, U232, and C233. Only deoxy-substitution of the 2'OH of U232 completely prevented the branching reaction. A strong effect of deoxy-substitution at A231 was ascribed to a structural effect (Section 12.5). Detailed characterization of the reaction mechanism is still in its initial phase.

The characteristic product of the branching reaction, the lariat cap, was first deduced from indirect analysis, such as primer extension, and resistance towards degradation by enzymes and alkali. Subsequently, the lariat was sequenced by enzymatic degradation and analysis by thin-layer chromatography (TLC).[3] Curiously, the structure of the lariat cap was already known from the literature. Several small lariats, including a 4 nt lariat, were analysed by nuclear magnetic resonance (NMR) imaging by Agbäck et al.[29] in a study attempting to describe the lariat products from spliceosomal splicing. The 4 nt lariat was found to have an unusual structure with the lariat ring locked in a rigid South-type conformation.

12.4.3 Biochemistry of GIR1

Optimization of the GIR1 reaction is mainly due to the early work of Jabri et al. on the minimal version (178.19) of *Naegleria andersoni* GIR1 (NanGIR1).[25] Notably, this ribozyme was reported to exclusively cleave by hydrolysis at the IPS and thus the branching reaction may not have been optimized. The basic setup for GIR1 cleavage studies is to pre-incubate the *in vitro* transcribed RNA in cleavage buffer at pH 5.5 (10 mM cacodylate or 10 mM acetate) for 5–10 min and then start the reaction using a pH jump by addition of cleavage buffer containing Hepes-KOH at pH 7.5. This setup eliminates a lag phase required for folding of the RNA and permits kinetic analysis. Based on this, the optimal conditions for NanGIR1 were found to be 1 M KCl, 25 mM MgCl$_2$ at 45 °C. CaCl$_2$ did not substitute for MgCl$_2$, but MnCl$_2$ could replace MgCl$_2$ and was even found to be more effective at lower concentrations. Cleavage in NaCl was less efficient than with KCl and LiCl inhibited the reaction. The polyamines spermine and spermidine could not replace the monovalent ion at concentrations used in group I introns. The activity of the NanGIR1 ribozyme under the above conditions increased linearly with temperature between 30 and 45 °C, and reduced again above 47 °C. The reaction rate was found to be first order in hydroxide ion concentration independent of the type of buffer in the range pH 4–8.5. *In vitro* selection was employed to select for NanGIR1 variants with a faster hydrolysis rate and less salt dependence. Variants were characterized that cleaved at 300-fold greater rates in 100 mM KCl.[30] A systematic analysis of the requirement for the *Didymium* ribozyme, and in particular in relation to the branching activity, has not been performed. Generally speaking, the conditions found by Jabri *et al.*[25] appear to apply to the *Didymium* ribozymes as well, but the mutations that relieved the NanGIR1 ribozyme of the high salt requirement did not have a similar effect in the DiGIR1 context.[1]

12.5 Modelling the Structure of GIR1

Most of the base-pairing scheme of GIR1 was based on the close similarity to group I introns (Chapter 10). In the first published 3D model,[10] known features of group I introns were supplemented with *in vitro* mutagenesis structure probing data. The most notable observations were the lack of P1, the presence of a novel P15 pseudoknot, and the unusual structure of J5/4. The P2-P2.1 domain was not included in the original model but was subsequently verified.[31] Parallel work with the *Naegleria* ribozyme corroborated the overall base-pairing scheme, except that P2.1 is absent in this species.[11,25]

The publication of an X-ray crystal structure at 3.1 Å resolution of the *Azoarcus* sp. tRNA^{Ile} intron (*Azo*)[32,33] prompted a revision of the base-pairing scheme of DiGIR1. The *Azo* intron was crystallized in a version that represents the second step of splicing, the structure that most closely resembles DiGIR1. The similarity suggested an extended P15 corresponding to the P1-P2 stack in *Azo* at the second step of splicing. The next development was the unexpected finding that the reaction catalysed by GIR1 is a branching reaction rather than sequential hydrolytic cleavages.[3] The unique product of the reaction, a 4 nt lariat, had previously been characterized by NMR in Chattopadhyaya's group in an attempt to describe the structure of the lariat resulting from spliceosomal splicing.[29] With this at hand, we could construct a model[34] that incorporated features from recent X-ray crystal structures of group I introns, including the *Azo*,[32,33] *Tetrahymena thermophila* (*Tet*),[35] and the *Staphylococcus aureus* bacteriophage *Twort* (*Two*)[36] introns, together with the unique features of

GIR1 like the P15 pseudoknot, the lariat fold and a type of three-way junction (3WJ) originally found in rRNA.[37] Two different base-pairing schemes were considered for the GU pair at the active site. Based on mutation analysis, G109:U207 was favoured over G109:U232 (Section 12.5.4). The P2-P2.1 domain presented an additional problem because the docking of this domain onto the remainder of the structure was not evident. The P2-P2.1 domain was thus excluded from the model.

12.5.1 Overall Structure

The model of the DiGIR1 is compact with three aligned helical stacks ($85 \times 52 \times 40$ Å without the P2-P2.1 domain). The P3-P7-P8-P9 domain (hereafter P3-P9 domain) and P10-P15 domains run in parallel and the P4-P5-P6 domain (hereafter P4-P6 domain) is slightly tilted with the L5 end pointing away from the rest (Figure 12.7). This helical arrangement is roughly similar to that of *Azo*.[33] The helical organization in DiGIR1 is mainly brought about by two structural features. First, the 3′ strand of P2 (G199-C204) has apparently been fused with nucleotides from J8/7 (A205-U207), thus forming the P15 pseudoknot. Since P3 is already embedded in the pseudoknot characteristic of group I introns, the resulting structure is a double pseudoknot. The formation of P15 shortens the canonical group I intron J8/7 from 7 (IC1 introns) or 6 nt (all other subgroups) to only 3 nt and in this way tightens the structure. Second, J15/3 organizes the three-way junction between P15, P3 and P8, resulting in a side-by-side parallel orientation of the P3-P9 and P10-P15 domains (Figure 12.8A).

12.5.2 Coaxially Stacked Helices

The P4-P6 domain in DiGIR1 is quite small. In the *Tet* intron this domain folds at an early stage and functions as a scaffold that facilitates the folding of the core.[38,39] This is unlikely to be the case in DiGIR1 and is consistent with the notion that DiGIR1 initially folds into an inactive conformation from which it is activated (Section 12.3.3). Substitution of L5 with a UUCG tetraloop has only a moderate effect on the cleavage kinetics, whereas substitution of the two As in L6 has a more pronounced effect (unpublished), probably due to an interaction with P3. Compared to group I introns, the interface between P5 and P4 is quite different in that J4/5 is missing. J5/4 and the architecture around it, including the bulged U156, are critical for the activity (Section 12.5.4). The *Naegleria* ribozyme differs considerably in the structure of the P4-P6 domain. First, P6 is extended and includes an internal loop that makes the tertiary interaction with P3 similar to what is known from the *Azo* intron. As a consequence, L6 is not involved in this tertiary interaction and the sequence is variable. Second, J5/4 is highly variable among *Naegleria* GIR1's, in many cases including an additional helix (P5.1) in some sequence variants.

The GIR1 Branching Ribozyme   243

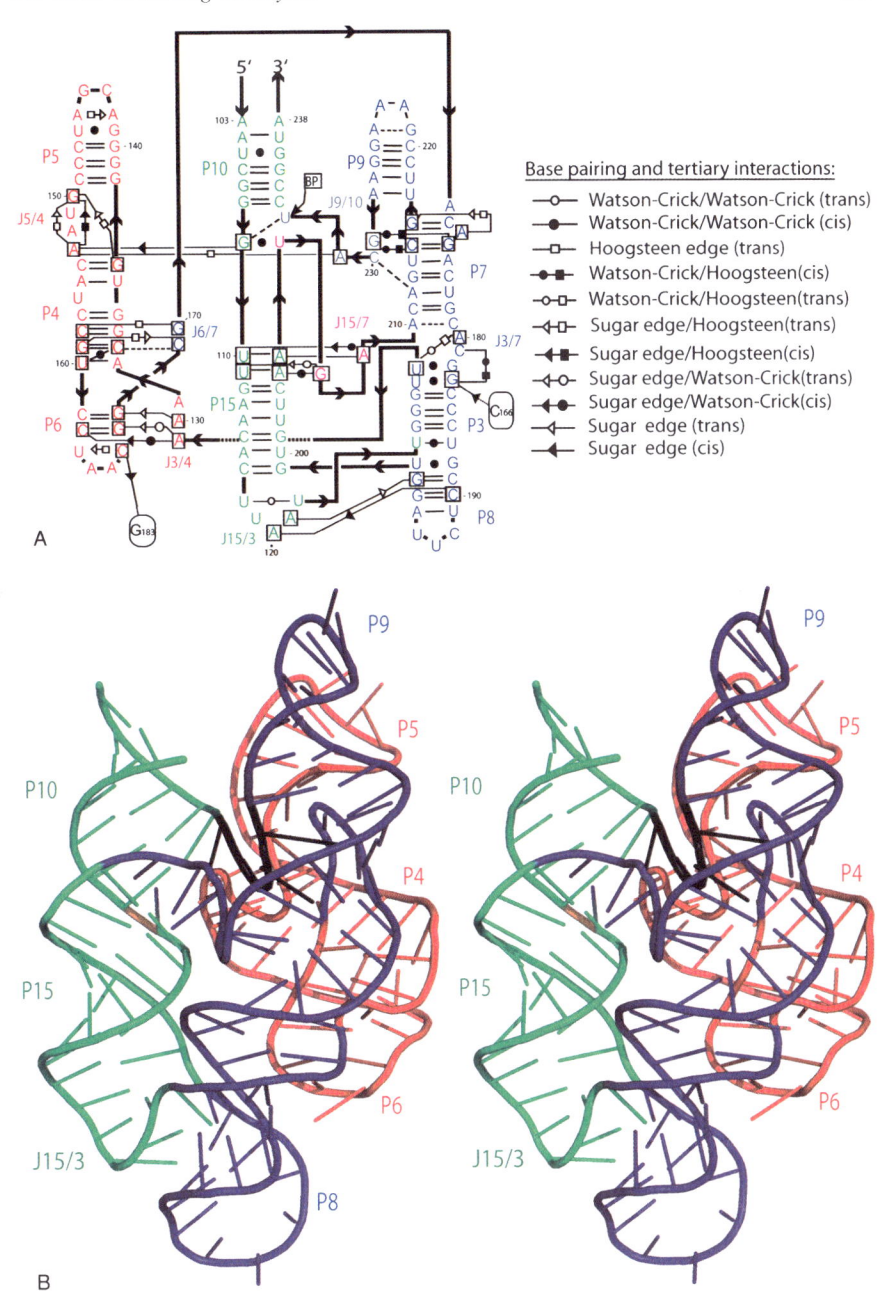

Figure 12.7 Model of the structure of the GIR1 ribozyme. (*A*) Secondary structure and proposed tertiary interactions of DiGIR1 showing Watson–Crick as well as non-Watson–Crick base-pairings. (*B*) 3D model of the overall structure of DiGIR1.[34] The paired stems (P) and one of the joining segments (J) are numbered.

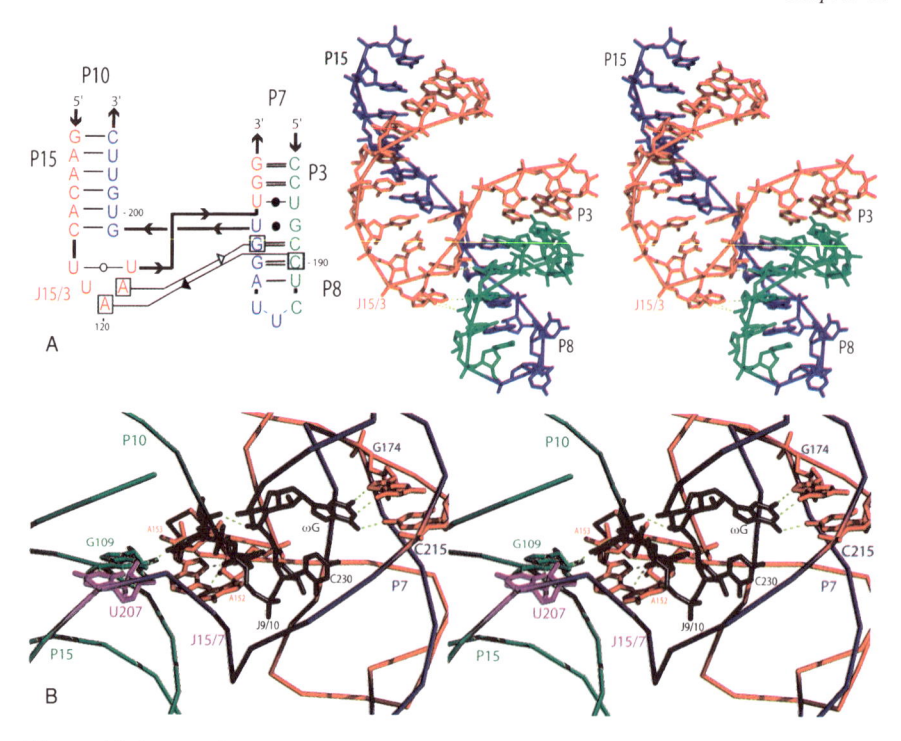

Figure 12.8 Details from the structural model of DiGIR1. (*A*) Base pairing in the three-way junction J15/3 and 3D model.[34] The boxed nucleotides are A120, A121, C190, and G197. (*B*) Modelled structure of the active site with the key nucleotides indicated. Colouring corresponds to that in Figure 12.7.

The P3-P9 domain contains a P7 guanosine binding site that is highly conserved among all group I introns. Here, an exogenous guanosine factor (exoG) binds during the first step of splicing and the intron terminal guanosine (ωG) binds during the second step of splicing and during the first step of the circularization pathway. GIR1 P7 binds only G229, the equivalent of ωG. Addition of GTP to a GIR1 cleavage reaction has no effect (unpublished), suggesting that added guanosine is unable to compete with G229 for binding to P7. P9 stacks upon P7 without any intervening nucleotides. Both the length and the sequence of the GAAA tetraloop that caps P9 are important for activity (unpublished), but the tetraloop receptor remains elusive. Experiments and modelling have excluded a receptor in the P4-P6 domain. Rather, we favour an interaction within the P2-P2.1 domain. P2 is involved in a conformational switch (Section 12.3) that results in the formation of the alternative HEG P1, and a receptor for L9 has been suggested in this structure (unpublished). However, this does not rule out the existence of an alternative receptor in the catalytically active conformation. NaGIR1 differs from DiGIR1 in that L9 is a 7–13 nt loop. Part of the loop sequence is complementary to the sequence downstream P2″, an interaction that could very well be a functional replacement of the L9:HEG P1 interaction in DiGIR1.

The P10-P15 domain consists of a 5 base-pair P10 stacked directly upon P15. A 6 base-pair P15 was proposed in our original model, but the analogy to *Azo* suggested an extension by 3 base-pair that was subsequently supported by mutagenesis analysis of the U110-A206 base-pair. The structure of this domain is very similar in DiGIR1 and NaGIR1. In fact, both P10 and P15 from *Naegleria* function in the DiGIR1 context[10] (unpublished). The P10-P15 stack is equivalent to the P10-P1-P2 stack formed at the second step of group I introns. Thus, the apparent absence of a P1 from the GIR1 ribozyme is related to the fact that GIR1 resembles a group I intron at the second, not the first, step (Section 12.6).

The P2-P2.1 top domain, not incorporated into the current model, forms a three-way junction with P10. P2 is the connection to the splicing group I intron component (GIR2) of the twin-ribozyme introns (Section 12.2). NaGIR1 has a loop of unknown structure in place of P2.1. One possible function of P2-P2.1 in DiGIR1 is to bridge the P4-P6 and P3-P9 domains and in this way stabilize the folding of the core in a manner similar to the direct interactions between these two domains known in group I introns.

12.5.3 Junctions and Tertiary Interactions Involving Peripheral Elements

The junctions between the P4-P6 and P9-P3 domains resemble those seen in group I introns. The interface between the two domains is modelled with the involvement of base-triple interactions between J3/4 and J6/7 in the shallow/minor groove of P6 and the deep/major groove of P4, respectively. J15/3 organizes a three-way junction of family C[37] (Figure 12.8A). U118 base-pairs with U122, and U123 base-pairs with U187 (P3). A120 and A121 form A-minor interactions in the minor groove of P8. Two other junctions, J15/7 and J9/10, are part of the active site and are discussed below.

Group I introns usually have peripheral structural elements that branch out from the core. These elements interact and contribute stabilization of the core but are few in GIR1, at least in the model that does not involve the P2-P2.1 domain. The two As in the L6 tetraloop make A-minor interactions in the shallow/minor groove of two consecutive G=C pairs in P3. In the *Naegleria* ribozyme the interaction is between As in J6/6a and P3, as observed recurrently in group I introns. As mentioned in Section 12.5.2, L9 is a candidate for an additional peripheral interaction, but an interaction partner has not been identified.

12.5.4 The Active Site

The key residues for the branching reaction interact in a pocket formed at the interface of P10, P7, J5/4, and J9/10 (Figure 12.8B). The architecture of the guanosine binding site is similar to that of group I introns (Chapter 10) and all substitutions of G229 are inactive.[1] A GU wobble pair is presented at the interface of P10 and P15 equivalent to the position at the interface of P10 and P1 at the second step in group I introns. In our original model, G109 forms a GU wobble base-pair with U232. Substitutions of U232 reduce cleavage by a factor of 3–4 but are still permissive for the branching reaction.[34] In contrast, mutations of G109 and U207 both result in poor branching in addition to a 2–4-fold reduction in the cleavage rate.[34] The G109:U207 pairing was thus favoured in the model. The GU wobble in *Azo* is recognized by the wobble receptor motif in the symmetric J5/4 internal loop. Here, A58 and A87 belonging to two consecutive AA pairs are involved in an intricate hydrogen bonding network and recognize the ribose 2'OH and the exocyclic amine, respectively, of the G in the GU pair. In DiGIR1, J4/5 is lacking, and the recognition of the GU pair by J5/4 appears to be quite different. A231 of the lariat fold appears oriented towards A153 of J5/4 due to hydrogen bonding between the 2'OH and a phosphate oxygen at C230. These two adenosines are among the most critical residues in DiGIR1 and are proposed to form a *trans*-Hoogsteen-Hoogsteen pair. The pairing places A153 in a position to hydrogen bond to the N2 of G109. The nucleotide carrying the nucleophile (U232) lies in the shallow/minor groove of the GU pair and is close enough to hydrogen bond to the N2 amino group of G109. Perhaps this interaction can explain the preference for a U at this position. Mutational analysis in NaGIR1[30] indicates that GU recognition in this ribozyme could be slightly different, perhaps related to the variation in the structure of J5/4.

J8/7 is a highly conserved sequence among group I introns. In *Azo*, it pirouettes through the active site, making contacts to all three helical domains as well as participating in the binding of structural and catalytic metal ions.[33] The sequence of J8/7 is clearly recognized in GIR1 (4/6 nucleotides are conserved), but its organization appears dramatically altered. The first three nucleotides are involved in P15 base-pairings. Thus, the recognition of P1-P2 in *Azo* by the three 5' nucleotides in J8/7 has been replaced by participation in the analogous stem structure by contributing base-paired residues. The third base-pair (involving U207) is a GU wobble base-pair that replaces the GU found in P1 in *Azo* at almost the same position. In *Azo*, the next three nucleotides make contacts with J7/3, P4, and P7, respectively.[33] However, in GIR1 the base-pairing of the 5' part of J8/7 restricts the course of the backbone and the next two nucleotides (G208 and A209) contact the shallow/minor groove of P15. Then, the last nucleotide of J8/7 (A210) mimics the fourth nucleotide in *Azo* (G170) in the sense that it makes the same hydrogen bonding and stacking interaction with J7/3. This leaves GIR1 short of the two nucleotides (C171 and A172) that contact P4 and P7. One interesting possibility is that the two added nucleotides from the lariat fold (C230 and A231) might take on some of these functions. At least, C230 can be stacked under the guanosine binding site in P7 in the same way that A172 is stacked in *Azo*. In *Azo*, J8/7 is critical in the binding of both of the catalytic metal ions. A particularly well-studied aspect of the catalytic site of the *Azoarcus* intron is the coordination of catalytic metal ions.[40] The structural differences observed in GIR1 could influence metal ion binding but this has not yet been addressed.

In the active site, the nucleotide carrying the nucleophilic 2'OH (U232) is unpaired and occupies a pocket that will accommodate any base. This is consistent with *in vitro* mutagenesis results that show that branching can occur with any nucleotide at this position (unpublished). In the *wt* sequence, a single hydrogen bond between U232 and A231 bond could help orient the 2'OH for nucleophilic attack. The distance U232 O2'–C230 O1P in the model is 2.80 Å, and in agreement with the proposed mechanism.

12.6 Phylogenetic Considerations

The similarities between GIR1 and *Azo* suggest a common ancestor or, alternatively, that one of the ribozymes was derived from the other. As previously noted, GIR1 may have originated from eubacterial group IC3 introns.[1] In this section we discuss the possibility that GIR1 originated by a misaligned reverse splicing event and that this was followed by structural transformations. Reversal of the splicing reaction is a well-established reaction suggested to be in part responsible for the observed phylogenetic distribution of group I introns.[41] G229 in GIR1 is equivalent to the ωG residue of a group I intron. The sequence that follows, 5'-CAU, is identical to the sequence of the 5'-exon immediately upstream of the *Azo* intron. In fact, this is the anticodon sequence of the tRNA^{Ile} in which the *Azo* intron resides. The suggestion is that GIR1 originated by reverse splicing of a eubacterial tRNA-like intron into a tRNA-like molecule in a situation where the exons were misaligned by three nucleotides. This resulted in reverse splicing on the 5' side of the anticodon instead of the 3' side. As a consequence, the intron became linked at its 3'-end (G229) to three nucleotides of 5'-exon (5'-C230AU) followed by the 3'-exon (C233 and onwards). The inclusion of the three nucleotides of 5'-exon is postulated to have blocked any further steps in reverse splicing, thus trapping the intron in a conformation that is reminiscent of a group I intron prior to the second step of splicing. We now consider several predictions that follows from this evolutionary model.

The first prediction is that GIR1 has lost the features characteristic of the first step of splicing. Here, a P1 hairpin including the 5' splice site is absent. Perhaps related to this, GIR1 is unable to bind the cofactor of the first step, GTP, well enough to compete with binding of the ωG analogue, G229. In *Azo*, an A residue links P7 and P9. This nucleotide has been hypothesized to sequester ωG in a G-A pair during the first step of splicing.[42] Consistent with the idea that such a function is redundant in GIR1, P7 and P9 are linked without intervening nucleotides.

The second prediction is a loss of conformational flexibility in GIR1 related to the conformational switch that occurs in *Azo* between the first and the second step of splicing. There are two important examples in GIR1 of replacement of tertiary interactions with base-pairing. This results in a loss in flexibility in the sense that the interaction is a one-step rather than a two-step interaction. The tethering of the (P10)-P1-P2 helical stack to the P3-P9 domain by an L2 tetraloop: P8 tetraloop receptor interaction in *Azo* is replaced by the lower 6 base-pairs of the P15 pseudoknot in GIR1. The docking of P1 into the core of a group I ribozyme is mediated by a minor groove interaction by the three 5' nucleotides of J8/7. This interaction has to accommodate both the P1 hairpin at the first step, and the lower part of the hairpin sandwiched between P10 and P2 at the second step. This interaction is replaced by the three upper base-pairs in GIR1 P15.

The third prediction is that GIR1 has undergone a structural reduction while maintaining the essentials of the catalytic core. The IC3 introns present in eubacterial tRNA rank among the smallest group I introns known. GIR1, and in particular DiGIR1, is even smaller (Table 12.1). The integrations of P1, P2, and parts of J8/7 into the P15 pseudoknot are characteristics of the GIR1 ribozyme. In DiGIR1, the reduction in size of the P4-P6 domain is another prominent feature. The internal loop that is involved in a minor groove interaction with P3 is lost and the interaction is replaced by a similar interaction using nucleotides in L6. The reduction in size of the P4-P6 domain correlates with the loss of the function of this domain as a scaffolding domain during folding. The preservation of the core is mainly seen in the conservation of the P7 architecture and the maintenance of an interaction between J5/4 and a GU wobble base-pair. A final consequence of the proposed origin of GIR1 is that the flanking sequences could have retained tRNA features. In this view, P10 is regarded as a fusion with the anticodon stem and P2.1 as the D arm. Further similarities in the 3' flanking sequences that depend on alternative folding are currently being analysed.

Perhaps the most stunning observation based on the model is that the topological differences between the GIR1 and the group I intron catalytic cores can be explained by relatively simple alterations at the sequence level. For example, a single transposition of six nucleotides (5' GUGUUC) from P15" of GIR1 to a position within J15/3 would shuffle some of the critical junctions within the core and alter the structural organization of J8/7 at the RNA level. The resulting topology would be that of a group I intron. Such a transposition event can be incorporated into the reverse splicing model or can it be envisaged without such an event.[34] Importantly, the feasibility of these evolutionary models can be experimentally addressed.

12.7 Concluding Remarks

The GIR1 branching ribozyme shows a clear structural resemblance to group I introns. In terms of the reaction catalysed, it is more reminiscent of the group II introns (Chapter 11) (or spliceosomal introns; Chapter 13). However, the catalytic core appears to constitute a unique fold and the product, although essentially being a branched RNA, appears to be unique in structure. For these reasons, we suggest that the GIR1 branching ribozyme is listed as an independent class of naturally occurring ribozymes, as originally suggested by Cech and Golden.[43] Besides being an addition to the shortlist of naturally occurring ribozymes, GIR1 is a new example of 2′,5′-phosphodiester bond-forming activity, of which around ten are known.[44] Compared with most other ribozymes, the characterization of the structure and reaction mechanism of GIR1 is still in its initial phase. In contrast, a substantial amount of information of the biological context has been gathered. A particularly interesting aspect of GIR1 is that a plausible hypothesis can be made on its evolutionary history. There are very few examples of this, the most recent being the suggestion that the HDV ribozyme originated from the human transcriptome.[45] The proposed development of GIR1 is more radical because it involves a structural and a functional derivation. We suggest that two events, a misaligned reverse splicing, and a transposition of six nucleotides are sufficient to account for the derivation of the basic GIR1 structure from a group I intron.

References

1. S. Johansen, C. Einvik and H. Nielsen, DiGIR1 and NaGIR1: Naturally occurring group I-like ribozymes with unique core organization and evolved biological role, *Biochimie*, 2002, **84**, 905–912.
2. S. Johansen and V.M. Vogt, An intron in the nuclear ribosomal DNA of *Didymium iridis* codes for a group I ribozyme and a novel ribozyme that cooperate in self-splicing, *Cell*, 1994, **76**, 725–734.
3. H. Nielsen, E. Westhof and S. Johansen, An mRNA is capped by a 2′, 5′ lariat catalyzed by a group I-like ribozyme, *Science*, 2005, **309**, 1584–1587.
4. W.A. Decatur, C. Einvik, S. Johansen and V.M. Vogt, Two group I ribozymes with different functions in a nuclear rDNA intron, *EMBO J*, 1995, **14**, 4558–4568.
5. S. Johansen and P. Haugen, A new nomenclature of group I introns in ribosomal DNA, *RNA*, 2001, **7**, 935–936.
6. F. Michel and E. Westhof, Modelling of the three-dimensional architecture of group I catalytic introns based on comparative sequence analysis, *J. Mol. Biol.*, 1990, **216**, 585–610.
7. S.O. Suh, K.G. Jones and M. Blackwell, A group I intron in the nuclear small subunit rRNA gene of *Cryptendoxyla hypophloia*, an ascomycetous fungus: Evidence for a new major class of Group I introns, *J. Mol. Evol.*, 1999, **48**, 493–500.
8. M. Belfort and R.J. Roberts, Homing endonucleases: Keeping the house in order, *Nucleic Acids Res.*, 1997, **25**, 3379–3388.
9. R.J. Roberts, M. Belfort, T. Bestor, A.S. Bhagwat, T.A. Bickle, J. Bitinaite, R.M. Blumenthal, S.K. Degtyarev, D.T. Dryden and K. Dybvig *et al.*, A nomenclature for restriction enzymes, DNA methyltransferases, homing endonucleases and their genes, *Nucleic Acids Res.*, 2003, **31**, 1805–1812.

10. C. Einvik, H. Nielsen, E. Westhof, F. Michel and S. Johansen, Group I-like ribozymes with a novel core organization perform obligate sequential hydrolytic cleavages at two processing sites, *RNA*, 1998, **4**, 530–541.
11. C. Einvik, W.A. Decatur, T.M. Embley, V.M. Vogt and S. Johansen, *Naegleria* nucleolar introns contain two group I ribozymes with different functions in RNA splicing and processing, *RNA*, 1997, **3**, 710–720.
12. P. Haugen, O.G. Wikmark, A. Vader, D.H. Coucheron, E. Sjøttem and S.D. Johansen, The recent transfer of a homing endonuclease gene, *Nucleic Acids Res.*, 2005, **33**, 2734–2741.
13. S.D. Johansen, A. Vader, E. Sjøttem and H. Nielsen, In vivo expression of a group I intron HEG from the antisense strand of *Didymium* ribosomal DNA, *RNA Biol.*, 2006, **3**, 157–162.
14. O.G. Wikmark, C. Einvik, J.F. De Jonckheere and S.D. Johansen, Short-term sequence evolution and vertical inheritance of the *Naegleria* twin-ribozyme group I intron, *BMC. Evol. Biol.*, 2006, **6**, 39.
15. A.B. Birgisdottir and S. Johansen, Site-specific reverse splicing of a HEG-containing group I intron in ribosomal RNA, *Nucleic Acids Res.*, 2005, **33**, 2042–2051.
16. S. Johansen, M. Elde, A. Vader, P. Haugen, K. Haugli and F. Haugli, In vivo mobility of a group I twintron in nuclear ribosomal DNA of the myxomycete *Didymium iridis*, *Mol. Microbiol.*, 1997, **24**, 737–745.
17. M. Elde, N.P. Willassen and S. Johansen, Functional characterization of isoschizomeric His-Cys box homing endonucleases from *Naegleria*, *Eur. J. Biochem.*, 2000, **267**, 7257–7266.
18. A. Vader, H. Nielsen and S. Johansen, In vivo expression of the nucleolar group I intron-encoded I-*Dir*I homing endonuclease involves the removal of a spliceosomal intron, *EMBO J.*, 1999, **18**, 1003–1013.
19. H. Nielsen, T. Fiskaa, A.B. Birgisdottir, P. Haugen, C. Einvik and S. Johansen, The ability to form full-length intron RNA circles is a general property of nuclear group I introns, *RNA*, 2003, **9**, 1464–1475.
20. A. Vader, S. Johansen and H. Nielsen, The group I-like ribozyme DiGIR1 mediates alternative processing of pre-rRNA transcripts in *Didymium iridis*, *Eur. J. Biochem.*, 2002, **269**, 5804–5812.
21. S.D. Johansen, P. Haugen and H. Nielsen, Expression of protein-coding genes embedded in ribosomal DNA, *Biol. Chem.*, 2007, **388**, 679–686.
22. W.A. Decatur, S. Johansen and V.M. Vogt, Expression of the *Naegleria* intron endonuclease is dependent on a functional group I self-cleaving ribozyme, *RNA*, 2000, **6**, 616–627.
23. S.L. Brehm and T.R. Cech, Fate of an intervening sequence ribonucleic acid: Excision and cyclization of the *Tetrahymena* ribosomal ribonucleic acid intervening sequence in vivo, *Biochemistry*, 1983, **22**, 2390–2397.
24. H. Nielsen, C. Einvik, T.E. Lentz, M.M. Hedegaard and S.D. Johansen, A conformational switch in the DiGIR1 ribozyme involved in release and folding of the downstream I-*Dir*I mRNA, 2007, in preparation.
25. E. Jabri, S. Aigner and T.R. Cech, Kinetic and secondary structure analysis of *Naegleria andersoni* GIR1, a group I ribozyme whose putative

biological function is site-specific hydrolysis, *Biochemistry*, 1997, **36**, 16345–16354.
26. M.J. Fedor, Structure and function of the hairpin ribozyme, *J. Mol. Biol.*, 2000, **297**, 269–291.
27. D.A. Lafontaine, T.J. Wilson, D.G. Norman and D.M. Lilley, The A730 loop is an important component of the active site of the VS ribozyme, *J. Mol. Biol.*, 2001, **312**, 663–674.
28. B. Clouet-d'Orval and O.C. Uhlenbeck, Hammerhead ribozymes with a faster cleavage rate, *Biochemistry*, 1997, **36**, 9087–9092.
29. P. Agbäck, A. Sandstrom, S. Yamakage, C. Sund, C. Glemarec and J. Chattopadhyaya, Solution structure of lariat RNA by 500 MHz NMR spectroscopy and molecular dynamics studies in water, *J. Biochem. Biophys. Methods*, 1993, **27**, 229–259.
30. E. Jabri and T.R. Cech, In vitro selection of the *Naegleria* GIR1 ribozyme identifies three base changes that dramatically improve activity, *RNA*, 1998, **4**, 1481–1492.
31. C. Einvik, H. Nielsen, R. Nour and S. Johansen, Flanking sequences with an essential role in hydrolysis of a self-cleaving group I-like ribozyme, *Nucleic Acids Res.*, 2000, **28**, 2194–2200.
32. P.L. Adams, M.R. Stahley, A.B. Kosek, J. Wang and S.A. Strobel, Crystal structure of a self-splicing group I intron with both exons, *Nature*, 2004, **430**, 45–50.
33. P.L. Adams, M.R. Stahley, M.L. Gill, A.B. Kosek, J. Wang and S.A. Strobel, Crystal structure of a group I intron splicing intermediate, *RNA*, 2004, **10**, 1867–1887.
34. B. Beckert, H. Nielsen, C. Einvik, S.D. Johansen, E. Westhof and B. Masquida, Structure and evolution of the GIR1 branching ribozyme, in preparation.
35. F. Guo, A.R. Gooding and T.R. Cech, Structure of the *Tetrahymena* ribozyme: Base triple sandwich and metal ion at the active site, *Mol. Cell*, 2004, **16**, 351–362.
36. B.L. Golden, H. Kim and E. Chase, Crystal structure of a phage *Twort* group I ribozyme-product complex, *Nat. Struct. Mol. Biol.*, 2005, **12**, 82–89.
37. A. Lescoute and E. Westhof, Topology of three-way junctions in folded RNAs, *RNA*, 2006, **12**, 83–93.
38. P.P. Zarrinkar and J.R. Williamson, Kinetic intermediates in RNA folding, *Science*, 1994, **265**, 918–924.
39. W.D. Downs and T.R. Cech, Kinetic pathway for folding of the *Tetrahymena* ribozyme revealed by three UV-inducible crosslinks, *RNA*, 1996, **2**, 718–732.
40. M.R. Stahley and S.A. Strobel, Structural evidence for a two-metal-ion mechanism of group I intron splicing, *Science*, 2005, **309**, 1587–1590.
41. D. Bhattacharya, V. Reeb, D.M. Simon and F. Lutzoni, Phylogenetic analyses suggest reverse splicing spread of group I introns in fungal ribosomal DNA, *BMC. Evol. Biol.*, 2005, **5**, 68.

42. P. Rangan, B. Masquida, E. Westhof and S.A. Woodson, Architecture and folding mechanism of the *Azoarcus* Group I Pre-tRNA, *J. Mol. Biol.*, 2004, **339**, 41–51.
43. T.R. Cech and B.L. Golden, Building a catalytic active site using only RNA, in: *The RNA World*, ed. R.F. Gesteland, T.R. Cech and J.F. Atkins, CSHL Press, Cold Spring, Harbor New York, 1999, pp. 321–349.
44. H. Nielsen and S.D. Johansen, A new RNA branching activity: The GIR1 ribozyme, *Blood Cells Mol. Dis.*, 2007, **38**, 102–109.
45. K. Salehi-Ashtiani, A. Luptak, A. Litovchick and J.W. Szostak, A genome-wide search for ribozymes reveals an HDV-like sequence in the human CPEB3 gene, *Science*, 2006, **313**, 1788–1792.
46. J.F. De Jonckheere, Evidence for the ancestral origin of group I introns in the SSUrDNA of *Naegleria* spp, *J. Eukaryot. Microbiol.*, 1994, **41**, 457–463.
47. J.F. De Jonckheere and S. Brown, Loss of the ORF in the SSUrDNA group I intron of one *Naegleria* lineage, *Nucleic Acids Res.*, 1994, **22** 3925–3927.
48. I. Dykova, I. Kyselova, H. Peckova, M. Obornik and J. Lukes, Identity of *Naegleria* strains isolated from organs of freshwater fishes, *Dis. Aquat. Organ*, 2001, **46**, 115–121.
49. I. Dykova, H. Peckova, I. Fiala and H. Dvorakova, Fish-isolated *Naegleria* strains and their phylogeny inferred from ITS and SSU rDNA sequences, *Folia Parasitol. (Praha)*, 2006, **53**, 172–180.
50. J.F. De Jonckheere and S. Brown, Three different group I introns in the nuclear large subunit ribosomal DNA of the amoeboflagellate *Naegleria*, *Nucleic Acids Res.*, 1998, **26**, 456–461.

CHAPTER 13
Is the Spliceosome a Ribozyme?

DIPALI G. SASHITAL AND SAMUEL E. BUTCHER

Department of Biochemistry, University of Wisconsin-Madison, 433 Babcock Drive, Madison, Wisconsin 53706, USA

13.1 Introduction

Precursor messenger RNA (pre-mRNA) splicing is the process by which introns are removed and exons are spliced together to form a mature mRNA transcript, and is an essential RNA processing step required for eukaryotic gene expression. Pre-mRNA splicing is catalyzed by the spliceosome, a massive megadalton RNA–protein complex assembled from five small nuclear ribonucleoprotein particles, or snRNPs.[1,2] Each snRNP contains one of the five small nuclear RNAs (snRNAs, denoted U1, U2, U4, U5 and U6) and many proteins. The total number of proteins in the spliceosome ranges from approximately 70 in *Saccharomyces cerevisiae* to over 300 in human.[1,3,4] Despite this abundance of proteins, a growing body of data suggests that the spliceosome is a ribozyme. Specifically, a complex that forms between U2 and U6 snRNAs is thought to be the RNA component responsible for catalysis. Owing to the complexity of the spliceosome, the ultimate determination of whether the spliceosome is a ribozyme will likely require high-resolution crystal structures, as was the case for the ribosome.[5,6] Despite the absence of an atomic level "picture" of the spliceosomal active site, the cumulative biochemical and genetic data clearly indicate that the spliceosome is controlled by an RNA center. In this chapter, we review the evidence that suggests the spliceosome is a ribozyme.

13.2 Similarity to Group II Self-splicing Introns

The spliceosome as a ribozyme hypothesis was first proposed over 20 years ago, based on the requirement for spliceosomal RNA and the mechanistic similarity to group II self-splicing introns.[7–9] The chemistry catalyzed by the spliceosome is identical to that of group II intron ribozymes (Chapter 11), involving two transesterification reactions within the pre-mRNA substrate (Figure 13.1). In

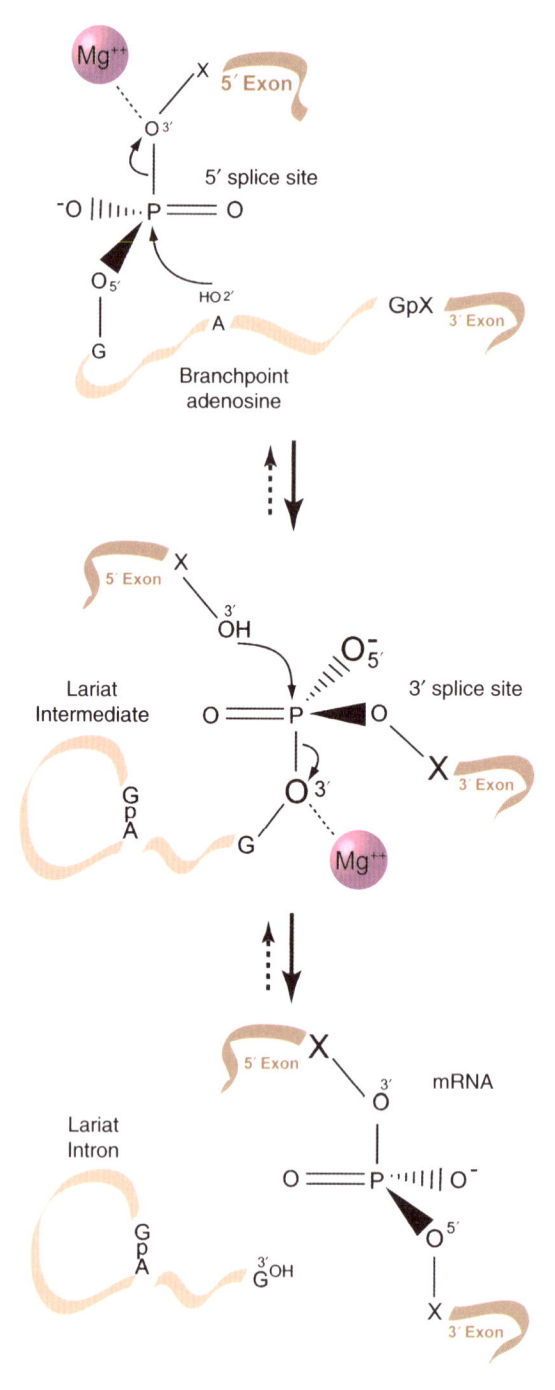

Figure 13.1 Splicing catalysis via two transesterification reactions. The 2′ oxygen of the branchpoint adenosine attacks at the 5′ splice site, freeing the 3′-OH of the 5′ exon for attack at the 3′ exon. The exons are ligated, forming mature mRNA, and the lariat intron is released. The 3′ oxygen leaving groups are stabilized by Mg^{2+} ions.

the first step, a 2′-OH of the "branchpoint" adenosine within the intron attacks at the 5′ splice site, cleaving the phosphodiester backbone and creating a 2′-5′ branched "lariat" intermediate. Cleavage at the 5′ end of the intron liberates the 3′-OH at the end of the 5′ exon, which then attacks the 3′ splice site in the second step, resulting in ligation of the exons and release of the lariat intron.

The splicing reactions catalyzed by group II self-splicing introns and the spliceosome are stereochemically identical.[10,11] Both the first and second steps of splicing result in complete inversion of stereochemistry (Figure 13.1), as is typical for an in-line S_N2 displacement reaction mechanism.[10] The fact that both steps of splicing are inhibited by R_P phosphorothioate strongly suggests that the two steps of splicing are catalyzed by two different active sites. If there were a single active site with inversion of phosphate stereochemistry, one would expect the two steps to be inhibited by different diastereomers (e.g., in a single active site, R_P inhibition of step 1 followed by inversion of stereochemistry should result in S_P inhibition of step 2). For both pre-mRNA splicing and group II intron self-splicing, the first active site catalyzes 5′ splice site cleavage, after which the active site must rearrange to allow the juxtaposition of the 5′ and 3′ exons for exon ligation in the second step. Given the different reactions catalyzed in the first and second steps of splicing, it is not surprising that more than one active site should be required. Although the data indicate that the two active sites are stereochemically distinct, they may be spatially very close or even share common functional groups.[10]

Additionally, the splicing reactions catalyzed by the spliceosome and group II self-splicing introns have highly analogous metal ion requirements. For both systems, a metal ion coordinates to the 3′ oxygen leaving groups during both the first and second steps of splicing (Figure 13.1).[12–15] This is a common catalytic strategy, in which metal ion coordination to the 3′ oxyanion leaving group shields the charge build-up in the transition state and thereby lowers the activation energy of the reaction.[16] Sulfur substitutions at the 3′ oxygen block splicing, but can be rescued by Mn^{2+}, which coordinates to sulfur more readily than Mg^{2+}.[14,15] For both the spliceosome and group II self-splicing introns, sulfur substitution at the 3′ splice site only blocks exon ligation when the 3′ splice site substrate is added in trans after completion of the first step, suggesting that in cis splicing a conformational change between the two steps of splicing is rate-limiting.[12,13]

The metal ion coordinated to the 3′ oxyanion leaving group is analogous to one out of the two metal ions involved in a general two-metal-ion catalytic mechanism.[16] Interestingly, the two-metal-ion catalytic mechanism has been observed by X-ray crystallography for a group I self-splicing intron.[17] Therefore, it is possible that the group II self-splicing introns and the spliceosome, as well as group I self-splicing introns, share similar mechanisms for metal ion catalysis.[18]

13.3 Role of snRNA in the Spliceosome Active Site

Although its catalytic mechanism is relatively simple, the spliceosome is one of the largest and most complex molecular machines in the cell, in part because splicing is a highly regulated process. Pre-mRNA introns can vary greatly in size, with the branchpoint adenosine typically located 18 to 37 nucleotides upstream of the 3′ splice site, and the 5′ splice site located anywhere from tens to hundreds of thousands of bases upstream of the other reactive regions, depending on the length of the intron.[19] The spliceosome identifies and juxtaposes these distant regions by directly binding the 5′ splice site and branchpoint sequence, primarily through base-pairing interactions with U1 and U2 snRNAs, respectively. U1 is subsequently replaced by U6 snRNA at the 5′ splice site, and both U1 and U4 snRNPs are released from the complex before formation of the catalytically active spliceosome.[1,2] As evidenced by these early events in the splicing cycle, RNA–RNA interactions are critical in substrate recognition and coordination.

The major consideration in the spliceosome as a ribozyme hypothesis is the composition of its active site. Though high-resolution structures of the catalytically active spliceosome have yet to be determined, a great deal of genetic and biochemical information has revealed a complex RNA network at the spliceosomal catalytic core. Substrate binding within this RNA framework is largely achieved by a complex of U2 and U6 snRNAs, which brings together the reactive regions of the pre-mRNA for the first catalytic step (Figures 13.2 and 13.3A). The U2-branchpoint sequence helix that forms during substrate recognition serves to position the 2′-OH of the branchpoint adenosine for nucleophilic attack,[20–24] while the 5′ splice site is coordinated through a base-pairing interaction between U6 and nucleotides 4–6 of the intron[25–27] (Figures 13.2 and 13.3A). Two components of the U5 snRNP are also important in positioning substrates within the active site. The 5′ splice site is partially tethered to the active site by an interaction between the 5′ exon and the invariant nine nucleotide loop 1 in U5 snRNA (Figure 13.3A).[26,28–32] Additionally, the U5 snRNP

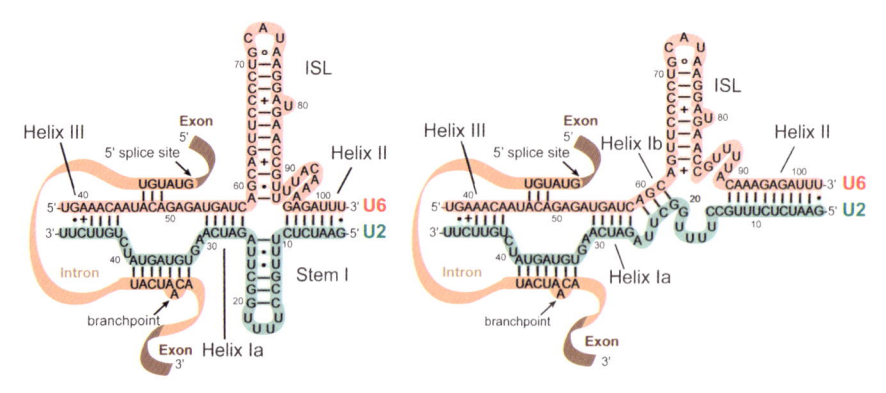

Figure 13.2 U2-U6 complex from *S. cerevisiae* bound to pre-mRNA substrate. In the four-helix junction conformation (left), U2 stem I is present and the U6 ISL is extended to include the AGC triad (U6:59–61). The AGC triad base-pairs with U2 residues in helix Ib, contained in the three-helix junction conformation (right). The ACAGAGA sequence (U6:47–53) and the branchpoint recognition sequence (U2:32–38) help coordinate the substrates in the active site.

Is the Spliceosome a Ribozyme?

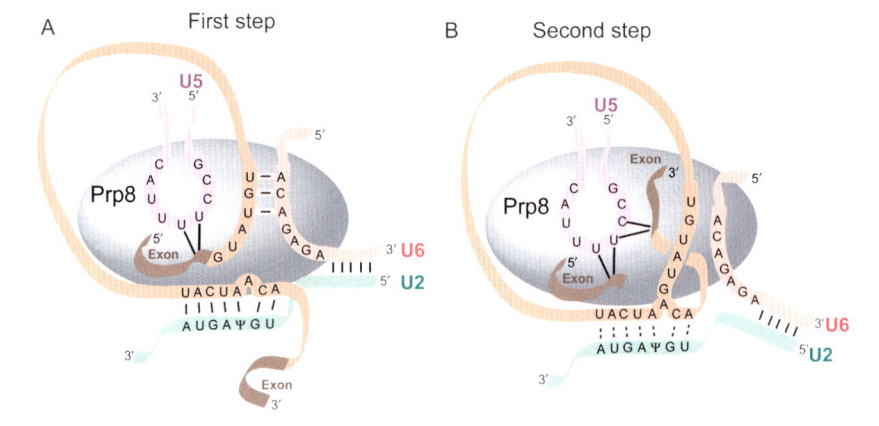

Figure 13.3 Interactions in the substrate binding center. (*A*) In the first step active site, the 5′ splice site is coordinated by the U6 ACAGAGA sequence and U5 loop 1, interactions that are stabilized by Prp8. The branchpoint adenosine is unpaired within a helix formed between the surrounding sequence and U2. (*B*) In the second step active site, the lariat intermediate may be displaced from the catalytic center, and both the 5′ and 3′ exons are coordinated by U5 loop 1 and Prp8.

component Prp8 stabilizes both the U5 and U6 interactions with the 5′ splice site (Figure 13.3A).[33,34] Of all the spliceosomal protein factors, Prp8 is the most intimately associated with the active site, involved in securing many of the substrate–snRNA interactions.[8,35]

Prp8 assists in positioning both substrates during exon ligation (Figure 13.3B),[36] although little is known about specific RNA–RNA and RNA–protein interactions in the second step active site. The U6-5′ splice site helix is disrupted, likely in favor of an interaction between the U6 ACAGAGA sequence and the 5′ and 3′ exons (Figure 13.3B).[37,38] It is unknown whether the lariat intermediate is displaced completely from the catalytic center, or if the U2-branchpoint sequence interaction remains intact. The interaction between U5 loop 1 and the 5′ exon that is present during the first step persists in the second step conformation, and loop 1 also forms an essential interaction with the 3′ exon (Figure 13.3B).[29,36,39,40]

Congruent with its essential role in substrate coordination in the spliceosome active site, a growing body of evidence indicates that the U2-U6 complex is the RNA component responsible for catalysis. The importance of U2 and U6 snRNAs is apparent from phylogenetic investigation of their sequences. The regions of U2 and U6 that form the core of the active site range between 80–85% sequence identity from yeast to mammals,[41] and U6 is the most highly conserved spliceosomal RNA.[42] Two regions of U6 are both invariant and essential for splicing function: the ACAGAGA sequence, which positions the 5′ splice site during the first step of splicing, and the AGC triad.[43] Several point mutations and backbone atomic substitutions within these regions block splicing.[44–48]

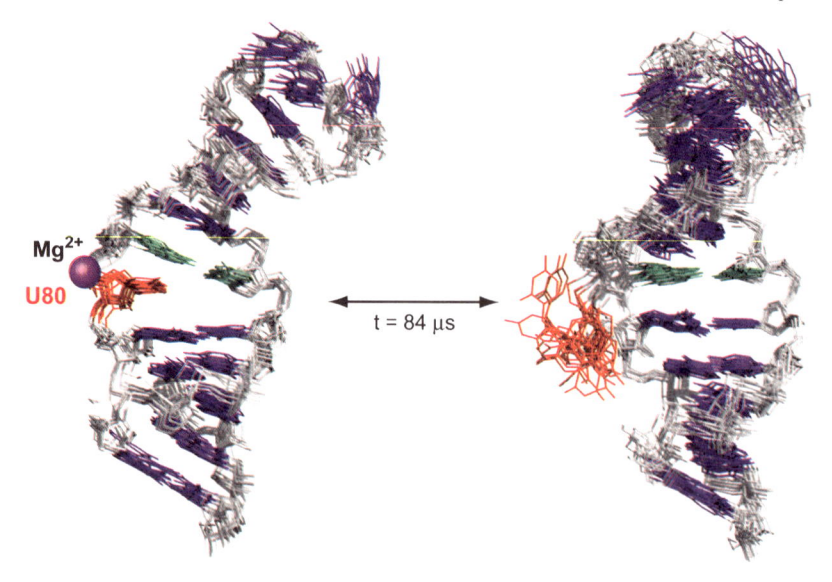

Figure 13.4 Base-flipping of U80 (red) in the U6 ISL internal loop occurs upon protonation of the adjacent C-A wobble (green) on an 84-μs time-scale. Metal ion binds at the *pro*-S$_P$ phosphate oxygen (magenta sphere) of U80 when the base is stacked in the helix.

Indeed, many phosphate oxygens along the U6 backbone are required, suggesting potential metal-binding locations.[47–49] Among these backbone locations, the most intriguing is at U80 within the intramolecular stem-loop (ISL) (Figures 13.2 and 13.4).[47,49] Splicing is blocked by S$_P$ phosphorothioate substitutions at U80 (Figure 13.4), but can be rescued by addition of the thiophilic ion Cd^{2+}, or Mn^{2+}.[49] The fact that the metallochemistry of a megadalton complex can be completely altered by substitution of a single oxygen atom with sulfur is striking, and interesting because the U80 site in U6 resides within the highly conserved U6 ISL helix. Structural and metal-binding studies of the isolated U6 ISL RNA indicate that U80 is unpaired within an internal loop and stereospecifically binds divalent metal, even in the absence of protein, which is consistent with the spliceosome as a ribozyme hypothesis.[50–53] The U80 metal binding site in the U6 ISL helix is relatively distant from the substrate binding region of the U2-U6 complex, suggesting that the ISL may fold down onto the ACAGAGA sequence, bringing the metal ion into position for a role in splicing catalysis. This hypothesis is supported by hydroxyl radical footprinting of catalytically active spliceosomes, in which the ISL is revealed to be in close proximity to the intron during the first step of splicing.[54] A U80G mutation in the internal loop of the U6 ISL is lethal, further implicating it in splicing catalysis.[45,55] Similar mutational sensitivity has been observed for the corresponding residue, U74, in human.[56] However, it is still possible that the U80 *pro*-S$_P$ phosphate oxygen site coordinates a structurally essential metal ion rather than an active site metal ion.

Metal ion binding at the U80 site appears to be regulated by the conformation of the U6 ISL internal loop, which undergoes a structural rearrangement in response to the protonation state of the adjacent A79, which has a pK_a near neutral pH.[52,53] In response to protonation, the U80 base undergoes a base flipping motion on the 84 ± 10 ms timescale (Figure 13.4).[52,53] Of potential significance to the splicing mechanism, only the non-protonated conformation, with U80 stacked-in the helix, can bind metal ion (Figure 13.4). Thus, base flipping of U80 is a proton-driven conformational switch that modulates the binding of an essential metal ion. Since metal binding at U80 is required for splicing, the observed conformational change has the potential to regulate the pre-mRNA splicing mechanism.[52,53] In this regard, interestingly, the angle between the upper and lower stems changes by 25° in the two structures, resulting in a 9–12 Å movement of the helix. This suggests that a change in protonation state and metal-binding at the U6 ISL could result in large-scale movement of RNA domains in the spliceosome.

Perhaps the most intriguing evidence of ribozyme activity in the spliceosome is the discovery of a catalytically active minimal human U2-U6 complex that can assemble *in vitro* and bind an RNA oligonucleotide containing the branchpoint sequence.[57,58] The minimal U2-U6 complex slowly catalyzes the formation of an unusual phosphotriester linkage when the 2′-OH of the branchpoint adenosine attacks along the backbone of the U6 AGC triad, between A59 and G60 (human numbering). This chemistry is intriguing, as it is similar, but not identical, to the first step of splicing. Owing to the lack of a 5′ splice site substrate within the minimal construct, it would be impossible for the minimal U2-U6 complex to catalyze the first step reaction. However, it can still activate the branchpoint adenosine as a nucleophile, and the AGC triad is thought to be juxtaposed with the putative active site based on *in vitro* crosslinking studies of the minimal U2-U6 construct.[59] A similar tertiary interaction has also been detected *in vivo* by genetic suppression in yeast,[60] and supports a potential location for the AGC triad in the active site of the spliceosome. Both A53 and G54 in the triad contain important backbone oxygen atoms *in vivo*.[47,48]

The other unusual aspect of the chemistry catalyzed by the minimal U2-U6 complex is the resultant formation of a phosphotriester, rather than a phosphodiester, in the reaction product. Phosphotriester formation is unfavorable, as it requires the protonation of an extra non-bridging phosphate oxygen, presumably by an activated water molecule, which would likely occur quite slowly at neutral pH. This is one possible reason why the reaction kinetics are extremely slow ($k_{obs} = 0.002$ min^{-1}).[57] Another explanation for the slow reaction kinetics could be that the minimal construct only rarely adopts the catalytic conformation. Significantly, the efficiency of the reaction increases when a highly conserved pseudouridine within the U2 branch site helix is incorporated into the minimal construct.[58] The conserved pseudouridine has been implicated in helping to position the branchpoint adenosine for nucleophilic attack.[24] Despite the differences between the minimal U2-U6 reaction and canonical splicing, the revelation that U2 and U6 snRNAs can base-pair, retain structural features found in the spliceosome active site, bind substrate, and catalyze a splicing-like reaction greatly strengthens the argument that the spliceosome is a ribozyme.

13.4 Conformation of the U2-U6 Complex and Parallels to Group II Intron Structures

Because of the importance of the U2-U6 complex in splicing catalysis, the conformation of the complex has been the target of extensive investigation using crosslinking, genetic suppression, and structural studies for both the *S. cerevisiae* and human sequences. The U2-U6 interaction is established toward the end of spliceosome assembly, following the disruption of a base-pairing interaction between U6 and U4 snRNA that is present during early assembly stages (Figure 13.5). The intermolecular stems present in the U4-U6 complex are mutually exclusive with the U6 ISL and U2-U6 helix Ia, both of which form in the U2-U6 complex (Figure 13.5).[44,56,61,62] The exact timing of U2-U6 association is unknown: U6 may be completely unwound from U4 prior to U2-U6 formation;[63] or U2 and U6 base-pairing may be concomitant with U4-U6 unwinding.[64]

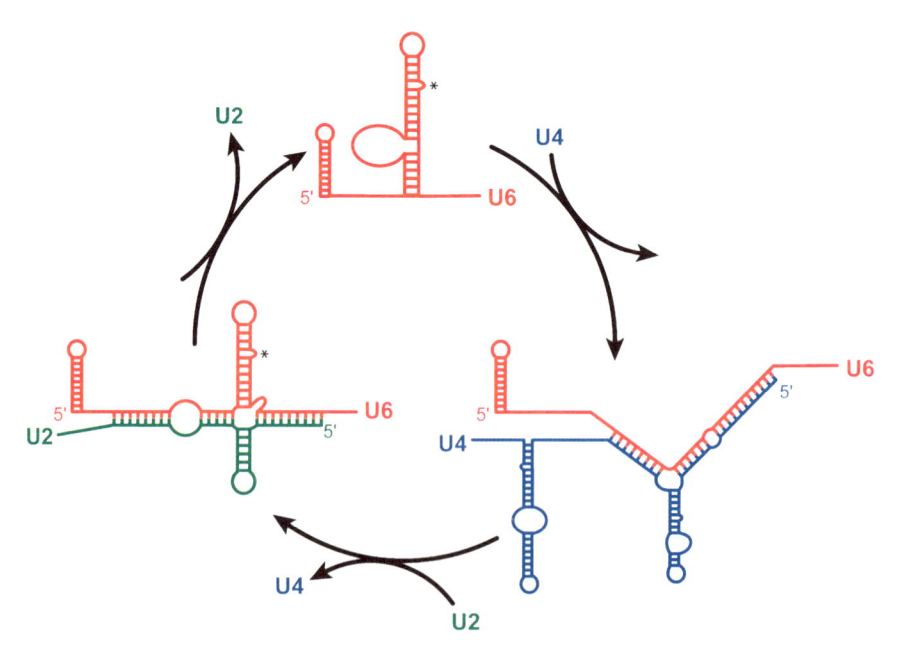

Figure 13.5 Conformations of U6 snRNA. Free U6 must form an extensive base-pairing interaction with U4 prior to incorporation into the spliceosome. The U4-U6 complex is mutually exclusive with the U6 ISL and helix I of the U2-U6 complex, and therefore must be unwound before activation of the spliceosome. An asterisk marks the U80 metal ion binding site.

In yeast, the AGC triad base-pairs with U2 to form helix Ib, which is essential for the second step of splicing (Figure 13.2).[44,65] This interaction has not been detected in human U2-U6; instead the last two residues of the AGC triad may form base-pairs within U6, extending the U6 ISL.[61] In addition, the U2 residues of helix Ib participate in an important intramolecular helix (U2 stem I) in both human and yeast (Figure 13.2).[61,66] Finally, two other intermolecular helices, helices II and III, are essential in human U2-U6,[61,67–69] but not in yeast.[70,71]

The folding of the yeast U2-U6 complex has also been investigated by NMR.[72] Interestingly, the secondary structure revealed through these studies differs from the helix Ib-containing three-helix junction detected by genetic studies in yeast. NMR detection of hydrogen bonds within the U2-U6 complex revealed that it adopts a four-helix junction fold, in which the U6 ISL is extended to include the AGC triad (Figure 13.2). Additionally, U2 stem I is observed within this four helix junction conformation (Figure 13.2). The observation that the intrinsic folding of the U2-U6 complex includes U2 stem I provides further evidence that this stem-loop plays an important role within the spliceosome active site. The extended U6 ISL and U2 stem I are both mutually exclusive with helix Ib, suggesting that the four-helix junction conformation may be present prior to the second step, during which helix Ib is essential.[44,65,72] Since a conformational change within the active site is required to transition between the first and second steps of splicing,[10,15] it is possible that the transformation from the four-helix to three-helix junction may be part of this structural rearrangement. The four-helix junction conformation is predicted to be slightly more stable than the three-helix junction conformation, such that relatively little free energy would be required to transition between the two secondary structures.[73]

The extension of the U6 ISL in the four-helix junction draws interesting parallels between the ISL and domain 5 of group II introns (Figure 13.6) (Chapter 11). Like the U6 ISL, domain 5 is a metal ion-binding stem-loop that is structurally and catalytically important, and the two stem-loops have long been thought to be evolutionarily related.[74] Both stem-loops contain an internal loop that binds a metal ion,[49,75] and inclusion of the AGC triad at the base of the U6 ISL extends the helix so that the two internal loops are located at the same central location within the stem-loop (Figure 13.6A). The structures of the U6 ISL and domain 5[76–78] can be superimposed with a backbone r.m.s.d. of 2 Å (Figure 13.6B). The base-pairs formed with the AGC triad are identical in the U6 ISL and domain 5, including a G-U wobble at the center of the triad. These similarities suggest that the AGC triad may have a common function in both the spliceosome and group II introns. One potential shared function could be in metal ion-binding, as the domain-5 AGC triad interacts with a metal ion that is likely essential for RNA folding and packing at the junction of several helices in the intron,[79] and phosphate oxygens in the same region of U6 are also required for splicing catalysis.[47,48] Intriguingly, domain 5 and the U6 ISL are functionally interchangeable in at least one case. Domain 5 can functionally replace the U6 ISL in a minor form of the spliceosome that splices a rare class of introns and contains an alternate form of U6 (known as U6atac).[80] This

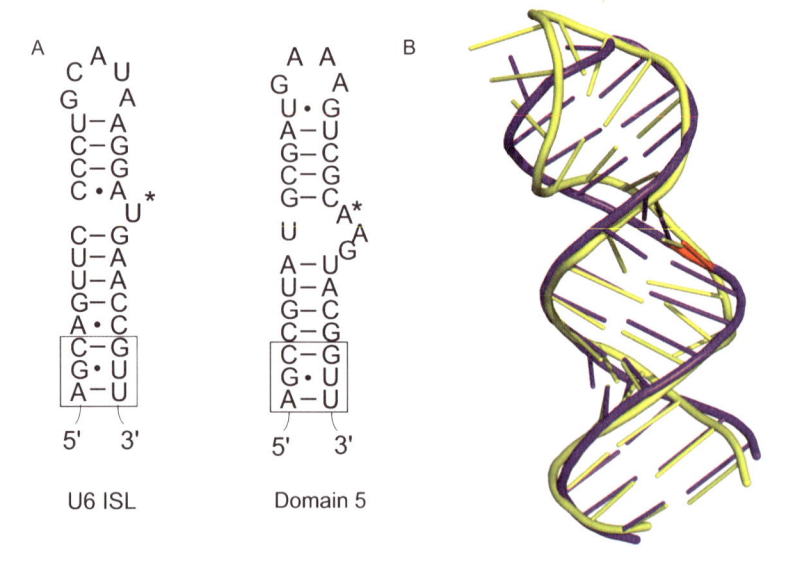

Figure 13.6 The U6 ISL and domain 5 are structurally and functionally related. (*A*) The secondary structure of extended yeast U6 ISL is strikingly similar to domain 5, including conserved base-pairing at the base of the helix. Metal ion binding occurs at internal loops (denoted by asterisks) in both stem-loops. (*B*) Overlay of extended U6 ISL (magenta) and domain 5 (yellow) structures. U80 is highlighted in red in the ISL structure.

observation strengthens the hypothesis that the U6 ISL and domain 5 of group II introns are functionally analogous structures.

Another shared structural feature between group II introns and the U2-U6 complex is the branchpoint sequence helix. In group II introns, the branchpoint adenosine is located in a stem-loop called domain 6,[81] a structure that is analogous to the U2-branchpoint sequence helix in the spliceosome active site. Both domain 6 and the U2-branchpoint helix serve to position the branchpoint adenosine for its role as the nucleophile during lariat formation. In the isolated spliceosomal branchpoint helix, the branchpoint adenosine is unpaired and bulged out of the helix.[24] The extrusion of the branchpoint adenosine is facilitated by the presence of a pseudouridine on the opposite side of the helix.[24] This pseudouridine is not present in domain 6, and NMR studies of the stem-loop indicate that the branchpoint adenosine is predominately stacked in the helix.[82] Nevertheless, domain 6 and the U2-branchpoint helix obviously serve similar functional roles in their respective systems, and represent another striking structural similarity between group II introns and the spliceosome.

13.5 RNA-mediated Regulation in the Spliceosome

Given the central role of pre-mRNA splicing in gene expression, the spliceosome must be precisely regulated during all stages of assembly and catalysis.

This regulation is achieved through a highly dynamic network of RNA–RNA, RNA–protein, and protein–protein interactions. Structural transitions within the RNA complexes are often required to proceed to the next step of assembly or catalysis, and are generally mediated by several "RNA chaperones",[83,84] including eight DExD/H-box RNA-dependent ATPases in yeast.[1,85] The spliceosomal snRNAs contain several regions that participate in more than one RNA–RNA interaction, creating competition between different structural elements. This competition indicates that the snRNAs are active participants not only in splicing catalysis, but also in the molecular switches that trigger spliceosome activity.

One example of competing RNA structures playing a role in regulation of spliceosome assembly is the sequestering of U6 snRNA in the catalytically inactive U4-U6 complex during spliceosome assembly (Figure 13.7). Surprisingly, the dissociation of human U4 from U6 can occur without external assistance from any protein, despite the extensive (25 base-pair) interaction between the two RNAs.[86] An alternate, extended U6 intramolecular helix, termed the telestem, competes with the U4-U6 interaction (Figure 13.7), providing enough folding free energy to promote the disassembly of the complex at physiologically relevant temperatures – presumably through a step-wise concerted mechanism similar to branch migration of DNA.[86] Although several proteins have been found to be involved in the disruption of the U4-U6 complex *in vivo* (including a helicase, Brr2, in yeast[87]), the intrinsic dissociation ability of human U4 and U6 RNA indicates that these spliceosomal snRNAs have some capacity to self-regulate.

Competing structures within the spliceosome active site may play an important role in regulating splicing activity, an idea described by a model in which the two conformations of the spliceosome active site required for each catalytic step are in kinetic competition with one another.[88] Destabilization of one conformation drives formation of the other, thereby promoting one catalytic step over the other.[88] Kinetic competition between the two active sites has been observed using mutants that destabilize specific RNA–RNA or RNA–protein

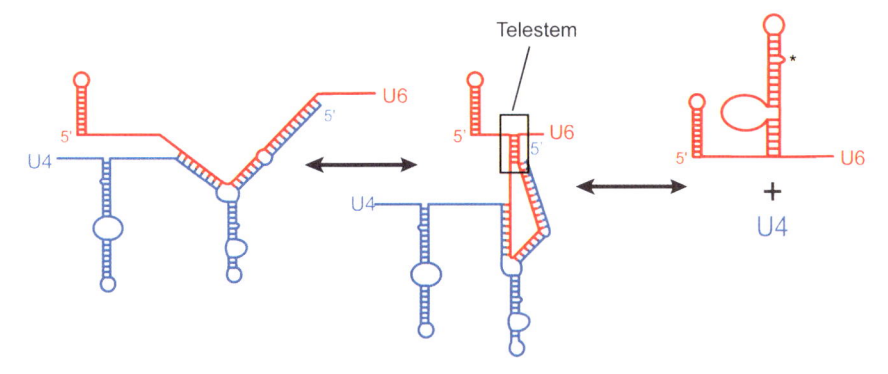

Figure 13.7 Intrinsic dissociation of U4 and U6 snRNAs. The U6 telestem assists in disrupting the U4–U6 interaction.

interactions within one conformation of the spliceosome active site.[37,89] For example, substitution of the branchpoint adenosine with guanosine normally blocks lariat formation; however, Prp8 mutations that destabilize the second step active site can rescue first step activity with the mutant substrate.[89] This observed kinetic competition suggests that competing RNA structures may play a role in regulating the transition from the first catalytic step to the second.

One potential regulatory RNA element in the spliceosome active site is the intramolecular U2 stem I, a structure that is important for efficient splicing.[61,66] U2 stem I is mutually exclusive with helix Ib and helix II (Figure 13.2), and therefore may play a role in regulating formation of helix Ib prior to the second step of splicing, a hypothesis that is supported by mutational evidence. Hyperstabilization of the internal loop at the center of stem I impairs splicing.[90] Evidence that U2 stem I and helix Ib are competing structures is provided by the observation that mutations that disrupt helix Ib inhibit the second step of splicing, but can be rescued by destabilization of U2 stem I.[65] These observations suggest that the relative stabilities of these RNA structures are precisely balanced in a manner that regulates splicing catalysis, which is consistent with the kinetic competition model. Interestingly, U2 stem I is significantly more stable in human than in yeast, owing to the presence of several post-translational modifications in the human stem-loop (Figure 13.8).[91] This divergence suggests that the importance of the stem I structure may vary between the two species. Indeed, stem I is thought to be completely unwound in the yeast spliceosome during exon ligation, when helix Ib is a required structural element of the U2-U6 framework (Figure 13.2).[44,65] In contrast, no

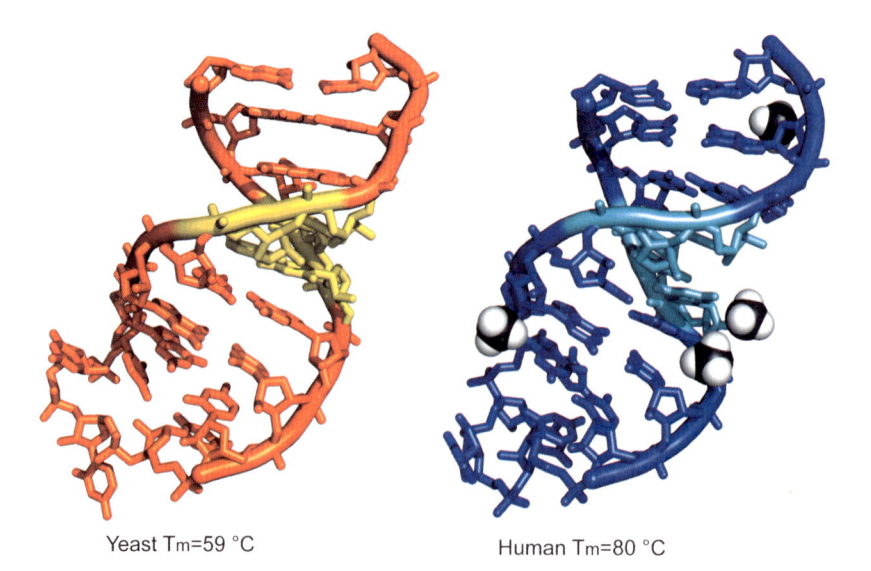

Yeast Tm=59 °C   Human Tm=80 °C

Figure 13.8 Yeast (left) and human (right) U2 stem I display disparate thermal stabilities, due to the presence of post-transcriptional modifications such as 2′-O-methyl (spheres) in human U2 snRNA. Non-Watson–Crick pairs are shown in yellow (yeast) and light blue (human).

Is the Spliceosome a Ribozyme? 265

evidence exists that helix Ib is required in humans.[61] This observation, combined with the high thermal stability of human U2 stem I,[91] suggests that U2 stem I never fully unwinds in the human spliceosome.

Another important structural rearrangement within U2 snRNA occurs in the stem II region just downstream of the branchpoint sequence (Figure 13.9).[92,93] Two competing structures, stem IIa and IIc, can form within this region, and each promotes different stages of the splicing cycle. Stem IIa is required for

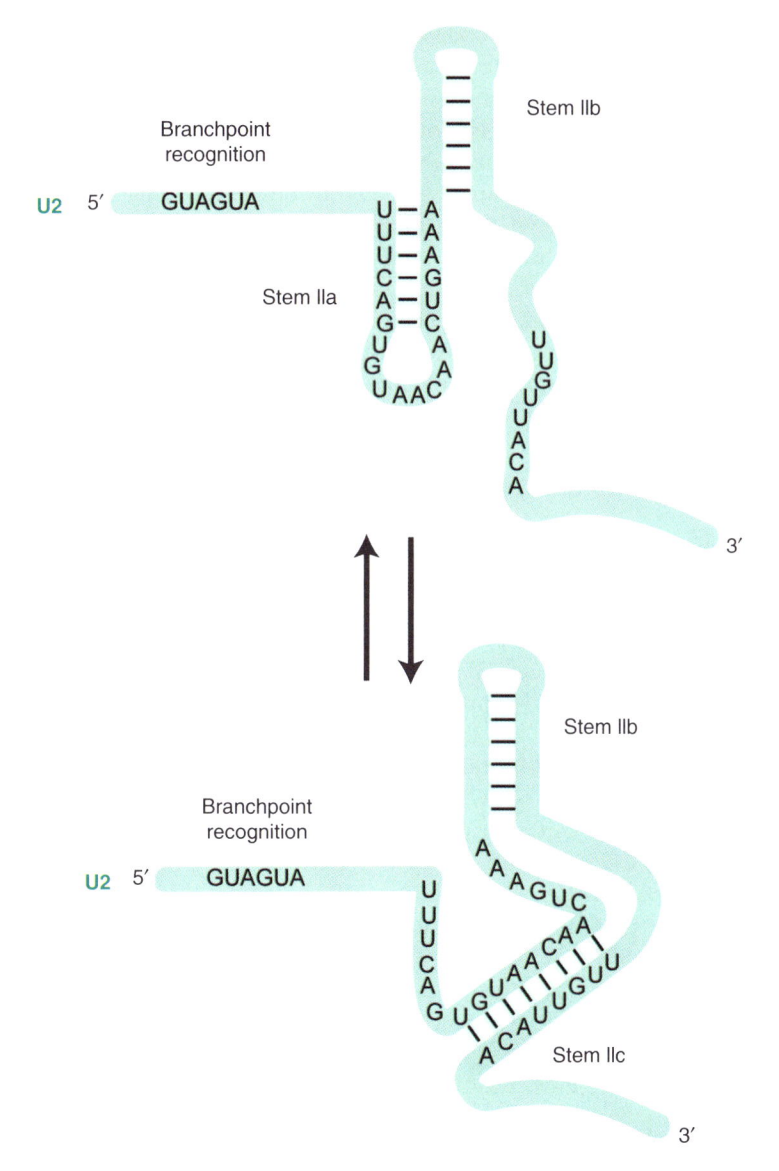

Figure 13.9 Stem II region of U2 snRNA contains competing stems IIa and IIc, which are important at different stages of the splicing cycle.

association of U2 snRNP with the pre-mRNA,[94,95] and is also present in an intermediate conformation between the two catalytic steps.[92,93] During both catalytic steps, stem IIa is unwound in favor of stem IIc.[92,93] This transition between stems IIa and IIc suggests that stem IIa is present in an "open" conformation that promotes substrate accommodation, while the "closed" stem IIc conformation promotes catalysis. Intriguingly, this switch between open and closed conformations is important for splicing fidelity: destabilization of the closed conformation heightens the requirements for substrate specificity, while destabilization of the open conformation relaxes the same requirements.[92,93] Thus, adopting the non-catalytic open conformation between the two catalytic steps may serve as a kinetic trap for spliceosomes bound to incorrect substrates, as the open conformation is more stable than the closed conformation when sub-optimal substrates are bound.

In summary, the spliceosome is an extraordinarily complex particle that clearly uses RNA to regulate its assembly and position substrates during splicing catalysis. It remains to be seen whether it is a ribozyme, with an all-RNA active site, or an "RNPzyme," with a composite RNA–protein active site. Nevertheless, the available biochemical data indicate that the spliceosome utilizes a complex network of RNA–RNA interactions to achieve a remarkable level of allosteric control during splicing.

References

1. D.A. Brow, *Annu. Rev. Genet.*, 2002, **36**, 333–360.
2. C.L. Will and R. Luhrmann, in: *The RNA World*, eds. R.F. Gesteland, T.R. Cech and J.F. Atkins, Cold Spring Harbor Laboratory Press, Woodbury, New York, 2006, pp. 369–400.
3. T.W. Nilsen, *Bioessays*, 2003, **25**, 1147–1149.
4. M.S. Jurica and M.J. Moore, *Mol. Cell*, 2003, **12**, 5–14.
5. T.R. Cech, *Science*, 2000, **289**, 905–920.
6. T.A. Steitz and P.B. Moore, *Trends Biochem. Sci.*, 2003, **28**, 411–418.
7. T.R. Cech, *Cell*, 1986, **44**, 207–210.
8. C.A. Collins and C. Guthrie, *Nat. Struct. Biol.*, 2000, **7**, 850–854.
9. P.A. Sharp, *Cell*, 1985, **42**, 397–400.
10. M.J. Moore and P.A. Sharp, *Nature*, 1993, **365**, 364–368.
11. R.A. Padgett, M. Podar, S.C. Boulanger and P.S. Perlman, *Science*, 1994, **266**, 1685–1688.
12. P.M. Gordon, E.J. Sontheimer and J.A. Piccirilli, *RNA*, 2000, **6**, 199–205.
13. P.M. Gordon, E.J. Sontheimer and J.A. Piccirilli, *Biochemistry*, 2000, **39**, 12939–12952.
14. E.J. Sontheimer, P.M. Gordon and J.A. Piccirilli, *Genes Dev.*, 1999, **13**, 1729–1741.
15. E.J. Sontheimer, S. Sun and J.A. Piccirilli, *Nature*, 1997, **388**, 801–805.
16. T.A. Steitz and J.A. Steitz, *Proc. Natl. Acad. Sci. U.S.A.*, 1993, **90**, 6498–6502.

17. M.R. Stahley and S.A. Strobel, *Science*, 2005, **309**, 1587–1590.
18. M. Yarus, *FASEB J.*, 1993, **7**, 31–39.
19. M.R. Green, *Annu. Rev. Genet.*, 1986, **20**, 671–708.
20. R. Parker, P.G. Siliciano and C. Guthrie, *Cell*, 1987, **49**, 229–239.
21. Y. Zhuang and A.M. Weiner, *Genes Dev.*, 1989, **3**, 1545–1552.
22. J. Wu and J.L. Manley, *Genes Dev.*, 1989, **3**, 1553–1561.
23. C.C. Query, M.J. Moore and P.A. Sharp, *Genes Dev.*, 1994, **8**, 587–597.
24. M.I. Newby and N.L. Greenbaum, *Nat. Struct. Biol.*, 2002, **9**, 958–965.
25. C.F. Lesser and C. Guthrie, *Science*, 1993, **262**, 1982–1988.
26. E.J. Sontheimer and J.A. Steitz, *Science*, 1993, **262**, 1989–1996.
27. S. Kandels-Lewis and B. Seraphin, *Science*, 1993, **262**, 2035–2039.
28. A. Newman and C. Norman, *Cell*, 1991, **65**, 115–123.
29. A.J. Newman and C. Norman, *Cell*, 1992, **68**, 743–754.
30. J.R. Wyatt, E.J. Sontheimer and J.A. Steitz, *Genes Dev.*, 1992, **6**, 2542–2553.
31. J.J. Cortes, E.J. Sontheimer, S.D. Seiwert and J.A. Steitz, *EMBO J.*, 1993, **12**, 5181–5189.
32. A.J. Newman, S. Teigelkamp and J.D. Beggs, *RNA*, 1995, **1**, 968–980.
33. I. Dix, C.S. Russell, R.T. O'Keefe, A.J. Newman and J.D. Beggs, *RNA*, 1998, **4**, 1675–1686.
34. V.P. Vidal, L. Verdone, A.E. Mayes and J.D. Beggs, *RNA*, 1999, **5**, 1470–1481.
35. R.J. Grainger and J.D. Beggs, *RNA*, 2005, **11**, 533–557.
36. C.A. Collins and C. Guthrie, *Genes Dev.*, 1999, **13**, 1970–1982.
37. M.M. Konarska, J. Vilardell and C.C. Query, *Mol. Cell*, 2006, **21**, 543–553.
38. C.A. Collins and C. Guthrie, *RNA*, 2001, **7**, 1845–1854.
39. R.T. O'Keefe and A.J. Newman, *EMBO J.*, 1998, **17**, 565–574.
40. R.T. O'Keefe, C. Norman and A.J. Newman, *Cell*, 1996, **86**, 679–689.
41. C. Guthrie and B. Patterson, *Annu. Rev. Genet.*, 1988, **22**, 387–419.
42. D.A. Brow and C. Guthrie, *Nature*, 1988, **334**, 213–218.
43. P. Fabrizio and J. Abelson, *Science*, 1990, **250**, 404–409.
44. H.D. Madhani and C. Guthrie, *Cell*, 1992, **71**, 803–817.
45. D.S. McPheeters, *RNA*, 1996, **2**, 1110–1123.
46. T. Wolff, R. Menssen, J. Hammel and A. Bindereif, *Proc. Natl. Acad. Sci. U.S.A.*, 1994, **91**, 903–907.
47. P. Fabrizio and J. Abelson, *Nucleic Acids Res.*, 1992, **20**, 3659–3664.
48. Y.T. Yu, P.A. Maroney, E. Darzynkiewicz and T.W. Nilsen, *RNA*, 1995, **1**, 46–54.
49. S.L. Yean, G. Wuenschell, J. Termini and R.J. Lin, *Nature*, 2000, **408**, 881–884.
50. A. Huppler, L.J. Nikstad, A.M. Allmann, D.A. Brow and S.E. Butcher, *Nat. Struct. Biol.*, 2002, **9**, 431–435.
51. N.J. Reiter, L.J. Nikstad, A.M. Allmann, R.J. Johnson and S.E. Butcher, *RNA*, 2003, **9**, 533–542.
52. H. Blad, N.J. Reiter, F. Abildgaard, J.L. Markley and S.E. Butcher, *J. Mol. Biol.*, 2005, **353**, 540–555.

53. N.J. Reiter, H. Blad, F. Abildgaard and S.E. Butcher, *Biochemistry*, 2004, **43**, 13739–13747.
54. B.M. Rhode, K. Hartmuth, E. Westhof and R. Luhrmann, *EMBO J.*, 2006, **25**, 2475–2486.
55. D.E. Ryan and J. Abelson, *RNA*, 2002, **8**, 997–1010.
56. J.S. Sun and J.L. Manley, *RNA*, 1997, **3**, 514–526.
57. S. Valadkhan and J.L. Manley, *Nature*, 2001, **413**, 701–707.
58. S. Valadkhan and J.L. Manley, *RNA*, 2003, **9**, 892–904.
59. S. Valadkhan and J.L. Manley, *RNA*, 2000, **6**, 206–219.
60. H.D. Madhani and C. Guthrie, *Genes Dev.*, 1994, **8**, 1071–1086.
61. J.S. Sun and J.L. Manley, *Genes Dev.*, 1995, **9**, 843–854.
62. T. Wolff and A. Bindereif, *Genes Dev.*, 1993, **7**, 1377–1389.
63. R.M. Vidaver, D.M. Fortner, L.S. Loos-Austin and D.A. Brow, *Genetics*, 1999, **153**, 1205–1218.
64. M.J. Frilander and J.A. Steitz, *Mol. Cell*, 2001, **7**, 217–226.
65. A.K. Hilliker and J.P. Staley, *RNA*, 2004, **10**, 921–928.
66. D.S. McPheeters and J. Abelson, *Cell*, 1992, **71**, 819–831.
67. B. Datta and A.M. Weiner, *Nature*, 1991, **352**, 821–824.
68. T.P. Hausner, L.M. Giglio and A.M. Weiner, *Genes Dev.*, 1990, **4**, 2146–2156.
69. J.A. Wu and J.L. Manley, *Nature*, 1991, **352**, 818–821.
70. D.J. Field and J.D. Friesen, *Genes Dev.*, 1996, **10**, 489–501.
71. D. Yan and M. Ares, Jr., *Mol. Cell Biol.*, 1996, **16**, 818–828.
72. D.G. Sashital, G. Cornilescu, C.J. McManus, D.A. Brow and S.E. Butcher, *Nat. Struct. Mol. Biol.*, 2004, **11**, 1237–1242.
73. S. Cao and S.J. Chen, *J. Mol. Biol.*, 2006, **357**, 292–312.
74. A. Jacquier, *Trends Biochem. Sci.*, 1990, **15**, 351–354.
75. R.K. Sigel, A. Vaidya and A.M. Pyle, *Nat. Struct. Biol.*, 2000, **7**, 1111–1116.
76. M. Seetharaman, N.V. Eldho, R.A. Padgett and K.T. Dayie, *RNA*, 2006, **12**, 235–247.
77. R.K. Sigel *et al.*, *Nat. Struct. Mol. Biol.*, 2004, **11**, 187–192.
78. L. Zhang and J.A. Doudna, *Science*, 2002, **295**, 2084–2088.
79. P.M. Gordon and J.A. Piccirilli, *Nat. Struct. Biol.*, 2001, **8**, 893–898.
80. G.C. Shukla and R.A. Padgett, *Mol. Cell*, 2002, **9**, 1145–1150.
81. C. Schmelzer and R.J. Schweyen, *Cell*, 1986, **46**, 557–565.
82. M.C. Erat, O. Zerbe, T. Fox and R.K. Sigel, *ChemBioChem*, 2007, **8**, 306–314.
83. R. Schroeder, A. Barta and K. Semrad, *Nat. Rev. Mol. Cell Biol.*, 2004, **5**, 908–919.
84. D. Herschlag, *J. Biol. Chem.*, 1995, **270**, 20871–20874.
85. J.P. Staley and C. Guthrie, *Cell*, 1998, **92**, 315–326.
86. D.A. Brow and R.M. Vidaver, *RNA*, 1995, **1**, 122–131.
87. D.H. Kim and J.J. Rossi, *RNA*, 1999, **5**, 959–971.
88. M.M. Konarska and C.C. Query, *Genes Dev.*, 2005, **19**, 2255–2260.
89. C.C. Query and M.M. Konarska, *Mol. Cell*, 2004, **14**, 343–354.
90. J. Wu and J.L. Manley, *Mol. Cell Biol.*, 1992, **12**, 5464–5473.

91. D.G. Sashital, V. Venditti, C.G. Angers, G. Cornilescu and S.E. Butcher, *RNA*, 2007, **13**, 328–338.
92. A.K. Hilliker, M.A. Mefford and J.P. Staley, *Genes Dev.*, 2007, **21**, 821–834.
93. R.J. Perriman and M. Ares, Jr., *Genes Dev.*, 2007, **21**, 811–820.
94. M.I. Zavanelli and M. Ares, Jr., *Genes Dev.*, 1991, **5**, 2521–2533.
95. M.I. Zavanelli, J.S. Britton, A.H. Igel and M. Ares, Jr., *Mol. Cell Biol.*, 1994, **14**, 1689–1697.

CHAPTER 14
Peptidyl Transferase Mechanism: The Ribosome as a Ribozyme

MARINA V. RODNINA

Institute of Physical Biochemistry, University of Witten/Herdecke, Witten, Germany

14.1 Introduction: Historical Background

Most natural ribozymes catalyze phosphoryl transfer reactions that require the activation of a ribose hydroxyl group (the nucleolytic ribozymes; self-splicing introns; perhaps the spliceosome) or a water molecule (RNase P) for nucleophilic attack of a phosphodiester bond.[1] However, the most abundant natural ribozyme is the ribosome, a molecular machine that synthesizes proteins and the only natural ribozyme that has polymerase activity. The bacterial ribosome is composed of three RNA molecules (rRNA) and over 50 proteins; it has two subunits, the larger of which has a sedimentation coefficient of 50S in prokaryotes (the 50S subunit), and the smaller which sediments at 30S (the 30S subunit); together they form 70S ribosomes. In principle, the catalytic center of the ribosome, the peptidyl transferase center, which is located on the 50S subunit, could contain both rRNA and ribosomal proteins. However, the pioneering work of Noller and colleagues suggested that 50S subunits that were largely depleted of proteins still exhibited catalytic activity, indicating that 23S rRNA had an important role in the reaction.[2] At that time, a contribution from the ribosomal proteins could not be excluded completely,[3,4] and protein-free 23S rRNA that promoted peptide bond formation could not be obtained. The determination of high-resolution crystal structures of the large ribosomal subunit by Steitz, Moore, and colleagues has revealed that the peptidyl transferase center, as localized by a transition-state analog, is composed of 23S rRNA.[5,6] This implied that synthesis of the peptide bond is catalyzed by RNA, and that the 23S rRNA components perform catalysis, while the proteins are the structural units that help to organize the ribozyme. This structural work provided a framework for recent biochemical, kinetic, genetic, and computational work that has resulted in a fairly consistent picture of the mechanism of peptide bond formation on the ribosome.

14.2 The Ribosome

The ribosome is a molecular machine that selects its substrates, aminoacyl-tRNAs (aa-tRNA), rapidly and accurately, and catalyzes the synthesis of peptides from amino acids. The ribosome has three tRNA-binding sites: A, P, and E sites (Figure 14.1). During the elongation cycle of protein synthesis, aa-tRNA is delivered to the A site of the ribosome in a ternary complex with elongation factor Tu (EF-Tu in bacteria, eEF1A in eukaryotes) and GTP. Following GTP hydrolysis and release from EF-Tu, aa-tRNA accommodates in the A site of the peptidyl transferase center and reacts with peptidyl-tRNA bound to the P site, yielding P-site deacylated tRNA and A-site peptidyl-tRNA that is extended by one amino acid residue. The subsequent movement of

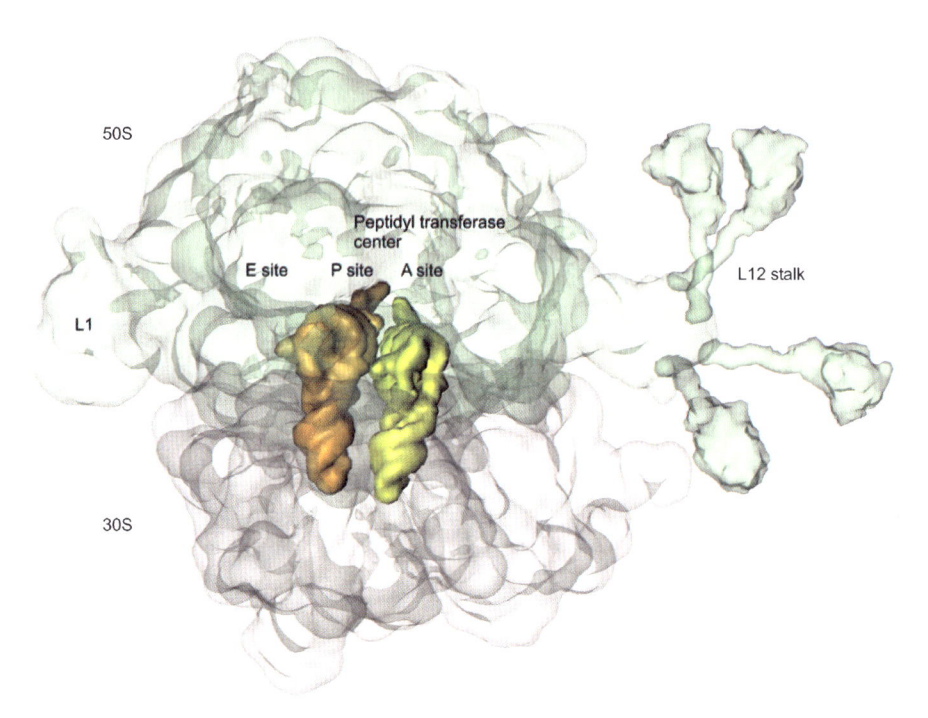

Figure 14.1 Model of the ribosome with bound tRNAs. The model is based on the crystal structure of E. coli ribosomes.[51,82] The tRNA positions in the P site (orange) and the A site (yellow) were adjusted.[25] The approximate location of the E site (shown without tRNA) is indicated.[25] Ribosomal protein L1 and the L12 stalk[83] are shown for orientation. (Reproduced from ref. 84 with permission from Elsevier.)

tRNAs and mRNA through the ribosome (translocation) is catalyzed by another elongation factor (EF-G in bacteria, eEF2 in eukaryotes). During translocation, peptidyl-tRNA and deacylated tRNA move to the P and E sites, respectively; a new codon is exposed in the A site for the interaction with the next aa-tRNA, and the deacylated tRNA is released from the E site. In addition to peptide bond formation, the second enzymatic activity that is associated with the peptidyl transferase center is the hydrolytic cleavage of the ester bond in peptidyl-tRNA during termination of protein synthesis. In contrast to peptide bond formation, which is an intrinsic activity of the ribosome and proceeds without the help of auxiliary factors, peptide release requires specialized release factors that recognize termination codons and promote the hydrolysis of the P-site peptidyl-tRNA, thereby liberating the translation product.

14.3 Peptidyl Transfer Reaction

The reaction catalyzed by the peptidyl transferase center of the ribosome is the aminolysis of an ester bond, with the nucleophilic α-amino group of A-site aa-tRNA attacking the carbonyl carbon of the ester bond linking the peptide moiety of P-site peptidyl-tRNA (Figure 14.2). The reactivity of esters with amines intrinsically is rather high, as the reaction in solution proceeds with a rate of 10^{-5}–10^{-4} $M^{-1} s^{-1}$ at room temperature. The ribosome increases the rate of peptide bond formation by 10^6–10^7-fold, and the exact rate of the reaction on the ribosome depends on the nature of both P- and A-site substrates. The maximum rate of peptide bond formation, measured with dipeptidyl-tRNA as P-site substrate and a minimal analog of the A-site substrate, puromycin (Pmn; O-methyl tyrosine linked to N6-dimethyl-adenosine *via* an amide bond), is about $50 s^{-1}$ at pH values >8.[7,8] The rate of the reaction varies about 10-fold

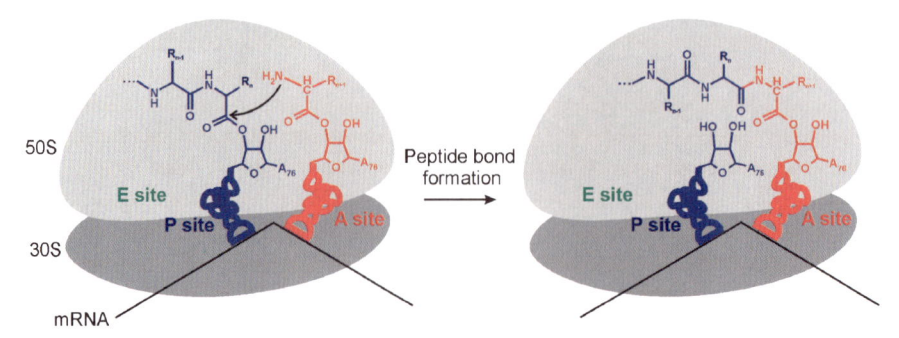

Figure 14.2 Schematic of peptide bond formation on the ribosome. The α-amino group of aminoacyl-tRNA in the A site (red) attacks the carbonyl carbon of the peptidyl-tRNA in the P site (blue) to produce a new, one amino acid longer peptidyl-tRNA in the A site and deacylated tRNA in the P site. The 50S subunit, where the peptidyl transferase center is located, is shown in light gray, the 30S subunit in dark gray. A, P, and E sites of the ribosome are indicated. (Reproduced from ref. 81 with permission from Elsevier.)

Table 14.1 Rate constants of peptide bond formation on the 70S ribosome with different substrates (at pH 7.5, 37 °C).

P-site substrate	A-site substrate	Rate constant (s^{-1})	Ref.
fMet-tRNA^{fMet}	Pmn	1	7,9,10
fMet-tRNA^{fMet}	C-Pmn	13	10
fMet-tRNA^{fMet}	CC-Pmn	4	10
fMet-tRNA^{fMet}	ACC-Pmn	2	10
fMetAla-tRNA^{Ala}	Pmn	57	a
fMetVal-tRNA^{Val}	Pmn	16	a
fMetSer-tRNA^{Ser}	Pmn	44	a
fMetArg-tRNA^{Arg}	Pmn	90	a
fMetLys-tRNA^{Lys}	Pmn	100	a
fMetPhe-tRNA^{Phe}	Pmn	12	7
fMetPhePhe-tRNA^{Phe}	Pmn	18	7
fMetAlaAsnMetPheAla-tRNA^{Ala}	Pmn	50	7
fMet-tRNA^{fMet}	Phe-tRNA^{Phe}	10	28

[a] I. Wohlgemuth, S. Brenner, M. Beringer, and M.V. Rodnina, unpublished data.

with different A-site substrates and 50–100-fold with the length of the peptidyl moiety of the P-site tRNA, the C-terminal amino acid of the peptidyl moiety, or the identity of the tRNA in the P site (Table 14.1).[7,9,10]

14.3.1 Characteristics of the Reaction off the Ribosome

Uncatalyzed peptide bond formation was studied using different model substrates. When amino acids and aminoacyl-adenylates were used as attacking nucleophile and reactive ester, respectively, reaction rates were about 10^{-5} M^{-1} s^{-1}, and varied 20-fold, depending on the identity of the amino acid, which probably reflected the differences in the pK_as of the attacking α-NH$_2$ group.[11]

Non-enzymatic ester aminolysis reactions are predicted to proceed through two tetrahedral intermediates (Figure 14.3). Early studies by Jencks and co-workers showed that aminolysis of methyl formate by different amines proceeds *via* general base-catalyzed attack by the unprotonated amine at high pH.[12,13] As the reaction pH is lowered, there is a break in the pH profile, suggesting a change in the rate-determining step of the reaction. At high pH, the attack on the ester carbonyl carbon by the amino nitrogen (step 1) is rapid and reversible, and trapping of the tetrahedral intermediate T$^\pm$ by general base-catalyzed proton removal is rate-determining (step 2). At low pH, the breakdown of the tetrahedral intermediate T$^-$ becomes rate-determining (step 3).[12] A late rate-limiting step is reflected in a high Brønsted coefficient, which is 0.8.[14] The reaction between the primary amine Tris(hydroxymethyl)aminomethane and organic esters such as *N*-formylglycine ethylene glycol ester (10^{-4} M^{-1} s^{-1}) showed a large unfavorable activation entropy ($T\Delta S^{\neq} = -13.1$ kcal mol^{-1}) and enthalpy ($\Delta H^{\neq} = 9.1$ kcal mol^{-1}).[15] Aminolysis of a natural substrate of the peptidyl transferase reaction, fMet-tRNA^{fMet}, had a more favorable entropic term

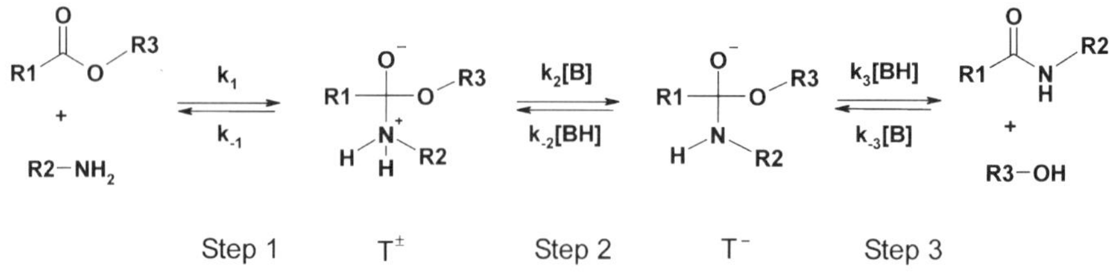

Figure 14.3 Scheme of the aminolysis reaction mechanism. Step 1, formation of the zwitterionic tetrahedral intermediate $T^{\pm}$. Step 2, deprotonation of the positively charged amino nitrogen, resulting in the second intermediate, T^{-}. Step 3, product formation. Proton transfers during the deprotonation of $T^{\pm}$ (step 2) and the formation the leaving group (step 3) can be catalyzed by base and acid, respectively. R1, R2, R3 are substituents.

Table 14.2 Activation parameters of uncatalyzed and catalyzed peptide bond formation (25 °C).[15,16] (Reproduced from ref. 81 with permission from Elsevier.)

			Activation parameters (kcal mol^{-1})		
Ester	Amine	Rate	$\Delta G^{\neq}$	$\Delta H^{\neq}$	$T\Delta S^{\neq}$
Uncatalyzed					
fGly-ethylene glycol	Tris	10^{-4} (M^{-1} s^{-1})	22.2	9.1	−13.1
fMet-tRNA^{fMet}	Tris	10^{-4} (M^{-1} s^{-1})	22.7	16.2	−6.5
Ribosome-catalyzed					
fMetPhe-tRNA^{Phe}	Pmn^a	10^{3} (M^{-1} s^{-1})	14.0	16.0	2.0
fMetPhe-tRNA^{Phe}	Pmn^b	5 (s^{-1})	16.5	17.2	0.7
fMet-tRNA^{fMet}	OH-Pmn^b	6×10^{-3} (s^{-1})	20.5	16.8	−3.7

fGly, N-formylglycine; fMet, N-formylmethionine.
^{a} Measured at limiting concentrations of Pmn; the rate obtained under these conditions (k_{cat}/K_M) is comparable to the second-order reaction of model substrates in solution.
^{b} Measured at saturating concentration of Pmn or OH-Pmn (k_{cat} conditions).

($T\Delta S^{\neq} = -6.5$ kcal mol^{-1}), and a somewhat less favorable activation enthalpy ($\Delta H^{\neq} = 16.2$ kcal mol^{-1})[16] (Table 14.2). Recent theoretical work by Warshel and colleagues[17] suggested that a significant part of the observed activation entropy of the solution reaction is due to the entropy of solvent reorganization.

14.3.2 Enzymology of the Peptidyl Transfer Reaction

14.3.2.1 Potential Mechanisms of Rate Acceleration by the Ribosome

Ribozymes possess a limited repertoire of groups that can take part in catalysis, compared to the chemically much more diverse protein enzymes. Nevertheless, ribozymes use various mechanisms, including general acid–base catalysis, metal ion-assisted catalysis, and substrate-alignment by base-pairing and other interactions, *i.e.*, they act in ways that are similar to conventional protein enzymes.[18] In principle, all these strategies might be used by the ribosome. Charged intermediates and/or transition states develop during the peptidyl transferase reaction, and the ribosome might catalyze the reaction by stabilizing them by electrostatic interactions or by abstracting/donating protons. The T$^\pm$ intermediate of the peptidyl transferase reaction contains a positively charged amino nitrogen and an oxyanion linked to the tetrahedral carbon, which would be stabilized by electrostatic interactions with the ribosome. After the nucleophilic attack, a proton from the positively charged amino nitrogen may be abstracted by a general base to generate an uncharged secondary amine. A proton is required to form the leaving group, the 3′ OH of the deacylated tRNA in the P site, which may require a general acid. For general acid–base catalysis (Chapter 2) to occur in an aqueous environment, the pK_as of the catalytic groups have to be close to neutrality to efficiently abstract or donate a proton during the reaction. The unperturbed pK_as of RNA bases are far from neutrality, *i.e.*, 3.5 and 4.2 for A and C and 9.2 for G and U, respectively.[19] Thus, if rRNA bases were to take part in chemical catalysis in an aqueous environment at neutral pH, their pK_as would have to be shifted quite significantly by a particular chemical environment. In addition, the efficiency of acid–base catalysis depends on Brønsted coefficients. Ionizing groups with pK_as outside the neutral pH may be efficient acid–base catalysts only when their Brønsted coefficients are sufficiently high. Thus, if acid–base catalysis contributed significantly to peptide bond formation, then the reaction rate should depend on the ionization of residues acting as general acid–base catalysts, and the Brønsted coefficient should be high. Furthermore, enzymes that employ general acid–base or covalent catalysis generally lower the activation enthalpy of the reactions they catalyze. The entropic contribution for the formation of the transition state in those cases is small and variable, with an average value close to zero.[20] Finally, the ribosome may promote peptide bond formation by a mechanism that differs in detail from the uncatalyzed aminolysis reaction in solution, as suggested by recent measurements of kinetic isotope effects for the peptidyl transferase reaction on isolated 50S subunits.[21] The reaction on the 50S subunit was described by the model in which the formation of the tetrahedral intermediate was the first irreversible step of the reaction and deprotonation of the amine occurred concurrently with intermediate formation.[21] A very low Brønsted coefficient ($\sim$0.2) suggested little charge development on the amine in the transition state.[21] On the other hand, molecular dynamics (MD) simulation suggested a late transition state for C–N bond formation and strong H-bonding between the nucleophile and the P-site O2′ before the proton is transferred, which implied that the geometry and charge distribution of the rate-limiting transition state may be similar to that of the high energy intermediate, irrespective of whether it actually is formed before or after the intermediate.[22]

14.3.2.2 *Experimental Approaches to Reaction on the Ribosome*

The elucidation of the catalytic mechanism of peptide bond formation required a complete reconstituted *in vitro* translation system in which parameters such as pH, temperature and ionic conditions could be changed. One critical issue in such experiments is to show that the rate of product formation reflects the chemistry step, rather than substrate binding or conformational rearrangements. In a complete translation system, this condition is unlikely to be fulfilled, because the overall rate of protein synthesis is determined by the time needed for aa-tRNA selection. The movement of aa-tRNA into the A site is a multistep process that requires structural rearrangements of the ribosome, EF-Tu and aa-tRNA.[23] Binding of aa-tRNA in complex with EF-Tu·GTP to the ribosome and codon recognition results in GTP hydrolysis by EF-Tu. Aa-tRNA is released from EF-Tu·GDP and moves through the ribosome into the peptidyl transferase center where its aminoacylated CCA terminus is engaged in multiple interactions with the rRNA.[24–27] The rate of accommodation of aa-tRNA in the A site is $\sim 10\,\mathrm{s}^{-1}$, and peptide bond formation follows instantaneously.[28,29] Because accommodation precedes peptide bond formation, it limits the rate of product formation as long as it is slower than peptidyl transfer.[29] Thus, studying the catalytic mechanism of peptide bond formation is possible only when the chemistry is uncoupled from accommodation. To overcome the accommodation problem, substrate analogs were used that mimic the aminoacylated 3′ terminus of aminoacyl-tRNA, such as Pmn and its derivatives containing additional one or two cytidines or adenine and two cytidines, C-Pmn, CC-Pmn, or ACC-Pmn, respectively. Such substrates bind to the A site on the 50S subunit very rapidly[7,15] and do not require accommodation. Early studies were performed with the so-called "fragment reaction" on 50S subunits utilizing N-blocked aminoacylated oligonucleotides, such as CCA-fMet, in the P site and Pmn in the A site.[30,31] Unfortunately, the fragment reaction required high concentrations of alcohol and the observed reaction rates appeared to be much slower than the rates of protein synthesis *in vivo*, raising the question whether such a decidedly non-physiological system was representative of the reaction in the cell. In particular, the possibility remained that the presence of the small subunit was necessary to induce a conformation of the active site favoring the reaction.[26]

One of the problems of the fragment reaction was the low affinity of small substrate analogs to their binding sites on the 50S subunit. To solve this problem, Steitz, Strobel and colleagues designed new, somewhat larger substrate analogs that could be used in a modified, alcohol-free version of the fragment reaction.[32–34] Furthermore, when full-length fMet-tRNA and Pmn or C-Pmn were used, the reaction rates on isolated 50S subunits were comparable to those measured on the 70S ribosomes[35] or observed *in vivo*. This strongly supports the conclusion that the 50S subunit alone possesses the full potential of catalyzing peptide bond formation.[35] The reactions catalyzed by the 70S ribosome and the 50S subunit are similar also in several other ways: they are susceptible to the same inhibitors,[36–39] and have similar pH profiles.[7,9,40,41]

Another possibility to overcome the accommodation problem is to decrease the rate of the chemistry reaction such that it becomes much slower than the tRNA accommodation step and thus amenable for biochemical analysis. This was accomplished by replacing the reactive nucleophilic α-NH$_2$ group by the much less reactive OH group.[28] The OH derivatives of either Pmn or aa-tRNA bind to the peptidyl transferase center in the same way as unmodified substrates,[34] and the reaction with Pmn-OH exhibited the same activation enthalpy as with unmodified Pmn (Table 14.2), suggesting that the change of the nucleophilic group did not considerably alter the reaction pathway. The hydroxyl derivative of Phe-tRNA was fully reactive with fMet-tRNA in the P site, but, because of the decreased nucleophilicity of the attacking group, the rate of reaction was much lower, $10^{-3}\,\mathrm{s}^{-1}$, *i.e.*, far below the rate of accommodation.

14.3.2.3 pH–Rate Profiles

For general acid–base catalysis to occur in an aqueous environment at physiological conditions, the pK_as of the catalytic groups have to be close to neutrality to efficiently abstract or donate a proton during the reaction (Chapter 2). Measuring the reaction rate at different pH values in the range 6–9 revealed that the rate of the reaction between OH-Phe-tRNA^{Phe} and fMet-tRNA was not affected by pH changes (Table 14.3). This result argues against an involvement of ionizing groups of the ribosome in the catalytic mechanism of the reaction.[28] Furthermore, peptide bond formation between full-length peptidyl-tRNA and unmodified aa-tRNAs did not show any pH dependence,[28] and – although the rate-limiting accommodation step probably would have masked part of a potential pH effect – these results are consistent with a small, if any, influence of ionizing groups of the ribosome on the reaction. Taking into account the ionization of the α-amino group of aa-tRNA ($pK_a = 8$) and a measured reaction rate of $6\,\mathrm{s}^{-1}$ observed at pH 6, the intrinsic rate of peptidyl transfer with unmodified full-length tRNA can be estimated to be $>300\,\mathrm{s}^{-1}$.[28]

Notably, the absence of a pH-dependence between pH 6 and 9 does not by itself eliminate acid–base catalysis, because a combination of an acid and a base with pK_as outside this range may produce the same effect. In such a case, the observed pH-independence is not due to the absence of deprotonation–protonation events, but instead is due to the decrease in the level of the functional form of the acid being offset by the increase in the functional form of the base.[42] In fact, there are examples of ribozymes using acid–base catalysis by nucleobases with pK_as outside the neutral range.[19,43] Notably, the reaction rates achieved by such catalysts are very moderate, usually $<1\,\mathrm{min}^{-1}$, for the following reason. Assuming the pK_as of the catalytic acid and base just

Table 14.3 Inhibition of peptide bond formation by protonation of ribosomal group(s). (Reproduced from ref. 81 with permission from Elsevier.)

A-site substrate	Rate decrease	Ref.
Pmn	150-fold	7
OH-Pmn	150-fold	7
C-Pmn	None	9
Phe-tRNA	None	28
OH-Phe-tRNA	None	28

outside the 6–9 range, e.g., pK_a = 5 and 10, respectively, the activity of such an enzyme at neutral pH (calculated as in ref. 42) is five orders of magnitude lower than the k_{max} value representing a bond-forming or -breaking event. Furthermore, k_{max} itself is a function of the Brønsted coefficient, which appears to be very low, β_{nuc} < 0.2, on the ribosome.[44] Thus, models assuming active acid–base catalysis by groups ionizing far outside the neutral pH range are inconsistent with the high rate of peptide bond formation on the ribosome.

In contrast to results obtained with full-length aminoacyl-tRNA, a pronounced pH dependence was observed for peptide bond formation with the minimal A-site substrate, Pmn (Table 14.3). Protonation of a ribosomal group with a pK_a around 7.5 reduced the rate of reaction about 150-fold.[7,9] This effect may, in principle, indicate an involvement of a ribosome residue as general base in reaction, although the magnitude of inhibition by protonation of a group with pK_a of 7.5 is much less than expected for an essential base. Alternatively, the pH effect may reflect a conformational rearrangement of active-site residues that impairs catalysis and does not take place with full-length aa-tRNA as A-site substrate.[7,28,45] This conclusion is corroborated by the observation that the reaction between C-Pmn in the A site and peptidyl-tRNA in the P site was not influenced by the ionization of ribosomal groups,[9] suggesting that the presence of the cytidine residue, which mimics C75 of the A-site tRNA and presumably its interaction with G2553, was sufficient to induce and stabilize the active conformation of the peptidyl transferase center. Generally, the reaction with full-length tRNA seems to be more robust than the reaction with Pmn, indicating the importance of more remote interactions for positioning the tRNA in the A site. Even when base pairs between C74 or C75 of P-site tRNA with rRNA residues G2252 or G2251, respectively, were disrupted by mutagenesis, there was no effect on the rate of peptide bond formation with aa-tRNA as A-site substrate.[46]

14.3.2.4 Activation Parameters

Crucial information about the mechanism of catalysis can be obtained by comparing the thermodynamic properties of the catalyzed and uncatalyzed reactions. Enzymes that employ general acid–base catalysis act by lowering the activation enthalpy of the reaction. If the ribosome acted as such a chemical catalyst, then the rate enhancement produced by the ribosome should result from a reduction of the activation enthalpy. If, however, the ribosome used other mechanisms of catalysis, such as substrate positioning in the active site, desolvation, or electrostatic shielding, then the rate enhancement produced by the ribosome should be largely entropic in origin. Comparing the rate of the ribosome-catalyzed reaction with a second-order model reaction in solution revealed that the acceleration is achieved by a major lowering of the activation entropy,[15] whereas the activation enthalpy was practically unchanged (Table 14.2). Consistent with the pH independence of the reaction with full-length tRNA, these findings suggest that general acid–base catalysis by ribosomal residues does not play a significant role in peptidyl transfer on the ribosome.

In complete agreement with the kinetic results, MD simulations of the peptidyl transfer reaction on the ribosome indicated that the catalytic effect was entirely of entropic origin and was mainly associated with lowering the free energy of solvent reorganization.[47] Interestingly, the contribution of substrate alignment or proximity at the active site was estimated to be quite small, about 3.4 kcal mol^{-1},[17,22] compared to the 8.5 kcal mol^{-1} reduction of the free energy of activation of the ribosome-catalyzed reaction.[15] In fact, activation entropies were similarly reduced in the reaction with pre-bound (k_{cat} conditions) and not pre-bound (k_{cat}/K_M conditions) substrates, indicating that the effect may not be associated with substrate binding.[15] It has been pointed out that the activation entropy is a macroscopic quantity that includes several microscopic contributions derived from solvation, molecular conformations, and phase-space configuration volumes, and that activation entropies are in many cases dominated by solvation effects.[17,22,47]

14.4 The Active Site

50S subunits are composed of two rRNA molecules, 23S rRNA and 5S rRNA, and more than 30 proteins (Figure 14.4A). The rRNA backbone in the peptidyl transferase center is found in very similar conformations in 50S subunits and 70S ribosomes,[33,34,48] or in ribosomes from different organisms.[5,6,24–26,32–34,48–51] However, small-scale reorientations can occur, and the movement of some

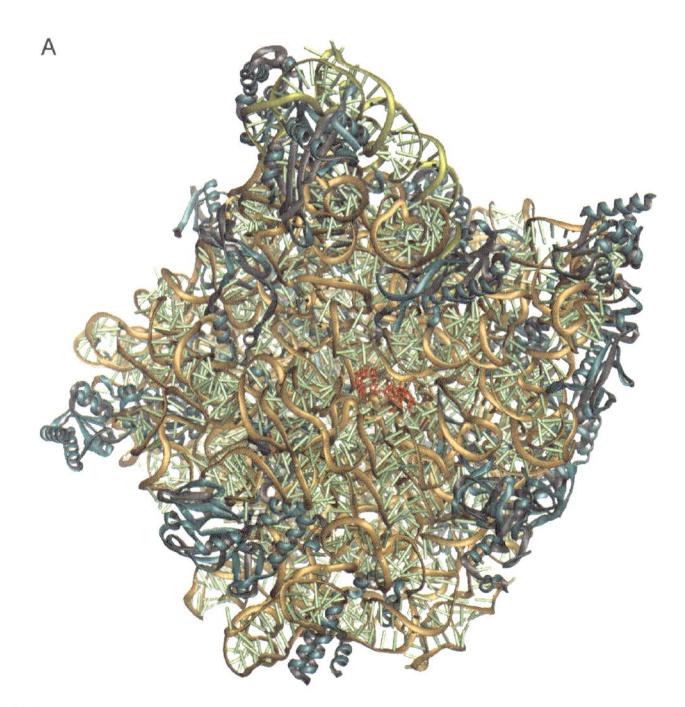

Figure 14.4 (*Continued*)

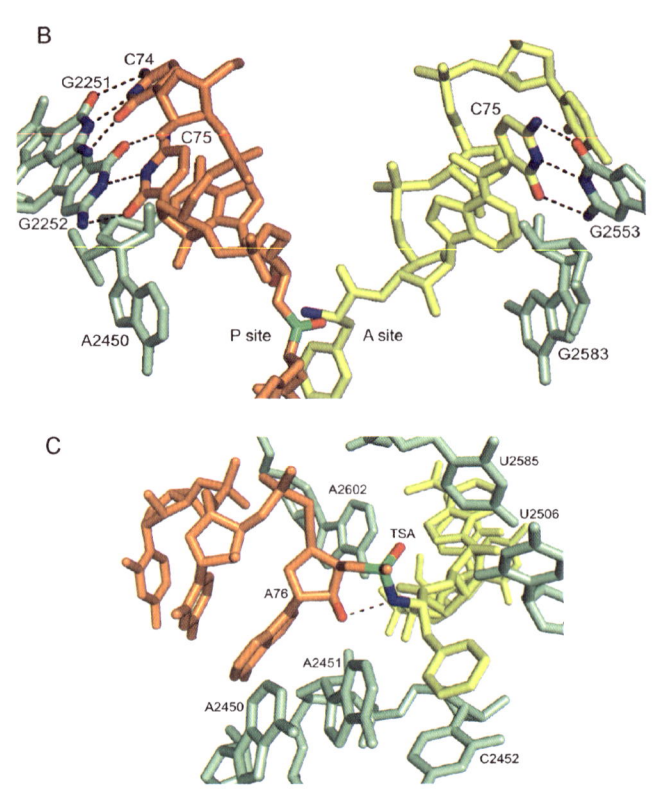

Figure 14.4 Structure of the peptidyl transferase center. (*A*) Crystal structure of the 50S subunit from *H. marismortui* with a transition state analog (red) bound to the active site (PDB entry 1VQP; ref. 33). Ribosomal proteins are blue, the 23S rRNA backbone is brown, the 5S rRNA backbone olive, and rRNA bases are pale green. The figure was generated with PyMol (http://www.pymol.org). (*B*) Substrate binding to the active site. Base pairs formed between cytosine residues of the tRNA analogs in the A site (yellow) and P site (orange) with 23S rRNA bases (green) are indicated (PDB entry 1VQN; ref. 34). The α-amino group of the A-site substrate (blue) is positioned for attack on the carbonyl carbon of the ester linking the peptide moiety of the P-site substrate (green). Inner-shell nucleotides are omitted for clarity. (*C*) Transition state analog (TSA) bound to the peptidyl transferase center (PDB entry 1VQP; ref. 33). The hydrogen bond between the nucleophilic nitrogen (blue) and the 2'-OH of P-site A76 is indicated. (Reproduced from refs. 81 and 84 with permission from Elsevier.)

particularly flexible nucleotides may have functional implications for the catalytic mechanism (see below).

Before the reaction, the tRNA substrates have to bind to their respective sites on the ribosome (Figure 14.4B).[5,52] The acceptor arms of the A and P-site tRNAs are located in a cleft of the 50S interface side.[25] Their universally conserved CCA ends are oriented and held in place by interactions with residues of 23S rRNA near the active site. The CCA ends of a full-length tRNA substrate or smaller substrate analogs are bound to the P site in essentially the same way.[34,48] In the P site, C74 and C75 of the tRNA are base-paired to G2251 and G2252 of the P loop of 23S rRNA.[6,33,53] The CCA end of the A-site tRNA is fixed by base-pairing of C75 with G2553 of the A loop of 23S rRNA.[6,26,27,33] The 3'-terminal A76 of both A and P-site tRNAs form interactions with residues G2583 and A2450, respectively. The conserved 23S rRNA bases A2451, U2506, U2585, C2452 and A2602 are located at the core of the peptidyl transferase center[6,26,33,34] (Figure 14.4C). The only protein located close to the peptidyl transferase center is L27.[48] The deletion of as few as three amino acids at the N terminus of L27 leads to an impaired activity of *Escherichia coli* ribosomes.[54] However, an *E. coli* strain lacking L27 is viable, and other organisms have ribosomes without protein L27 or any protein groups at the place where the N-terminus of L27 is located, suggesting that L27 is not part of an evolutionary conserved mechanism that is expected to employ identical residues in all organisms. Given the high degree of sequence conservation of rRNA, in particular at the peptidyl transferase center,[5,55,56] the active site is likely to consist of RNA in all organisms.

14.4.1 Structures of the Reaction Intermediates

High-resolution crystal structures of several ribosomal complexes have provided insight into the pre-reaction and post-reaction states, and the use of transition state analogs has revealed the conformation of the active site during peptidyl transfer. To visualize the pre-reaction state, Steitz and colleagues solved the crystal structure of a ribosomal complex containing an aa-tRNA analogue (CC-hydroxypuromycin, CC-hPmn) as A-site substrate in combination with a P-site substrate analog.[34] CC-hPmn represents a tRNA-like CCA end linked to an aminoacyl-like tyrosine in which the α-NH$_2$ group is replaced by an OH group. Because the OH group does not attack the P-site electrophile as efficiently as an α-NH$_2$ group would do, this approach prevented product formation from occurring during crystallization.[34] In this model for the pre-reaction state, the nucleophilic group (OH) forms hydrogen bonds with both N3 of A2451 and the 2'-OH of A76 of the P-site substrate. A somewhat different environment of the nucleophilic α-NH$_2$ group was reported for a substrate analog attached to a short RNA hairpin bound to *D. radiodurans* 50S subunits.[26] The reason for this discrepancy is not clear, and its solution awaits high-resolution structures of subunits with, preferably, full-size aminoacyl- and peptidyl-tRNAs.

The crystal structures of *H. marismortui* 50S subunits complexed with different transition state analogs revealed that the reaction proceeds through a tetrahedral intermediate with (S) chirality. The oxyanion of the tetrahedral intermediate is stabilized by a water molecule that is positioned by nucleotides A2602 and U2584.[33] The only atom within hydrogen-bonding distance of the α-NH$_2$ group mimic is the 2'-OH of A76 of the P-site portion of the transition state analog.[33] N3 of A2451, which is within hydrogen-bonding distance in the pre-reaction state, seems to lose this interaction with the attacking group during the reaction. The transition state analogs are held in place and stabilized by an intricate H-bond network composed of water molecules and ribosomal bases.[22,23]

After peptidyl transfer, the growing peptide chain is esterified to the A-site tRNA. In the crystal structure of a ribosome-product complex that represents the state directly after the reaction,[32] the peptide moiety points towards the exit tunnel without making specific contacts with the ribosome. The tRNA molecules are present in a very similar position and conformation as before peptide bond formation, and only some bases of 23S rRNA have adopted a different conformation compared with pre-reaction state, especially U2585. This structure indicates that peptide bond formation and the movement of the tRNA 3′ termini during hybrid state formation are not directly coupled.[32] In contrast, eventually the 3′ ends of both A- and P-site tRNA can move spontaneously from their initial binding sites on the 50S subunit to the P and E sites, respectively, forming the hybrid states.[57] Hybrid states are characterized by a very slow reaction of A/P bound peptidyl transferase-tRNA with Pmn.[58-60] This low, but significant reactivity of peptidyl-tRNA in the A/P state indicates that the active site can accommodate a different arrangement of the substrate. A rotational movement of the 3′ ends of the tRNAs was suggested to accompany peptide bond formation,[26] although the mechanism by which this movement would contribute to the catalysis of enzymatic reaction is unclear.

14.4.2 Conformational Rearrangements of the Active Site

14.4.2.1 Induced Fit

Active sites of many enzymes undergo conformational changes during substrate binding and catalysis, and there is some evidence for rearrangements at the active site of the ribosome that comes from comparison of the structures of 50S complexes with A-site substrate analogs of different lengths.[34] The smallest A-site substrate contained the terminal C75 and A76 of the tRNA and the aminoacyl moiety. The substrate was bound to the 50S subunit through the typical A site-binding contacts (see above) and the conformation of active site residues in the complex was similar to that with an empty A site. The P-site substrate, CCA-pcb, a CCA oligonucleotide linked to the residue mimicking the growing peptide chain, exhibited a conformation that was unfavorable for the peptidyl transfer reaction, as the carbonyl oxygen was pointing towards the α-NH_2 group of the A site-substrate mimic and the atoms that are required to react were about 4 Å apart.

Binding of an A-site substrate that additionally included C74 induced a conformational change in the catalytic center.[34] The stacking of the CCA bases positioned the substrate closer to the center of the active site and involved rearrangements of residues G2583, U2506, and U2585. These rearrangements resulted in a conformation of the active site in which the carbonyl carbon of peptidyl-tRNA was more appropriately oriented for attack by the nucleophile. In this pre-reaction ground state, the reactive α-NH$_2$ group formed hydrogen bonds with both N3 of A2451 and the 2′-OH of A76 of the P-site substrate. The conformation of the rRNA in this complex was nearly identical to that observed in the complex with transition state analogs, suggesting that rRNA rearrangements induced by substrate binding indeed reflect the reaction pathway. These structures suggest an induced-fit mechanism in which binding of the correct substrate to the A site induces repositioning of both substrates and active-site residues to promote peptide bond formation.[34]

One implication of the induced-fit model is that the rate of peptide bond formation with CC-Pmn as A site substrate should be higher than with C-Pmn. However, measurements under conditions of substrate limitation did not detect gross changes in the rates of catalysis.[9] Surprisingly, under conditions of substrate saturation the rate of peptide bond formation with CC-Pmn and ACC-Pmn was even lower than that with C-Pmn (Table 14.1).[10] MD simulations suggested that the U2585 base – which changes its position upon binding of CC-Pmn to the A site – is fairly mobile and spans a range of conformations between the two steps observed in the crystal structures.[22] In addition, the activation energies of the induced and uninduced reactions were found to be quite similar.[22] These results indicated that the effect of induced fit on the rate of peptidyl transfer is small.

Notably, the rate of reaction with CC-Pmn was independent of pH,[10] whereas the rate–pH profile of the reaction with C-Pmn showed a single pK_a close to 7, most probably reflecting the ionization of the α-NH$_2$ group of C-Pmn[9]. These results suggest that, depending of the substrate in the A site, the details of the reaction pathway may change and different elemental steps may determine the reaction rate. While the chemical step determines the rate of the reaction with Pmn (and possibly C-Pmn),[7,15] the reaction with the full-size aa-tRNA is limited by the preceding step of aa-tRNA accommodation in the A site.[28,29] The latter step is independent of pH and, because it is rate-limiting, masks the ionization of groups involved in peptidyl transfer, *i.e.*, the α-NH$_2$ group of amino acid [pK_a = 8 (ref. 61)]. Likewise, binding of CC-Pmn to the ribosome (or the rearrangements at the active site) may be sufficiently slow to limit the rate of the following steps; hence the reduced rate and lack of pH dependence of reaction. This suggests that interactions of C74 with the ribosome change the reaction pathway and, therefore, the effect of induced fit on the chemistry step of peptide bond formation cannot be assessed biochemically.

14.4.2.2 Role of the P-site Substrate

Binding of the correct P-site substrate, *i.e.*, full-length peptidyl-tRNA, results in a much faster peptidyl transferase reaction, compared to short substrate analogs.[35] The reaction on 50S subunits was faster when full-size fMet-tRNA^{fMet} (ref. 35) or Ala-tRNA^{Ala} (ref. 62), rather than a short fragment representing the aminoacylated 3' terminus of the tRNA, were used as P-site substrate. These findings suggest that full-size tRNA in the P site may be important for maintaining the active conformation of the peptidyl transferase center.[35] This raises the possibility that an interaction of a part of the tRNA molecule beyond the 3'-terminal sequence induces a conformational change of the peptidyl transferase center, thus enhancing the catalytic activity. Apart from the contacts in the peptidyl transferase center itself, tRNA interactions with the 50S subunit involve elements (helix 69 of 23S rRNA; protein L5) that contact the elbow region of the P-site tRNA.[25] These contacts may be formed in the complexes of the 50S subunit with fMet-tRNA^{fMet}, but not in those with oligonucleotide substrate analogs. The contact site is probably located in the acceptor domain of the tRNA, because Ala-tRNA^{Ala} and an Ala-tRNA minihelix were equally active on 50S subunits, whereas a 9-mer 3'-terminal fragment of the same tRNA had a much lower activity.[62] The only contact of the acceptor domain with the 50S subunit is through protein L5 which, therefore, may be involved in modulating the peptidyl transferase activity.

14.4.2.3 Conformational Flexibility of the Active Site

Three inner-shell residues of 23S rRNA seem to be particularly flexible. MD simulations indicate that U2506 can adopt energetically equivalent conformational orientations, allowing it to be flexible without the need for induced fit.[22] U2585 is seen in different conformations depending on the length of the A-site substrate (see above) but also seems to change position during the progression of the reaction.[32] Residue A2602, which lies between the A and P sites in the active site, can adopt different orientations and shows weak electron density in the crystal structures.[48] A2602 is a critical residue for the hydrolytic activity of the active site during protein release.[8,63] However, it is not clear whether the flexibility of any of these residues is functionally important for either peptidyl transfer or for the hydrolysis of peptidyl-tRNA.

In the crystal structure of the ribosome-tRNA complex, the non-canonical A2450-C2063 base pair was reported to move towards A76 of the P-site tRNA, bringing the A-C pair into hydrogen bonding distance of the 2'-OH group of A76.[50] The authors suggested that the A-C pair, which is expected to have a pK_a in the range 6.0–6.5,[64] may act as a proton donor/acceptor during the peptidyl transferase reaction.[7] If the pK_a of the A-C pair were shifted by the ribosome environment, this value could be compatible with the observed pH–rate dependence of the Pmn reaction showing an apparent pK_a of 7.5.[7] The relevance of the A2450-C2063 movement remains unclear, as it was not observed in the high-resolution crystal structure of a very similar ribosome-tRNA complex.[48] Replacement of the A2450-C2063 pair by an isosteric but uncharged G-U wobble pair results in a >200-fold decrease of the rate of peptide bond formation with either aa-tRNA or Pmn as A-site substrates; the effect of the replacement on the pH–rate profile could not be assessed due to impaired binding of Pmn to the A site.[65] As described above, ionizing groups of the ribosome contribute little if at all to catalysis (Table 14.3).[7–9,28] This argues against an important role of the ionization of the A2450-C2063 base pair in the chemistry step, but does not exclude that it is involved in the reaction in other ways, e.g., by providing additional stabilizing interactions for the P-site tRNA. Most probably, the detrimental effects of the mutation were due to a gross structural change in the peptidyl transferase center, rather than to a specific catalytic effect.[65]

14.4.3 Probing the Catalytic Mechanism: Effects of Base Substitutions

The results of the structural work provided the basis to probe the role of active-site residues for the catalysis of peptide bond formation. Mutation of any residue in the core of the active site leads to lethal phenotypes in *E. coli*,[66–68] which makes it impossible to express pure populations of mutant ribosomes. In some instances, the problem can be circumvented by expressing a mixture of mutant and wild-type ribosomes, measuring kinetics on a mixed population of ribosomes, and then deconvoluting the contributions of the wild-type and mutants ribosomes to the observed time course of product formation.[7] However, only after techniques had been developed to introduce affinity tags into rRNA, by which it was possible to purify mutant ribosomes, could more systematic studies be undertaken aimed at assessing the specific role of each of the active-site residues.[8,65,69] Several rRNA bases at the active site were mutated, and the effects of replacements were examined *in vivo* and *in vitro*.

N3 of the conserved base A2451 is located within 3–4 Å of the reactive α-NH$_2$ group, giving rise to the suggestion that A2451 might have a critical role during the formation of the tetrahedral intermediate.[6] However, spontaneous mutation of A2451U in rat mitochondrial rRNA conferred chloramphenicol resistance to cells in culture,[70] suggesting that in some organisms the mutation was not lethal and thus the replacement is unlikely to abolish the peptidyl transfer reaction. *E. coli* ribosomes carrying mutations of A2451 accumulated in polysomes *in vivo* and were active in peptide synthesis *in vitro*.[67,68] The Pmn reaction on A2451U mutant ribosomes was slowed down by a factor of 150 only.[7,8] Compared to the 10^6–10^7 rate acceleration by the ribosome, this is a too small effect for the replacement of an essential or important residue. Additionally, the same mutation had very little effect on peptide bond formation with full-size aa-tRNA, which argues against a catalytic function of A2451 in the reaction mechanism.[8] Further analysis of A2451U and G2447A mutations (the latter residue forming a part of the charge relay system postulated to bring about the catalytic function of A2451) in two organisms, *E. coli* and *M. smegmatis*, strongly argues against a critical role of A2451 in peptide bond formation.[8,45,71] Rather, the A2451U mutation alters the structure of the peptidyl transferase center and seems to function as a pivot point in stabilizing the ordered structure of the active site.[45] However, the 2′-OH of the ribose moiety of A2451 seems to be important, as it may take part in the intricate hydrogen bond network in the active site and interact directly with the critical

2'-OH group of the P-site tRNA.[22,47] Consistent with this suggestion, substitution of the 2'-OH of A2451 by hydrogen led to an impaired peptidyl transferase activity.[72]

Further mutagenesis in the so-called "inner shell" of the peptidyl transferase center (positions U2506, U2585 and A2602) showed that all but one of the nine mutations tested exhibited drastic reductions (30–9400-fold compared to wild-type ribosomes) in the rate of peptide bond formation with Pmn as A-site- and fMetPhe-tRNA as P-site substrates.[8] Unexpectedly, U2506 seemed to be the most critical residue (4700–9400-fold reduction). In striking contrast to the effects observed with Pmn, no large deficiency in the rate of peptidyl transfer was found for any of the A2451, U2506, U2585 and A2602 mutants when full-size aa-tRNA or C-Pmn were used as A-site substrates, suggesting that none of the rRNA bases in the peptidyl transferase center takes part in chemical catalysis.[8,9] The effects of mutations on cell viability were attributed to an impaired peptide release during the termination step of protein synthesis.[8,63]

14.4.4 Importance of the 2'-OH of A76 of the P-site tRNA

The only group found within hydrogen-bonding distance of the attacking nucleophile in the transition state analogs is the 2'-OH of A76 of peptidyl-tRNA in the P site.[33,34] The presence of the 2'-OH is crucial for the reaction, both on isolated 50S subunits[73] and 70S ribosomes.[74] Hydrogen bonding between the 2'-OH of A76 and the nucleophilic α-NH$_2$ group may help to position the nucleophile. Alternatively, the 2'-OH could act as a general acid, providing a proton for the leaving group, the 3'-oxygen of A76, and as a general base to deprotonate the nucleophile in the catalysis of the peptidyl transfer reaction.[74] For the 2'-OH to participate in proton transfer, its pK_a would need to be reduced towards neutrality. However, the large rate effect of the 2'-OH replacements – substitutions by hydrogen (2'-deoxy) or fluor (2'-fluoro) reduced the ribosome activity by six orders of magnitude[74] – is inconsistent with the limited effect of pH changes on peptide bond formation with Pmn (a factor of 150), C-Pmn (no pH dependence), or aa-tRNA (no pH dependence).[7,9,28] Alternatively, by analogy to group I intron splicing,[75,76] it was proposed that the 2'-OH may coordinate a catalytic metal ion, such as Mg^{2+} or K$^+$,[74] in line with earlier suggestions.[77] However, extensive studies aimed at the identification of metal binding sites in the 50S crystal structures did not reveal Mg^{2+} ions or monovalent metal ions in the active center that could directly promote catalysis or any other groups in the vicinity of the 2'-OH of the P-site tRNA that would be able to shift its pK_a.[33]

One attractive role for the 2'-OH of A76 of the P-site tRNA is to take part in a proton shuttle that bridges the attacking α-NH$_2$ group and the leaving 3' oxygen; several catalytic pathways can be envisaged.[33,47] The attack of the α-NH$_2$ group on the ester carbon may result in a six-membered transition state, where the 2'-OH group donates its proton to the adjacent 3' oxygen while simultaneously receiving one of the amino protons. Such a scenario would not require a pK_a shift of the 2′-OH group, owing to the concerted nature of the bond-forming and -breaking events. The structural details of the putative transition state and the involvement of an ordered water molecule in the proton shuttle remain to be tested.[17,22,33,47,78] The residues in the active site may provide a stable hydrogen-bond network that stabilizes the transition state.[47] The peptidyl transferase center can thus be viewed as a rather rigid environment of preorganized dipoles that do not need to undergo major rearrangements during the reaction. The most favorable mechanism would not involve any general acid–base catalysis by ribosomal groups. Rather, the catalytic effect is mainly associated with lowering the free energy of solvent reorganization.[47]

14.5 Conclusions and Evolutionary Considerations

Peptide bond formation on the ribosome is driven by a favorable entropy change. The A- and P-site substrates are precisely aligned in the active center by interactions of their CCA sequences and of the nucleophilic α-amino group with residues of 23S rRNA in the active site. The most favorable catalytic pathway involves a six-membered transition state (Figure 14.5) in which proton shuttling occurs *via* the 2′-OH of A76 of the P-site tRNA. The reaction does not

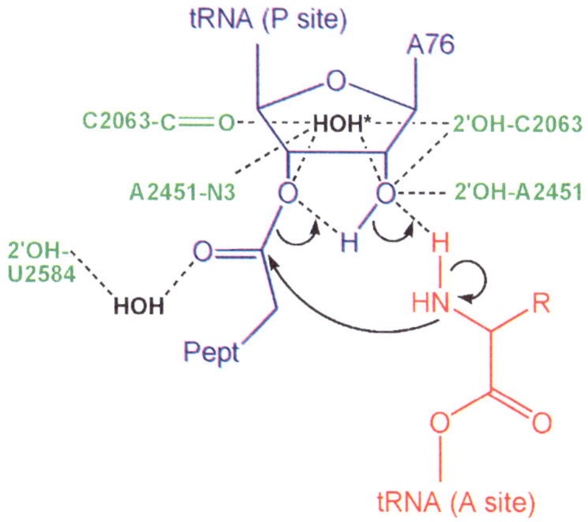

Figure 14.5 Concerted proton shuttle mechanism of peptide bond formation. Peptidyl-tRNA (P site) and aminoacyl-tRNA (A site) are blue and red, respectively, and ribosome residues are green, and ordered water molecules are gray. The attack of the α-NH$_2$ group on the ester carbonyl carbon results in a six-membered transition state in which the 2′-OH group of the A-site A76 ribose moiety donates its proton to the adjacent leaving 3′ oxygen and simultaneously receives a proton from the amino group.[33,47] Ribosomal residues are not involved in chemical catalysis but are part of the H-bond network that stabilizes the transition state. (Reproduced from refs. 81 and 84 with permission from Elsevier.)

involve chemical catalysis by ribosomal groups, but may be modulated by conformational changes at the active site that can be induced by protonation. In addition to placing the reactive groups into close proximity and precise orientation relative to each other, the ribosome appears to work by providing an electrostatic environment that reduces the free energy of forming the highly polar transition state, shielding the reaction against bulk water, helping the proton shuttle forming the leaving group, or a combination of these effects. With this preorganized network, the ribosome avoids the extensive solvent reorganization that is inevitable in the corresponding reaction in solution, resulting in significantly more favorable entropy of activation of the reaction on the ribosome.

In the early RNA world, where RNA provided both genetic information and catalytic function, the earliest protein synthesis would have had to be catalyzed by RNA, and the ribosome may provide insights as to how ancient RNA catalysts worked. The ribosome accelerates the peptidyl transfer reaction by a factor of about 10^7 and does not utilize chemical catalysis by the bases of rRNA. It is much less efficient than many protein enzymes, which use chemical catalysis and accelerate reactions by up to 10^{23}-fold.[79] Apparently, the evolutionary pressure had a much larger influence on increasing the speed and fidelity of the rate-limiting steps of protein synthesis that do not involve chemistry, such as substrate binding,[80] than on the chemistry step of peptide bond formation, allowing the ribosome to retain its catalytic strategy during the evolution of a prebiotic translational ribozyme into a modern ribosome. Later, the functions of the RNA-only catalysts may have been expanded by adding proteins. One particularly interesting example of how the protein factors modulate the activity of the ribosome is the hydrolysis of peptidyl-tRNA, which takes place in the peptidyl transferase center but needs to be induced by release factors. In the absence of release factor, the ribosome protects peptidyl-tRNA from hydrolysis by precluding the access of water to positions within the active site, from where it could attack the ester group, and the release factors may promote a conformational rearrangement at the active site to move the ester group of peptidyl-tRNA out of the protected conformation and allow the attack of the water molecule.[34] A conserved glutamine residue of the termination factor may help to position a hydrolytic water molecule for rapid reaction with the peptidyl-tRNA substrate. Thus, the function of the same active center is diversified by including protein groups into the catalytic center of the ribozyme, representing a step in evolution from the RNA to the protein world.

References

1. E.A. Doherty and J.A. Doudna, Ribozyme structures and mechanisms, *Annu. Rev. Biochem.*, 2000, **69**, 597.
2. H.F. Noller, V. Hoffarth and L. Zimniak, Unusual resistance of peptidyl transferase to protein extraction procedures, *Science*, 1992, **256**, 1416.

3. P. Khaitovich, A.S. Mankin, R. Green, L. Lancaster and H.F. Noller, Characterization of functionally active subribosomal particles from *Thermus aquaticus*, *Proc. Natl. Acad. Sci. U.S.A.*, 1999, **96**, 85.
4. G. Diedrich, C.M. Spahn, U. Stelzl, M.A. Schafer, T. Wooten, D.E. Bochkariov, B.S. Cooperman, R.R. Traut and K.H. Nierhaus, Ribosomal protein L2 is involved in the association of the ribosomal subunits, tRNA binding to A and P sites and peptidyl transfer, *EMBO J.*, 2000, **19**, 5241.
5. N. Ban, P. Nissen, J. Hansen, P.B. Moore and T.A. Steitz, The complete atomic structure of the large ribosomal subunit at 2.4 Å resolution, *Science*, 2000, **289**, 905.
6. P. Nissen, J. Hansen, N. Ban, P.B. Moore and T.A. Steitz, The structural basis of ribosome activity in peptide bond synthesis, *Science*, 2000, **289**, 920.
7. V.I. Katunin, G.W. Muth, S.A. Strobel, W. Wintermeyer and M.V. Rodnina, Important contribution to catalysis of peptide bond formation by a single ionizing group within the ribosome, *Mol. Cell*, 2002, **10**, 339.
8. E.M. Youngman, J.L. Brunelle, A.B. Kochaniak and R. Green, The active site of the ribosome is composed of two layers of conserved nucleotides with distinct roles in peptide bond formation and peptide release, *Cell*, 2004, **117**, 589.
9. J.L. Brunelle, E.M. Youngman, D. Sharma and R. Green, The interaction between C75 of tRNA and the A loop of the ribosome stimulates peptidyl transferase activity, *RNA*, 2006, **12**, 33.
10. M. Beringer and M.V. Rodnina, Importance of tRNA interactions with 23S rRNA for peptide bond formation on the ribosome: Studies with substrate analogs, *Biol. Chem.*, 2007, **388**, 687.
11. A.L. Weber and L.E. Orgel, The formation of dipeptides from amino acids and the 2'(3')-glycyl ester of an adenylate, *J. Mol. Evolution*, 1979, **13**, 185.
12. A.C. Satterthwait and W.P. Jencks, The mechanism of the aminolysis of acetate esters, *J. Am. Chem. Soc.*, 1974, **96**, 7018.
13. G.M. Blackburn and W.P. Jencks, The mechanism of the aminolysis of methyl formate, *J. Am. Chem. Soc.*, 1986, **90**, 2638.
14. W.P. Jencks and M. Gilchrist, Nonlinear structure-reactivity correlations. The reactivity of nucleophilic reagents toward esters, *J. Am. Chem. Soc.*, 1968, **90**, 2622.
15. A. Sievers, M. Beringer, M.V. Rodnina and R. Wolfenden, The ribosome as an entropy trap, *Proc. Natl. Acad. Sci. U.S.A.*, 2004, **101**, 7897.
16. M.V. Rodnina, M. Beringer and P. Bieling, Ten remarks on peptide bond formation on the ribosome, *Biochem. Soc. Trans.*, 2005, **33**, 493.
17. P.K. Sharma, Y. Xiang, M. Kato and A. Warshel, What are the roles of substrate-assisted catalysis and proximity effects in peptide bond formation by the ribosome?, *Biochemistry*, 2005, **44**, 11307.
18. J.A. Doudna and J.R. Lorsch, Ribozyme catalysis: Not different, just worse, *Nat. Struct. Mol. Biol.*, 2005, **12**, 395.

19. P.C. Bevilacqua, T.S. Brown, S. Nakano and R. Yajima, Catalytic roles for proton transfer and protonation in ribozymes, *Biopolymers*, 2004, **73**, 90.
20. R. Wolfenden, M. Snider, C. Ridgway and B. Miller, The temperature dependence of enzyme rate enhancements, *J. Am. Chem. Soc.*, 1999, **121**, 7419.
21. A.C. Seila, K. Okuda, S. Nunez, A.F. Seila and S.A. Strobel, Kinetic isotope effect analysis of the ribosomal peptidyl transferase reaction, *Biochemistry*, 2005, **44**, 4018.
22. S. Trobro and J. Åqvist, Analysis of predictions for the catalytic mechanism of ribosomal peptidyl transfer, *Biochemistry*, 2006, **45**, 7049.
23. M.V. Rodnina, K.B. Gromadski, U. Kothe and H.J. Wieden, Recognition and selection of tRNA in translation, *FEBS Lett.*, 2005, **579**, 938.
24. J.L. Hansen, T.M. Schmeing, P.B. Moore and T.A. Steitz, Structural insights into peptide bond formation, *Proc. Natl. Acad. Sci. U.S.A.*, 2002, **99**, 11670.
25. M.M. Yusupov, G.Z. Yusupova, A. Baucom, K. Lieberman, T.N. Earnest, J.H. Cate and H.F. Noller, Crystal structure of the ribosome at 5.5 Å resolution, *Science*, 2001, **292**, 883.
26. A. Bashan, I. Agmon, R. Zarivach, F. Schluenzen, J. Harms, R. Berisio, H. Bartels, F. Franceschi, T. Auerbach, H.A. Hansen, E. Kossoy, M. Kessler and A. Yonath, Structural basis of the ribosomal machinery for peptide bond formation, translocation, and nascent chain progression, *Mol. Cell*, 2003, **11**, 91.
27. D.F. Kim and R. Green, Base-pairing between 23S rRNA and tRNA in the ribosomal A site, *Mol. Cell*, 1999, **4**, 859.
28. P. Bieling, M. Beringer, S. Adio and M.V. Rodnina, Peptide bond formation does not involve acid-base catalysis by ribosomal residues, *Nat. Struct. Mol. Biol.*, 2006, **13**, 423.
29. T. Pape, W. Wintermeyer and M.V. Rodnina, Complete kinetic mechanism of elongation factor Tu-dependent binding of aminoacyl-tRNA to the A site of the *E. coli* ribosome, *EMBO J.*, 1998, **17**, 7490.
30. R.E. Monro, Catalysis of peptide bond formation by 50S ribosomal subunits from *Escherichia coli*, *J. Mol. Biol.*, 1967, **26**, 147.
31. R.E. Monro and K.A. Marcker, Ribosome-catalysed reaction of puromycin with a formylmethionine-containing oligonucleotide, *J. Mol. Biol.*, 1967, **25**, 347.
32. T.M. Schmeing, A.C. Seila, J.L. Hansen, B. Freeborn, J.K. Soukup, S.A. Scaringe, S.A. Strobel, P.B. Moore and T.A. Steitz, A pre-translocational intermediate in protein synthesis observed in crystals of enzymatically active 50S subunits, *Nat. Struct. Biol.*, 2002, **9**, 225.
33. T.M. Schmeing, K.S. Huang, D.E. Kitchen, S.A. Strobel and T.A. Steitz, Structural insights into the roles of water and the 2′ hydroxyl of the P site tRNA in the peptidyl transferase reaction, *Mol. Cell*, 2005, **20**, 437.
34. T.M. Schmeing, K.S. Huang, S.A. Strobel and T.A. Steitz, An induced-fit mechanism to promote peptide bond formation and exclude hydrolysis of peptidyl-tRNA, *Nature*, 2005, **438**, 520.

35. I. Wohlgemuth, M. Beringer and M.V. Rodnina, Rapid peptide bond formation on isolated 50S ribosomal subunits, *EMBO Rep.*, 2006, **7**, 669.
36. P.B. Moore and T.A. Steitz, The structural basis of large ribosomal subunit function, *Annu. Rev. Biochem.*, 2003, **72**, 813.
37. A. Yonath, Antibiotics targeting ribosomes: Resistance, selectivity, synergism and cellular regulation, *Annu. Rev. Biochem.*, 2005, **74**, 649.
38. J. Poehlsgaard and S. Douthwaite, The bacterial ribosome as a target for antibiotics, *Nat. Rev. Microbiol.*, 2005, **11**, 870.
39. N. Polacek and A.S. Mankin, The ribosomal peptidyl transferase center: Structure, function, evolution, inhibition, *Crit. Rev. Biochem. Mol. Biol.*, 2005, **40**, 285.
40. B.E. Maden and R.E. Monro, Ribosome-catalyzed peptidyl transfer. Effects of cations and pH value, *Eur. J. Biochem.*, 1968, **6**, 309.
41. K. Okuda, A.C. Seila and S.A. Strobel, Uncovering the enzymatic pK_a of the ribosomal peptidyl transferase reaction utilizing a fluorinated puromycin derivative, *Biochemistry*, 2005, **44**, 6675.
42. P.C. Bevilacqua, Mechanistic considerations for general acid-base catalysis by RNA: Revisiting the mechanism of the hairpin ribozyme, *Biochemistry*, 2003, **42**, 2259.
43. M.J. Fedor and J.R. Williamson, The catalytic diversity of RNAs, *Nat. Rev. Mol. Cell Biol.*, 2005, **6**, 399.
44. K. Okuda, A.C. Seila and S.A. Strobel, Uncovering the enzymatic pK_a of the ribosomal peptidyl transferase reaction utilizing a fluorinated puromycin derivative, *Biochemistry*, 2005, **44**, 6675.
45. M. Beringer, C. Bruell, L. Xiong, P. Pfister, P. Bieling, V.I. Katunin, A.S. Mankin, E. C. Bottger and M.V. Rodnina, Essential mechanisms in the catalysis of peptide bond formation on the ribosome, *J. Biol. Chem.*, 2005, **280**, 36065.
46. J.S. Feinberg and S. Joseph, A conserved base-pair between tRNA and 23S rRNA in the peptidyl transferase center is important for peptide release, *J. Mol. Biol.*, 2006, **364**, 1010.
47. S. Trobro and J. Åqvist, Mechanism of peptide bond synthesis on the ribosome, *Proc. Natl. Acad. Sci. U.S.A.*, 2005, **102**, 12395.
48. M. Selmer, C.M. Dunham, F.V.T. Murphy, A. Weixlbaumer, S. Petry, A.C. Kelley, J.R. Weir and V. Ramakrishnan, Structure of the 70S ribosome complexed with mRNA and tRNA, *Science*, 2006, **313**, 1935.
49. J. Harms, F. Schluenzen, R. Zarivach, A. Bashan, S. Gat, I. Agmon, H. Bartels, F. Franceschi and A. Yonath, High resolution structure of the large ribosomal subunit from a mesophilic eubacterium, *Cell*, 2001, **107**, 679.
50. A. Korostelev, S. Trakhanov, M. Laurberg and H.F. Noller, Crystal structure of a 70S ribosome-tRNA complex reveals functional interactions and rearrangements, *Cell*, 2006, **126**, 1065.
51. B.S. Schuwirth, M.A. Borovinskaya, C.W. Hau, W. Zhang, A. Vila-Sanjurjo, J.M. Holton and J. H. Cate, Structures of the bacterial ribosome at 3.5 Å resolution, *Science*, 2005, **310**, 827.

52. H.F. Noller, Ribosomal RNA and translation, *Annu. Rev. Biochem.*, 1991, **60**, 191.
53. R.R. Samaha, R. Green and H.F. Noller, A base pair between tRNA and 23S rRNA in the peptidyl transferase centre of the ribosome, *Nature*, 1995, **377**, 309.
54. B.A. Maguire, A.D. Beniaminov, H. Ramu, A.S. Mankin and R.A. Zimmermann, A protein component at the heart of an RNA machine: The importance of protein L27 for the function of the bacterial ribosome, *Mol. Cell*, 2005, **20**, 427.
55. R.R. Gutell, B. Weiser, C.R. Woese and H.F. Noller, Comparative anatomy of 16-S-like ribosomal RNA, *Prog. Nucl. Acid Res. Mol. Biol.*, 1985, **32**, 155.
56. H.F. Noller and C.R. Woese, Secondary structure of 16S ribosomal RNA, *Science*, 1981, **212**, 403.
57. D. Moazed and H.F. Noller, Intermediate states in the movement of transfer RNA in the ribosome, *Nature*, 1989, **342**, 142.
58. Y. Semenkov, T. Shapkina, V. Makhno and S. Kirillov, Puromycin reaction for the A site-bound peptidyl-tRNA, *FEBS Lett.*, 1992, **296**, 207.
59. Y.P. Semenkov, T.G. Shapkina and S.V. Kirillov, Puromycin reaction of the A-site bound peptidyl-tRNA, *Biochimie*, 1992, **74**, 411.
60. D. Sharma, D.R. Southworth and R. Green, EF-G-independent reactivity of a pre-translocation-state ribosome complex with the aminoacyl tRNA substrate puromycin supports an intermediate (hybrid) state of tRNA binding, *RNA*, 2004, **10**, 102.
61. R. Wolfenden, The mechanism of hydrolysis of amino acyl RNA, *Biochemistry*, 1963, **2**, 1090.
62. N.Y. Sardesai, R. Green and P. Schimmel, Efficient 50S ribosome-catalyzed peptide bond synthesis with an aminoacyl minihelix, *Biochemistry*, 1999, **38**, 12080.
63. N. Polacek, M.J. Gomez, K. Ito, L. Xiong, Y. Nakamura and A. Mankin, The critical role of the universally conserved A2602 of 23S ribosomal RNA in the release of the nascent peptide during translation termination, *Mol. Cell*, 2003, **11**, 103.
64. Z. Cai and I. Tinoco, Jr., Solution structure of loop A from the hairpin ribozyme from tobacco ringspot virus satellite, *Biochemistry*, 1996, **35**, 6026.
65. A.E. Hesslein, V.I. Katunin, M. Beringer, A.B. Kosek, M.V. Rodnina and S.A. Strobel, Exploration of the conserved A^+C wobble pair within the ribosomal peptidyl transferase center using affinity purified mutant ribosomes, *Nucl. Acids Res.*, 2004, **32**, 3760.
66. G.W. Muth, L. Ortoleva-Donnelly and S.A. Strobel, A single adenosine with a neutral pK_a in the ribosomal peptidyl transferase center, *Science*, 2000, **289**, 947.
67. J. Thompson, D.F. Kim, M. O'Connor, K.R. Lieberman, M.A. Bayfield, S.T. Gregory, R. Green, H.F. Noller and A.E. Dahlberg, Analysis of mutations at residues A2451 and G2447 of 23S rRNA in the

peptidyltransferase active site of the 50S ribosomal subunit, *Proc. Natl. Acad. Sci. U.S.A.*, 2001, **98**, 9002.
68. N. Polacek, M. Gaynor, A. Yassin and A.S. Mankin, Ribosomal peptidyl transferase can withstand mutations at the putative catalytic nucleotide, *Nature*, 2001, **411**, 498.
69. A.A. Leonov, P.V. Sergiev, A.A. Bogdanov, R. Brimacombe and O.A. Dontsova, Affinity purification of ribosomes with a lethal G2655C mutation in 23 S rRNA that affects the translocation, *J. Biol. Chem.*, 2003, **278**, 25664.
70. S.E. Kearsey and I.W. Craig, Altered ribosomal RNA genes in mitochondria from mammalian cells with chloramphenicol resistance, *Nature*, 1981, **290**, 607.
71. M. Beringer, S. Adio, W. Wintermeyer and M.V. Rodnina, The G2447A mutation does not affect ionization of a ribosomal group taking part in peptide bond formation, *RNA*, 2003, **9**, 919.
72. M.D. Erlacher, K. Lang, B. Wotzel, R. Rieder, R. Micura and N. Polacek, Efficient ribosomal peptidyl transfer critically relies on the presence of the ribose 2′-OH at A2451 of 23S rRNA, *J. Am. Chem. Soc.*, 2006, **128**, 4453.
73. A.A. Krayevsky and M.K. Kukhanova, The peptidyltransferase center of ribosomes, *Prog. Nucl. Acid Res. Mol. Biol.*, 1979, **23**, 1.
74. J.S. Weinger, K.M. Parnell, S. Dorner, R. Green and S.A. Strobel, Substrate-assisted catalysis of peptide bond formation by the ribosome, *Nat. Struct. Mol. Biol.*, 2004, **11**, 1101.
75. P.L. Adams, M.R. Stahley, A.B. Kosek, J. Wang and S.A. Strobel, Crystal structure of a self-splicing group I intron with both exons, *Nature*, 2004, **430**, 45.
76. S.O. Shan and D. Herschlag, Probing the role of metal ions in RNA catalysis: Kinetic and thermodynamic characterization of a metal ion interaction with the 2′-moiety of the guanosine nucleophile in the *Tetrahymena* group I ribozyme, *Biochemistry*, 1999, **38**, 10958.
77. A. Barta and I. Halama, in: *Ribosomal RNA and group I introns*, ed. R. Green and R. Schroeder, R.G. Landes Company, Austin, Texas, 1996, p. 35.
78. M.A. Rangelov, G.N. Vayssilov, V.M. Yomtova and D.D. Petkov, The syn-oriented 2′-OH provides a favorable proton transfer geometry in 1,2-diol monoester aminolysis: Implications for the ribosome mechanism, *J. Am. Chem. Soc.*, 2006, **128**, 4964.
79. A. Radzicka and R. Wolfenden, A proficient enzyme, *Science*, 1995, **267**, 90.
80. M.V. Rodnina and W. Wintermeyer, Fidelity of aminoacyl-tRNA selection on the ribosome: Kinetic and structural mechanisms, *Annu. Rev. Biochem.*, 2001, **70**, 415.
81. M. Beringer and M.V. Rodnina, The ribosomal peptidyl transferase, *Mol. Cell*, 2007, **26**, 311.

82. A. Vila-Sanjurjo, W.K. Ridgeway, V. Seymaner, W. Zhang, S. Santoso, K. Yu and J.H. Cate, X-ray crystal structures of the WT and a hyper-accurate ribosome from Escherichia coli, *Proc. Natl. Acad. Sci. U.S.A.*, 2003, **100**, 8682.
83. M. Diaconu, U. Kothe, F. Schlunzen, N. Fischer, J.M. Harms, A.G. Tonevitsky, H. Stark, M.V. Rodnina and M.C. Wahl, Structural basis for the function of the ribosomal L7/12 stalk in factor binding and GTPase activation, *Cell*, 2005, **121**, 991.
84. M.V. Rodnina, M. Beringer and W. Wintermeyer, How ribosomes make peptide bonds, *Trends Biochem. Sci.*, 2007, **32**, 20.

CHAPTER 15
Folding Mechanisms of Group I Ribozymes

SARAH A. WOODSON[a] AND PRASHANTH RANGAN[b]

[a] T.C. Jenkins Department of Biophysics, Johns Hopkins University, 3400 N. Charles St., Baltimore MD 21218-2685, USA; [b] Skirball Institute of Biomolecular Medicine, New York University Medical School, New York NY 10016, USA

15.1 Introduction

Recent discoveries have exposed the vast repertoire of cellular functions for RNA. No longer considered just a transducer of genetic information, RNA is known to function as a catalyst, genetic regulator and mediator of epigenetic information. These biological functions depend on dynamic structural changes in the RNA itself, in response to ligands, protein co-factors or other RNAs. The discovery of self-splicing introns and RNase P proved that RNA could function as a biological catalyst,[1,2] and stimulated interest in understanding how RNA catalysts select and activate their substrates. More recent discoveries of metabolite-sensing riboswitches[3] and small regulatory RNAs[4] illustrate how RNAs can directly control the expression of target genes. Increasingly, defects in RNA metabolism and regulation are linked to human genetic disease.[5–7] Thus, the parameters that affect the speed and faithfulness of RNA folding and assembly have become critical questions in RNA biology.

Many insights on how RNAs fold come from experiments on group I ribozymes and other catalytic RNAs, which continue to be important model systems for understanding RNA dynamics.[8] Group I ribozymes have a stable tertiary structure, the formation of which can be easily demonstrated by measuring the RNA's catalytic activity. This chapter focuses on what is known about the folding mechanism of group I ribozymes, and how the results have illuminated general principles of RNA folding.

15.2 Multi-domain Architecture of Group I Ribozymes

Group I ribozymes, which are derived from a family of self-splicing RNAs, are a good model system for folding studies, in part because they contain a well-folded and phylogenetically conserved core.[9,10] The ribozyme core can be divided into three helical domains, containing paired regions P3-P9, P4-P6, and P1-P2, respectively (Figure 15.1). As described in more detail in Chapter 10, interactions between core helices occur through a complex web of tertiary contacts, including a central triple helix at the junction of P4 and P6, a pseudoknot created by helices P3 and P7, and minor groove interactions between P3 and P6.[11,12]

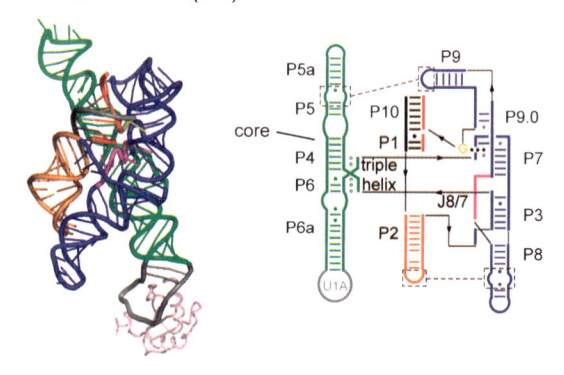

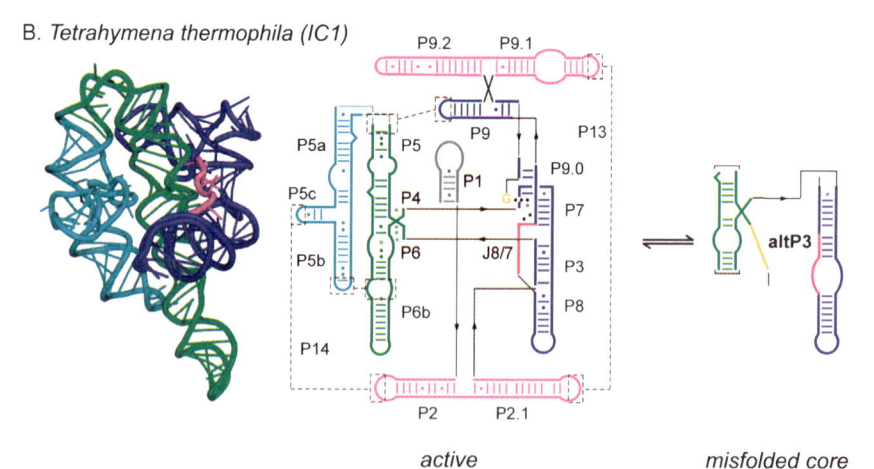

Figure 15.1 Structure of group I ribozymes. (*A*) *Azoarcus* bacterium (subgroup IC3), and (*B*) *Tetrahymena thermophila* (IC1), colored by domain. The catalytic core is shaded gray. Paired (P) and joining (J) regions are numbered as in Burke *et al.*[136] Ribbons from X-ray crystal structures of the *Azoarcus* splicing complex[137] and *Tetrahymena* ribozyme[138] were made with Pymol (Delano Scientific, http://pymol.sourceforge.net/).

The complex topology of the ribozyme core is intrinsically unstable. Consequently, additional tertiary contacts around the periphery are needed to reinforce the core.[13,14] These helices vary among subgroups, opening an opportunity to study the evolutionary constraints on folding and stability. In the smallest members of the family, such as the *Azoarcus* ribozyme, GAAA tetraloop–receptor interactions pin the ends of P2 and P9[15,16] (Figure 15.1). In larger family members, such as the *Tetrahymena* ribozyme, more elaborate helical extensions encircle the RNA and buttress the core tertiary interactions.[14] As we shall see, these peripheral helices can also determine the structures of the folding intermediates.

The active site for self-splicing is placed at the interface of all three major domains. Consequently, all three domains must fold correctly for splicing to occur, placing a premium on the fidelity and efficiency of the folding process. Within the active site, the non-bridging oxygen of the reactive phosphodiester and the oxygen of the leaving group are coordinated by two Mg^{2+} ions,[17,18] which also simultaneously coordinate nucleotides in all three helical domains. Thus, the reactivity of the RNA and the movement of substrates in and out of the active site are intimately linked to the structure of the catalytic core.

15.3 RNA Folding Problem

The RNA folding problem can be framed on lines similar to the Levinthal paradox for protein folding.[19] That is, how do RNA molecules fold into a unique structure without searching through all possible conformations available to them on a physiologically relevant timescale? Beginning with experiments on small hairpins and tRNA, the answers to these questions are gradually becoming clear.

15.3.1 Hierarchical Folding of tRNA

Early work on tRNA established that RNA folding was hierarchical, in that the secondary structure was independently stable and formed before the tertiary structure.[20] Base paired helices, which form the secondary structure, are stable under a wide range of conditions, while the tertiary interactions are only stable at low temperature and in salt or Mg^{2+}.[21] Unfolding experiments in Na^+ showed that the tertiary interactions and the D stem are first to melt followed by the other three stems of the tRNA molecule.[22] In the presence of Mg^{2+} ions, unfolding becomes a two-step transition going from the native state to a random coil state. This step is accompanied by the loss of bound Mg^{2+} ions in the D stem involved in tertiary interaction.[23]

The folding hierarchy also extends to the dynamics of tRNA. Using temperature jump and NMR experiments, it was shown that the secondary structure forms in 10 to 100 µs, whereas the L-shaped tertiary structure formed in 1–100 ms.[24,25] This relatively simple picture, in which the secondary structure forms first, followed by slower establishment of tertiary interactions, has been supported by many subsequent experiments on the dynamics of small hairpins and ribozymes.[26,27]

15.3.2 Coupling of Secondary and Tertiary Structure

In larger RNAs, exceptions to this simple hierarchy reveal that RNA folding mechanisms are, in reality, more complex. First, as we shall see below, non-native base pairs can also be captured by tertiary interactions, resulting in misfolded intermediates that must partly unfold to reach the native state.[8] Second, tertiary folding can be coupled to reorganization of the secondary structure, when the most stable conformation in the absence of Mg^{2+} is not the native one.[28] For example, folding of the P5abc domain in the *Tetrahymena* ribozyme is accompanied by a shift in the base pairing of the P5c helix.[29] Secondary structure rearrangements have been also proposed for a pseudoknot in the *E. coli* α operon mRNA,[30] the Hepatitis delta virus ribozyme,[31] and the Varkud satellite ribozyme.[32] Finally, RNA folding (milliseconds) is faster than synthesis by cellular polymerases (seconds). Thus, in the cell, folding of the RNA is vectorial and coupled to the kinetics of RNA polymerase elongation. Pausing at particular sites can influence the probability of forming certain intermediate conformations and thus the overall folding pathway of the RNA.[33]

15.4 Late Events: Formation of Tertiary Domains in the *Tetrahymena* Ribozyme

Given early results on tRNA dynamics, the next question is how multi-domain RNAs assemble, such as the *Tetrahymena* ribozyme and the ribozyme from *B. subtilis* RNase P. Early hydroxyl radical footprinting, chemical base modification and crosslinking experiments showed that the P4-P6 domain of the *Tetrahymena* ribozyme folded more readily and at lower Mg^{2+} concentrations than the P3-P9 domain (Figure 15.2).[34–36] These results were consistent with the fact that the P4-P6 domain can fold independently of the rest of the ribozyme,[37] and thus should be stable even under conditions in which the rest of the ribozyme is disordered.

Landmark studies using oligonucleotide hybridization and RNase H to monitor the formation of secondary structure showed that the P4-P6 domain of the *Tetrahymena* intron folds completely in less than a minute while the P3-P9 domain folded more slowly, at $\sim 0.7\,\text{min}^{-1}$.[38,39] From these results, it was initially proposed that the P4-P6 domain assembles first, and then provides a scaffold for assembly of the rest of the ribozyme.[39] The rate-limiting step was thought to be the assembly of the central triple helix that joins the two domains.[39]

15.4.1 Time-resolved Footprinting of Intermediates

Synchrotron hydroxyl radical footprinting using a synchrotron X-ray beam showed that the P4-P6 folds most rapidly, at a rate of $\sim 1\,\text{s}^{-1}$, followed by the

Folding Mechanisms of Group I Ribozymes 299

Tetrahymena ribozyme

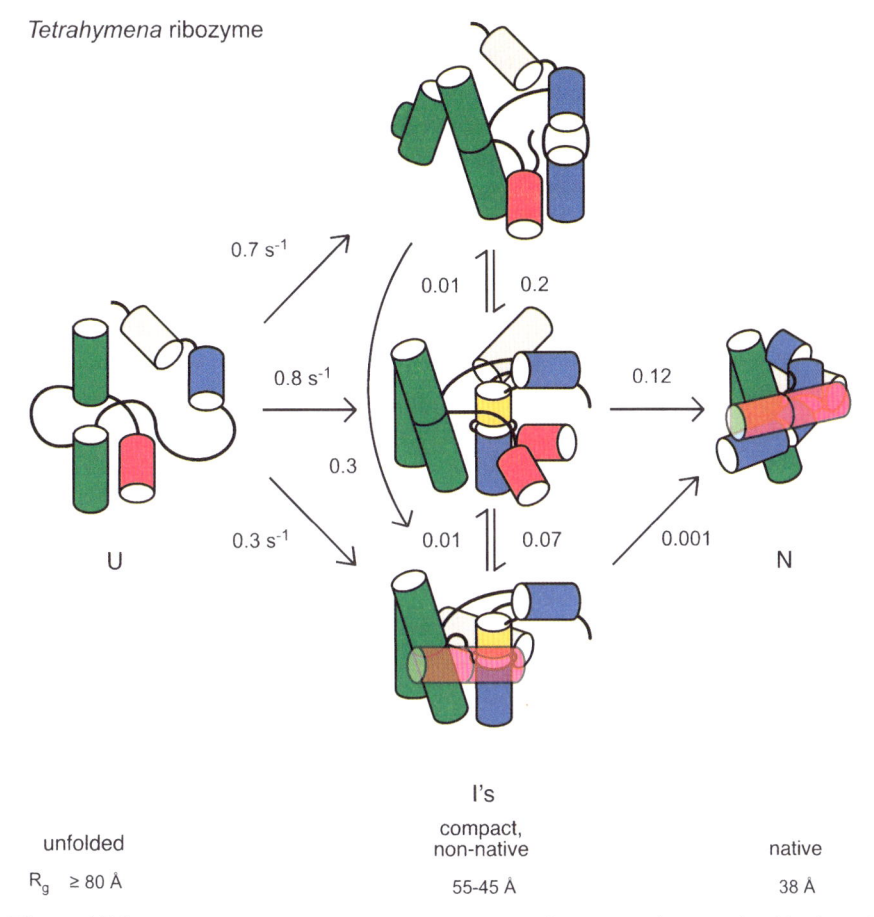

unfolded	I's compact, non-native	native
$R_g \geq 80$ Å	55-45 Å	38 Å

Figure 15.2 Folding pathway of the *Tetrahymena* ribozyme. Time-resolved hydroxyl radical footprinting[40,61] reveals three classes of tertiary folding intermediates in which P5abc (green; top), the entire P4-P6 domain (green; middle), or P4-P6 plus peripheral helices (pink; bottom) are folded. Slow organization of the core is due to mispairing of P3 (yellow).[50] Native gel analysis, however, shows that there are more than three intermediates.[46] Late stages of folding correlate with compaction of the RNA;[53,80] S. Moghaddam, R. Briber and S.A. Woodson (manuscript in preparation). Cartoon redrawn from ref. 61.

peripheral helices, which fold at a rate of $\sim 0.3\,\text{s}^{-1}$, then the intron core, which folds very slowly at a rate of $\sim 1\,\text{min}^{-1}$ at 42 °C (Figure 15.2).[40] This method allowed the backbone protection of individual nucleotides to be resolved with about 10 ms time resolution, providing a detailed look at the emergence of RNA tertiary interactions.[41]

These results initially suggested that the RNA structure forms sequentially, with each domain serving as the template for the next. However, experimental observations and theoretical considerations suggested that folding could not be strictly sequential. First, P5abc, which is the first part of the ribozyme to become fully folded, is not required for catalytic activity, although it stabilizes the active conformation of the ribozyme.[42,43] Second, mutations in the P5abc domain were found to speed up the overall folding rate of the RNA, suggesting that tertiary interactions in the P4-P6 domain must open up in order for other domains to fold.[44] Therefore, native tertiary interactions can kinetically trap the RNA by stabilizing the folding intermediates and raising the energy barrier that must be crossed on the way to the native state.

15.4.2 Misfolding of the Intron Core

If some parts of the *Tetrahymena* ribozyme can fold within seconds, why does the P3-P9 domain, which is in the interior of the ribozyme, require the longest time to fold completely? The explanation for this came from the presence of metastable folding intermediates of the *Tetrahymena* pre-rRNA and ribozyme that can be resolved from the native RNA on non-denaturing polyacrylamide gels.[45,46] The intermediates migrated as a broad smear that converted into the native RNA at a similar rate, suggesting the simultaneous refolding of many different intermediates.[45] The folding rate was increased by urea and high temperature, despite the fact that these conditions destabilize the folded RNA structure.[47] This observation was consistent with the presence of misfolded structures, which needed to partially unfold before being able to form the native structure. Urea was similarly found to increase the overall folding rate of the *B. subtilis* RNase P ribozyme, which was linked to misfolding of nucleotides joining the specificity and catalytic domains.[48,49]

Chemical modification interference and site-directed mutagenesis showed that folding intermediates of the *Tetrahymena* ribozyme are stabilized by mispairing of the native P3 pseudoknot, which is replaced by a non-native base pairing between the 3′ strand of P3 and J8/7 (altP3; Figure 15.1).[50] Thus, the prevalence of metastable folding intermediates in the *Tetrahymena* ribozyme is due in part to the propensity of the pseudoknot in the catalytic core to misfold. These results also explained why the P3-P9 domain forms more slowly than the P4-P6 domain. Remarkably, a single base substitution U273 to A, which stabilizes the native P3 helix relative to altP3, results in fast folding of the intron core, with an apparent rate constant of $1\,\mathrm{s}^{-1}$.[46]

15.4.3 Peripheral Stability Elements

The peripheral helices also stabilize altP3, increasing the stability (and lifetime) of the misfolded conformers. In addition to mutations that destabilize the P4-P6 domain,[44,51] mutations that destabilize the long-range P13 and P14 loop–loop interactions also increase the overall folding rate.[52] Thus, the peripheral stability elements, which are present in many RNAs, are a double-edged sword. On the one hand, they enhance the stability of the intrinsically unstable structural motifs that make up the catalytic core.[14] On the other hand, premature formation of peripheral interactions, before the core is correctly folded, can kinetically trap the RNA in misfolded conformations that refold slowly, if at all.[52,53]

15.5 Kinetic Partitioning among Parallel Folding Pathways

15.5.1 Theory and Experiment

The complexity of the ribozyme folding pathway and the presence of misfolded intermediates can be explained by free energy landscape models, which were first used to describe protein folding.[54,55] In such models, metastable conformers correspond to minima in the free energy surface, and the folding trajectory is a stochastic search along the energy surface for the lowest free energy state. The propensity of the *Tetrahymena* ribozyme to become kinetically trapped in misfolded intermediates is consistent with a rough free energy landscape, in which small changes in sequence or conformation result in large changes in stability.[51,56] Many local minima with similar free energies as the native state are separated by high energy barriers, resulting in slow transitions between them. This rough free energy landscape for RNA folding arises from the low sequence complexity of RNA[57,58] and the stability of RNA secondary structure compared with the tertiary interactions.[21]

An important prediction of free energy landscape models is that the folding population will kinetically partition among competing folding pathways toward the native state.[56] In a typical folding experiment, the initial "unfolded" ensemble occupies many thermally accessible conformations. When experimental conditions are shifted to favor the native state, a subset of the population directly reaches the native state without populating intermediate states. The rest becomes kinetically trapped in local energy minima that represent partially misfolded intermediates.[56] These fast and slow folding routes exist simultaneously in the population, because individual molecules stochastically begin folding from different points in the conformational landscape. As predicted by the kinetic partitioning model, 5–10% of the *Tetrahymena* pre-rRNA was able to fold in a much shorter time than predicted from the observed folding rate.[47] These rates are consistent with a small fraction of RNA folding directly to the native conformation.

15.5.2 Single Molecule Folding Studies

Direct evidence for parallel folding pathways came from single-molecule FRET experiments conducted on the *Tetrahymena* ribozyme,[59] in which fluorescent donor and acceptor dyes were conjugated to the 5′ and 3′ ends of the RNA. As the RNA folds, the donor and acceptor dyes are brought closer together, increasing the FRET signal compared to the unfolded state. A histogram of folding times showed that the ribozyme folded with two distinct rate constants: about 12% of the population folded at the rate of $\sim 1\,\mathrm{s}^{-1}$ and the rest folded at a rate of $\sim 0.02\,\mathrm{s}^{-1}$.[59] These rates are consistent with the results of solution ensemble studies.[46,47] The mutation U273A or preincubation in salt increased the proportion of RNA that folded rapidly,[53] also as predicted by ensemble experiments.[46,51,60] Thus, the partitioning of the population among different pathways depends on the sequence of the RNA, the distribution of molecules in the initial unfolded ensemble, and the relative stability of the transition state ensemble.[56]

15.5.3 Estimating the Flux through Footprinting Intermediates

Recently, a more detailed analysis of time-resolved footprinting data has shown that the progression of tertiary structure from P4-P6 to P3-P9 is also best explained by partitioning of the population among parallel folding pathways (Figure 15.2).[61] The footprinting data could be reasonably modeled by a folding mechanism involving three different intermediates, although the actual number of intermediate states is likely to be much higher. Relatively few molecules were found to fold directly from U to N in 10 mM Mg^{2+} under the conditions of the footprinting experiments.[61] By contrast, the direct pathway accounted for 18% of the flux when the RNA was refolded in 1.5 M NaCl,[61] which is consistent with more rapid organization of the RNA in monovalent salts (see below).[60]

15.5.4 Kinetic Partitioning *In Vivo*

The *Tetrahymena* ribozyme or intron self-splices 20–50 times more efficiently in the cell than *in vitro*.[62,63] Nonetheless, mutations that increase the propensity of the RNA to misfold *in vitro* also lower the fraction of spliced RNA in *E. coli* and in yeast, indicating that the RNA folds through similar intermediates in the cell as in the test tube.[64,65] Unexpectedly, the half-life of the unspliced pre-RNA remaining in the cell was too long to account for the accumulation of spliced products.[65,66] This led to the realization that the unspliced pre-RNA likely represents a residue of misfolded RNA that is incompetent to splice and that is degraded before it has a chance to refold.[66] Thus, the intracellular activity of the RNA is best explained by partitioning of new transcripts along parallel folding routes. How misfolded transcripts are identified and discarded is an important question that is just beginning to be addressed.

15.6 Early Events: Counterion-dependent RNA Collapse

If the folding kinetics are determined by the flux through alternative pathways to the native state, what interactions determine which pathways are preferred? This is a critical question, as the reliability of the folding process can determine the fidelity of RNA processing reactions and the assembly of RNA–protein complexes. Many experiments from different laboratories have shown that the earliest steps of tertiary folding are driven by the association of counterions with the RNA, which in turn induces a collapse into more compact intermediates.[67] These compact intermediates undergo conformational rearrangements to form stable tertiary folding intermediates and the native RNA.

At low ionic strength, charge repulsion drives the double helices apart, keeping the RNA chain in an extended state. Thus, one of the main requirements for folding is the neutralization of the phosphate backbone by cations.[58,68,69] We estimate that condensation of counterions around the RNA neutralizes more than 90% of the phosphate charge.[69] This charge neutralization triggers the collapse of the RNA into compact intermediates.[70–72] Following collapse, subsequent conformational rearrangements lead to stable tertiary folding intermediates and ultimately the native structure.

15.6.1 Compact Non-native Form of bI5 Ribozyme

Evidence for compact yet non-native states of RNA come from various experiments on group I ribozymes. Based on size-exclusion chromatography, native gel electrophoresis, footprinting and other biochemical assays, the yeast mitochondrial bI5 group I intron was proposed to form a collapsed state early in the folding pathway at low to moderate Mg^{2+} concentrations.[70,73] This collapsed state, which is not yet catalytically active, can form the active structure in higher Mg^{2+}, or be captured by the CPB2 protein in 7 mM $MgCl_2$ to form the native RNA–protein complex.[74] Photo-crosslinking of bI5 RNA in the compact and native states showed that the compact state already contains native-like structure, despite the fact that the RNA is not catalytically active under these conditions.[75] Similarly, the group I intron from *Candida albicans* was found to form the compact state early in the folding pathway, before the appearance of tertiary interactions.[76]

15.6.2 Small Angle X-ray Scattering of *Tetrahymena* Ribozyme

Through small angle X-ray scattering (SAXS) it was shown that the *Tetrahymena* intron goes from an extended structure with a radius of gyration (R_g) of 74 Å, to a compact globular intermediate with an R_g of 51–55 Å in the presence of counterions, to a folded structure with an R_g of 45–47 Å in 10–15 mM $MgCl_2$.[77] A similar picture was obtained from SAXS experiments on tRNA and RNase P,[78] which again showed that the expanded RNA in low salt buffer collapsed in a cooperative transition to a more compact globule in salt and $MgCl_2$.

Time-resolved SAXS experiments on the *Tetrahymena* ribozyme showed that the collapse occurred in two kinetic phases, with time constants on the order of 7 and 140 ms.[72] The model favored by the authors was that of an unfolded ensemble that gives rise to a partially collapsed state, which then rearranges to a more compact intermediate. Another explanation for the two kinetic phases is presence of multiple parallel folding pathways. A quadruple mutant of the *Tetrahymena* ribozyme in which key tertiary contacts throughout the RNA were destabilized was able to go through the initial transition to R_g of 55 Å but not the second transition to 45 Å,[79] suggesting that the initial phase is driven primarily by non-specific interactions but that further compaction requires specific folding of the RNA chain.

To better delineate which interactions contribute to the multiple collapse transitions in the *Tetrahymena* ribozymes, time-resolved SAXS and time-resolved hydroxyl radical footprinting experiments were carried out under the same conditions.[80] No backbone protections were observed during the initial collapse transition (<10 ms; $R_g = 55$ Å), but tertiary contacts within the P4-P6 domain appeared during or just after the second transition (100 ms; $R_g = 45$ Å). As expected, tertiary contacts within the other domains of the ribozyme appeared more slowly. Together, these studies suggest that the initial collapse transition precedes the stable formation of tertiary contacts, and may be driven primarily by counterion-dependent relaxation of the charge repulsion of the phosphates. The formation of stable and well-folded structures in RNA, however, correlates with the nucleation and maintenance of specific tertiary interactions.

15.6.3 Native-like Folding Intermediates in the *Azoarcus* Ribozyme

Further insights into the interactions that drive early folding transitions came from experiments on the *Azoarcus* group I ribozyme. This small and very stable group I ribozyme folds in two macroscopic transitions, as the concentration of Mg^{2+} is increased from zero to 15 mM $MgCl_2$ (Figure 15.3).[81] At low ionic strength, some double helices are present, but the tertiary structure is unfolded (U). Sub-millimolar Mg^{2+} induces the assembly of the core helices and the formation of compact native-like intermediates (I_C), while tenfold higher Mg^{2+} concentrations are needed to form the native tertiary structure (N), which is

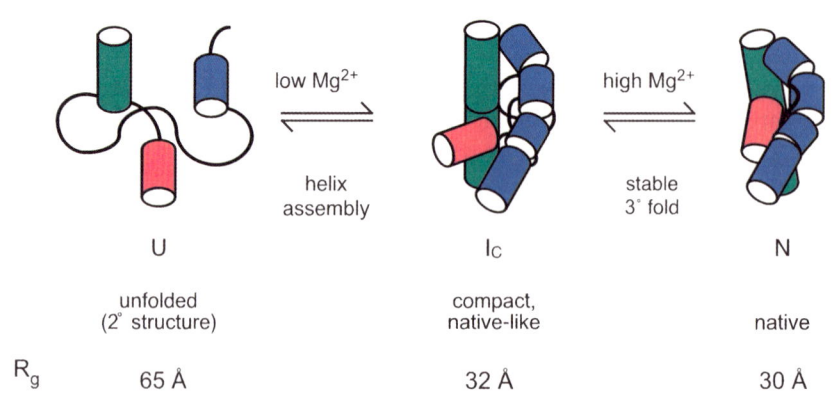

Figure 15.3 Equilibrium folding pathway of the *Azoarcus* ribozyme. Low (0.2 mM) Mg^{2+} ion concentration induces cooperative collapse to compact intermediates that are folded but dynamic.[83] In 2 mM Mg^{2+} the tertiary structure becomes stably folded and catalytic activity ensues. The overall folding time at 37 °C is <50 ms.[81] Redrawn from ref. 81.

detected by catalytic activity and protection of the RNA backbone from hydroxyl radical cleavage. Unlike the *Tetrahymena* ribozyme, the *Azoarcus* ribozyme folds in less than 50 ms, suggesting that a large fraction of the RNA population can fold directly to the native state without becoming trapped in misfolded intermediates.[81]

Small-angle neutron and X-ray scattering showed that assembly of core helices coincides with cooperative collapse of the RNA into more compact structures.[82,83] Surprisingly, the intermediates were nearly as compact as the native RNA (Figure 15.3). Site-directed mutagenesis of nucleotides involved in key tertiary interactions destabilized the compact intermediate state (I_C), as well as the native state, showing that tertiary interactions have already formed in the intermediate.[83] This was unexpected, because the RNA backbone remains fully accessible to the solvent, as judged from cleavage by hydroxyl radical. One possibility is that the tertiary interactions in the intermediate state are dynamic, allowing solvent to enter the interior of the RNA.

15.6.4 Early Folding Intermediates of the P4-P6 RNA

Experiments on the P4-P6 domain of the *Tetrahymena* ribozyme further illustrate the interplay of counterions and RNA tertiary interactions in the formation of RNA tertiary structure. The P4-P6 domain is formed by a 150° bend in J5/5a, which brings the P4, P5 and P6 helices into proximity with the P5abc subdomain.[37,84] Using a pyrene fluorophore conjugated to P4, Silverman *et al.*[85,86] showed that the P4-P6 domain folded through more than one mechanism, and that the folding transition occurred before all of the tertiary contacts are established.[87] Using time-resolved footprinting, Deras *et al.*[88] found that the folding rate of the P4-P6 RNA increased several hundred fold when the RNA was preincubated in 100–200 mM NaCl before the addition of Mg^{2+}. This increase in the folding rate correlated with "electrostatic relaxation" of the RNA conformation, which was observed as an early folding transition in NaCl by SAXS.[89]

Interestingly, folding of the P5abc subdomain was delayed by interactions with P4-P6; mutations that prevented interactions between the two halves of the P4-P6 domain decreased the apparent folding time of P5abc from 0.5 s to <50 ms.[88] This result suggested that transient interactions between helices also slow down the conformational search that ultimately produces the complete 3D fold. This kind of search through transient yet populated compact states has been proposed to explain the slow folding of a large group II ribozyme[90] (Chapter 11).

15.7 Counterions and Folding of Group I Ribozymes

15.7.1 Metal Ions and RNA Folding

Because of their negative charge, RNAs require cations to stabilize their tertiary structure and for catalysis.[67,91] Thus, the free-energy landscape and the folding kinetics are expected to be highly sensitive to the presence of counterions. Most metal ions associate non-specifically with the negative charge of the RNA.[92] Counterions are strongly attracted to the RNA due to the polyelectrolyte effect, which arises from the close spacing of negative charges.[93] In the counterion condensation model, condensed ions are confined to a volume surrounding the RNA, but remain hydrated and can diffuse freely with respect to the RNA. The local concentration of condensed ions is sufficient to neutralize 80–95% of the RNA's negative charge.[94]

When the RNA folds, the phosphates are brought into greater proximity, intensifying the negative charge potential of the RNA. The thermodynamic stabilization of the folded RNA by Mg^{2+} can be explained by the preferential association of cations with the folded RNA compared with the unfolded RNA.[92,95,96] The strength of this interaction is greater for multivalent cations than monovalent cations, in part explaining why RNAs are more stably folded in Mg^{2+} than in Na^+ or K^+.[97,98] The distribution of metal ions around the RNA depends on the RNA's electrostatic potential, which has been successfully predicted by mean field theories such as the nonlinear Poisson–Boltzmann equation[95,99] and Brownian dynamics simulations.[100]

15.7.2 Valence and Size of Counterions Matter

Because most of the cations associated with RNA remain fully or partly hydrated, this non-specific charge screening can be satisfied by a wide range of cations. Not surprisingly, the *Tetrahymena* and *Azoarcus* ribozymes were found to form near-native structures in almost all cations tested, including polyamines.[69,101] As expected, the folded RNA is stabilized more efficiently by multivalent cations than by monovalent cations.[69] However, the folding free energy increased linearly with the size or charge density of group IIA metal ions (Figure 15.4).[102] As the same trend was observed among polyamines[103] and in

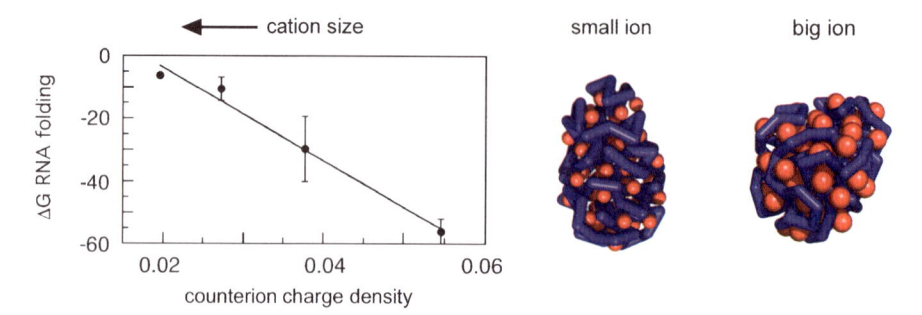

Figure 15.4 RNA folding and charge density of metal ions. Left: the stability of the folded *Tetrahymena* ribozyme in group IIA metal ions decreases with the size of the hydrated ion, or increases with the counterion charge density. Right: compact globules formed by a model polyelectrolyte (blue) with a small ion the size of Mg^{2+} or a larger ion the size of Ba^{2+}. (Reprinted from ref. 102 with permission.)

simulations of a model polyelectrolyte,[102] this effect is likely due to the excluded volume of the condensed counterions, rather than changes in hydration energy or other chemical properties. Non-denaturing gel electrophoresis, size-exclusion chromatography and SAXS all suggest that low charge density counterions produce a more dynamic ensemble of folded RNAs than high charge density counterions.[102]

Although the native RNA is less stable in large, monovalent ions than in small, multivalent ions, the folding kinetics of the *Tetrahymena* ribozyme are faster. This is because the folding intermediates are less stable in monovalent ions than in divalent or trivalent ions.[60,104] The change in the folding kinetics was consistent with a movement of the folding transition state ensemble toward the native state in monovalent ions, and a broader or more plastic transition state ensemble in monovalent ions compared with divalent ions.[104] This agrees with the result that an increased fraction of the ribozyme population folds directly from U to N in NaCl, compared with $MgCl_2$.[61] Overall, the folding kinetics of the *Tetrahymena* ribozyme were optimal in a mixture of monovalent and divalent ions that approximates physiological conditions.[60,105]

15.7.3 Specific Metal Ion Coordination and Folding

Group I ribozymes require Mg^{2+} (or Mn^{2+}) for catalytic activity.[106,107] Footprinting studies showed that while the *Azoarcus* ribozyme forms 3D structure in any counterion, the active site requires Mg^{2+} to fold.[101] Tb^{3+} cleavage assays identified two high affinity Mg^{2+} binding sites in the active site,[101] which were subsequently found to agree with two active site metal ions observed by X-ray diffraction.[18] Direct coordination of partially dehydrated metal ions is expected to be selective for specific types of ions, because they depend on the coordination geometry and hydration energies of the metal.[108]

Site bound metal ions can influence folding pathways by stabilizing a particular conformation of the RNA. Additional examples of coordinated metal ions in group I ribozymes are a cluster of five Mg^{2+} ions in the P5abc domain of the *Tetrahymena* ribozyme,[109,110] and the selective association of a K^+ ion with the AA platform in tetraloop receptors.[111] The latter may explain why the *Azoarcus* ribozyme, which is stabilized by two tetraloop–receptor motifs, can fold readily in Na^+ while the *Tetrahymena* ribozyme cannot.[101] On the other hand, because dehydration of the metal ion and specific coordination require a highly organized and favorable binding site within the RNA, such interactions are likely to occur relatively late in the folding process.

15.8 Protein-dependent Folding of Group I Ribozymes

The genetics of gene expression in yeast mitochondria revealed that many group I and group II introns require proteins to splice *in vivo*.[112,113] Some of these proteins are encoded by ORFS within the intron itself, while others, such as the *Neurospora* CYT-18 tyrosyl tRNA synthetase and the yeast CBP2 protein, are encoded by nuclear genes. Given the weakness of RNA tertiary interactions and the tendency of the catalytic core to misfold, how might proteins promote folding of the RNA and the assembly of an active RNP? As we shall see below, proteins can assist the folding of group I ribozymes and other RNAs by stabilizing the native fold and by stimulating reorganization of folding intermediates.[114]

15.8.1 Stabilization of RNA Tertiary Structure

There are now many examples in which proteins stabilize RNA tertiary structure by specifically binding the correctly folded RNA. For example, the *Neurospora* CYT-18 protein binds to the P4 and P6 helices of group I introns,[115] functionally replacing the P5abc element found in the *Tetrahymena* ribozyme and other members of its subfamily.[116] Chemical probing and mutagenesis experiments show that stabilization of the P4-P6 domain by binding of CYT-18 also stabilizes the conformation of the catalytic core.[74,117–119]

The yeast mitochondrial bI5 ribozyme is catalytically active in 40 mM $MgCl_2$, demonstrating that the RNA is able to function as a catalyst independently of its cognate proteins.[120] Genetic and biochemical experiments showed that bI5 requires the CBP2 protein to function under physiological conditions,[120,121] and chemical base modification and hydroxyl radical footprinting experiments confirmed that binding of CBP2 to the RNA stabilizes the tertiary structure of the RNA under physiological conditions.[122,123] CBP2 also promotes interactions between the core of the ribozyme and a separate domain containing the 5′ splice site helix, ensuring the formation of an active RNP.[123]

Recent single-molecule FRET studies showed that CBP2 can bind the compact form of bI5 RNA in both non-specific and specific modes.[124] Non-specific binding results in large dynamic fluctuations in the RNA, which ultimately lead to the native complex. Studies of the *Aspergillus nidulans* I-*Ani*I protein and its cognate group I ribozyme also found that assembly of the native RNP goes through several stages. Binding of the protein to the partly disordered RNA stimulates a structural reorganization of the RNA and the protein required to achieve the mature complex.[125]

15.8.2 Stimulation of Refolding by RNA Chaperones

Although RNA binding proteins such as CYT-18 and CBP2 stabilize the native structure of the RNA, they do not necessarily ensure the accuracy of folding. In fact, both proteins have been found to also stabilize misfolded RNA conformations.[126,127] Thus, just like the peripheral RNA stability elements they replace, RNA binding proteins can stabilize RNA tertiary structure at the expense of rapid refolding kinetics. Interestingly, splicing of yeast and *Neurospora* mitochondrial introns also requires DEAD-box RNA-dependent ATPases, which have the potential to unwind the RNA.[128,129] Using some clever RNA engineering, Lambowitz and colleagues showed that the *Neurospora* CYT-19 can accelerate the assembly of active RNA-CYT-18 complex, and can stimulate refolding of the P3 helix within the catalytic core.[128] In addition to DEAD-box proteins, RNA binding proteins that passively bind single-stranded RNA such as StpA can also help refold RNAs *in vivo*, and may help overcome the propensity of the catalytic core of group I ribozymes to misfold.[130–132] Refolding of RNA intermediates can also be triggered by the passage of ribosomes along the mRNA, directly coupling the structure of the RNA to translation.[133–135]

15.9 Conclusion

The stability and reactivity of group I introns have made them invaluable model systems for understanding how RNAs fold and how they function within the environment of the cell. Critical issues that are beginning to be addressed are which interactions nucleate the folding reaction itself, folding fidelity and kinetic partitioning among parallel routes, the extent of cooperativity among RNA tertiary interactions, and the recognition of compact yet dynamic intermediate states by RNA-binding proteins. These questions are likely to be addressed by increasingly sophisticated tools for probing the dynamics of RNA *in vitro* and within cells.

References

1. K. Kruger, P.J. Grabowski, A.J. Zaug, J. Sands, D.E. Gottschling and T.R. Cech, *Cell*, 1982, **31**, 147.
2. C. Guerrier-Takada, K. Gardiner, T. Marsh, N. Pace and S. Altman, *Cell*, 1983, **35**, 849.
3. M. Mandal and R.R. Breaker, *Nat. Rev. Mol. Cell Biol.*, 2004, **5**, 451.
4. G. Storz, J.A. Opdyke and A. Zhang, *Curr. Opin. Microbiol.*, 2004, **7**, 140.
5. N.A. Faustino and T.A. Cooper, *Genes Dev.*, 2003, **17**, 419.
6. L.P. Ranum and T.A. Cooper, *Annu. Rev. Neurosci.*, 2006, **29**, 259.
7. G.A. Calin and C.M. Croce, *Cancer Res.*, 2006, **66**, 7390.
8. S.A. Woodson, *Cell Mol. Life Sci.*, 2000, **57**, 796.
9. F. Michel, A. Jacquier and B. Dujon, *Biochimie*, 1982, **64**, 867.
10. T.R. Cech, *Annu. Rev. Biochem.*, 1990, **59**, 543.
11. P.L. Adams, M.R. Stahley, M.L. Gill, A.B. Kosek, J. Wang and S.A. Strobel, *RNA*, 2004, **10**, 1867.
12. Q. Vicens and T.R. Cech, *Trends Biochem. Sci.*, 2006, **31**, 41.
13. A.A. Beaudry and G.F. Joyce, *Biochemistry*, 1990, **29**, 6534.
14. V. Lehnert, L. Jaeger, F. Michel and E. Westhof, *Chem. Biol.*, 1996, **3**, 993.
15. M. Tanner and T. Cech, *RNA*, 1996, **2**, 74.
16. Y. Ikawa, K. Nohmi, S. Atsumi, H. Shiraishi and T. Inoue, *J. Biochem. (Tokyo)*, 2001, **130**, 251.

17. J.A. Piccirilli, J.S. Vyle, M.H. Caruthers and T.R. Cech, *Nature*, 1993, **361**, 85.
18. M.R. Stahley and S.A. Strobel, *Science*, 2005, **309**, 1587.
19. C. Levinthal, in: *Mossbauer Spectroscopy in Biological Systems: Proceedings of a Meeting held at Allerton House, Monticello, Illinois*, ed. J.T.P. DeBrunner and E. Munck, University of Illinois Press, Champaign-Urbana IL, 1969, p. 22.
20. J.R. Fresco, A. Adams, R. Ascione, D. Henley and T. Lindahl, *Cold Spring Harbor Symp. Quant. Biol.*, 1966, **31**, 527.
21. P. Brion and E. Westhof, *Annu. Rev. Biophys. Biomol. Struct.*, 1997, **26**, 113.
22. A. Stein and D.M. Crothers, *Biochemistry*, 1976, **15**, 160.
23. A. Stein and D.M. Crothers, *Biochemistry*, 1976, **15**, 157.
24. D.M. Crothers, P.E. Cole, C.W. Hilbers and R.G. Shulman, *J. Mol. Biol.*, 1974, **87**, 63.
25. D.C. Lynch and P.R. Schimmel, *Biochemistry*, 1974, **13**, 1841.
26. I.J. Tinoco and C. Bustamante, *J. Mol. Biol.*, 1999, **293**, 271.
27. D.M. Crothers, in: *RNA*, ed. D. Söll, S. Nishimura and P. Moore, Elsevier, Oxford UK, 2001, p. 61.
28. D. Thirumalai, *Proc. Natl. Acad. Sci. U.S.A.*, 1998, **95**, 11506.
29. M. Wu and I. Tinoco, Jr., *Proc. Natl. Acad. Sci. U.S.A.*, 1998, **95**, 11555.
30. T.C. Gluick and D.E. Draper, *J. Mol. Biol.*, 1994, **241**, 246.
31. T.S. Brown, D.M. Chadalavada and P.C. Bevilacqua, *J. Mol. Biol.*, 2004, **341**, 695.
32. A.A. Andersen and R.A. Collins, *Proc. Natl. Acad. Sci. U.S.A.*, 2001, **98**, 7730.
33. T. Pan and T. Sosnick, *Annu. Rev. Biophys. Biomol. Struct.*, 2006, **35**, 161.
34. D.W. Celander and T.R. Cech, *Science*, 1991, **251**, 401.
35. A.R. Banerjee and D.H. Turner, *Biochemistry*, 1995, **34**, 6504.
36. W.D. Downs and T.R. Cech, *RNA*, 1996, **2**, 718.
37. F.L. Murphy and T.R. Cech, *Biochemistry*, 1993, **32**, 5291.
38. P.P. Zarrinkar and J.R. Williamson, *Science*, 1994, **265**, 918.
39. P.P. Zarrinkar and J.R. Williamson, *Nat. Struct. Biol.*, 1996, **3**, 432.
40. B. Sclavi, M. Sullivan, M.R. Chance, M. Brenowitz and S.A. Woodson, *Science*, 1998, **279**, 1940.
41. B. Sclavi, S. Woodson, M. Sullivan, M.R. Chance and M. Brenowitz, *J. Mol. Biol.*, 1997, **266**, 144.
42. R. Russell and D. Herschlag, *J. Mol. Biol.*, 1999, **291**, 1155.
43. T.H. Johnson, P. Tijerina, A.B. Chadee, D. Herschlag and R. Russell, *Proc. Natl. Acad. Sci. U.S.A.*, 2005, **102**, 10176.
44. D.K. Treiber, M.S. Rook, P.P. Zarrinkar and J.R. Williamson, *Science*, 1998, **279**, 1943.
45. V.L. Emerick and S.A. Woodson, *Proc. Natl. Acad. Sci. U.S.A.*, 1994, **91**, 9675.
46. J. Pan, M.L. Deras and S.A. Woodson, *J. Mol. Biol.*, 2000, **296**, 133.

47. J. Pan, D. Thirumalai and S.A. Woodson, *J. Mol. Biol.*, 1997, **273**, 7.
48. T. Pan and T.R. Sosnick, *Nat. Struct. Biol.*, 1997, **4**, 931.
49. T. Pan, X. Fang and T. Sosnick, *J. Mol. Biol.*, 1999, **286**, 721.
50. J. Pan and S.A. Woodson, *J. Mol. Biol.*, 1998, **280**, 597.
51. M.S. Rook, D.K. Treiber and J.R. Williamson, *J. Mol. Biol.*, 1998, **281**, 609.
52. J. Pan and S.A. Woodson, *J. Mol. Biol.*, 1999, **294**, 955.
53. R. Russell, X. Zhuang, H.P. Babcock, I.S. Millett, S. Doniach, S. Chu and D. Herschlag, *Proc. Natl. Acad. Sci. U.S.A.*, 2002, **99**, 155.
54. J.D. Bryngelson, J.N. Onuchic, N.D. Socci and P.G. Wolynes, *Proteins*, 1995, **21**, 167.
55. K.A. Dill and H.S. Chan, *Nat. Struct. Biol.*, 1997, **4**, 10.
56. D. Thirumalai and S.A. Woodson, *Acc. Chem. Res.*, 1996, **29**, 433.
57. D. Herschlag, *J. Biol. Chem.*, 1995, **270**, 20871.
58. D. Thirumalai, N. Lee, S.A. Woodson and D. Klimov, *Annu. Rev. Phys. Chem.*, 2001, **52**, 751.
59. X. Zhuang, L.E. Bartley, H.P. Babcock, R. Russell, T. Ha, D. Herschlag and S. Chu, *Science*, 2000, **288**, 2048.
60. S.L. Heilman-Miller, J. Pan, D. Thirumalai and S.A. Woodson, *J. Mol. Biol.*, 2001, **309**, 57.
61. A. Laederach, I. Shcherbakova, M.P. Liang, M. Brenowitz and R.B. Altman, *J. Mol. Biol.*, 2006, **358**, 1179.
62. S.L. Brehm and T.R. Cech, *Biochemistry*, 1983, **22**, 2390.
63. F. Zhang, E.S. Ramsay and S.A. Woodson, *RNA*, 1995, **1**, 284.
64. T. Nikolcheva and S.A. Woodson, *J. Mol. Biol.*, 1999, **292**, 557.
65. S.A. Jackson, S. Koduvayur and S.A. Woodson, *RNA*, 2006, **12**, 2149.
66. S.P. Koduvayur and S.A. Woodson, *RNA*, 2004, **10**, 1526.
67. S.A. Woodson, *Curr. Opin. Chem. Biol.*, 2005, **9**, 104.
68. J.L. Leroy, M. Gueron, G. Thomas and A. Favre, *Eur. J. Biochem.*, 1977, **74**, 567.
69. S.L. Heilman-Miller, D. Thirumalai and S.A. Woodson, *J. Mol. Biol.*, 2001, **306**, 1157.
70. K.L. Buchmueller, A.E. Webb, D.A. Richardson and K.M. Weeks, *Nat. Struct. Biol.*, 2000, **7**, 362.
71. X.W. Fang, T. Pan and T.R. Sosnick, *Nat. Struct. Biol.*, 1999, **6**, 1091.
72. R. Russell, I.S. Millett, M.W. Tate, L.W. Kwok, B. Nakatani, S.M. Gruner, S.G. Mochrie, V. Pande, S. Doniach, D. Herschlag and L. Pollack, *Proc. Natl. Acad. Sci. U.S.A.*, 2002, **99**, 4266.
73. A.E. Webb and K.M. Weeks, *Nat. Struct. Biol.*, 2001, **8**, 135.
74. A.E. Webb, M.A. Rose, E. Westhof and K.M. Weeks, *J. Mol. Biol.*, 2001, **309**, 1087.
75. K.L. Buchmueller and K.M. Weeks, *Biochemistry*, 2003, **42**, 13869.
76. M. Xiao, M.J. Leibowitz and Y. Zhang, *Nucleic Acids Res.*, 2003, **31**, 3901.
77. R. Russell, I.S. Millett, S. Doniach and D. Herschlag, *Nat. Struct. Biol.*, 2000, **7**, 367.

78. X. Fang, K. Littrell, X.J. Yang, S.J. Henderson, S. Siefert, P. Thiyagarajan, T. Pan and T.R. Sosnick, *Biochemistry*, 2000, **39**, 11107.
79. R. Das, L.W. Kwok, I.S. Millett, Y. Bai, T.T. Mills, J. Jacob, G.S. Maskel, S. Seifert, S.G. Mochrie, P. Thiyagarajan, S. Doniach, L. Pollack and D. Herschlag, *J. Mol. Biol.*, 2003, **332**, 311.
80. L.W. Kwok, I. Shcherbakova, J.S. Lamb, H.Y. Park, K. Andresen, H. Smith, M. Brenowitz and L. Pollack, *J. Mol. Biol.*, 2005.
81. P. Rangan, B. Masquida, E. Westhof and S.A. Woodson, *Proc. Natl. Acad. Sci. U.S.A.*, 2003, **100**, 1574.
82. U.A. Perez-Salas, P. Rangan, S. Krueger, R.M. Briber, D. Thirumalai and S.A. Woodson, *Biochemistry*, 2004, **43**, 1746.
83. S. Chauhan, G. Caliskan, R.M. Briber, U. Perez-Salas, P. Rangan, D. Thirumalai and S.A. Woodson, *J. Mol. Biol.*, 2005, **353**, 1199.
84. Y.H. Wang, F.L. Murphy, T.R. Cech and J.D. Griffith, *J. Mol. Biol.*, 1994, **236**, 64.
85. S.K. Silverman and T.R. Cech, *Biochemistry*, 1999, **38**, 14224.
86. S.K. Silverman, M.L. Deras, S.A. Woodson, S.A. Scaringe and T.R. Cech, *Biochemistry*, 2000, **39**, 12465.
87. S.K. Silverman and T.R. Cech, *RNA*, 2001, **7**, 161.
88. M.L. Deras, M. Brenowitz, C.Y. Ralston, M.R. Chance and S.A. Woodson, *Biochemistry*, 2000, **39**, 10975.
89. K. Takamoto, R. Das, Q. He, S. Doniach, M. Brenowitz, D. Herschlag and M.R. Chance, *J. Mol. Biol.*, 2004, **343**, 1195.
90. L.J. Su, M. Brenowitz and A.M. Pyle, *J. Mol. Biol.*, 2003, **334**, 639.
91. D.E. Draper, D. Grilley and A.M. Soto, *Annu. Rev. Biophys. Biomol. Struct.*, 2005, **34**, 221.
92. L.G. Laing, T.C. Gluick and D.E. Draper, *J. Mol. Biol.*, 1994, **237**, 577.
93. G.S. Manning, *Biophys. Chem.*, 1977, **7**, 189.
94. G.S. Manning, *Q. Rev. Biophys.*, 1978, **11**, 179.
95. V.K. Misra and D.E. Draper, *Biopolymers*, 1998, **48**, 113.
96. D. Grilley, A.M. Soto and D.E. Draper, *Proc. Natl. Acad. Sci. U.S.A.*, 2006, **103**, 14003.
97. R. Shiman and D.E. Draper, *J. Mol. Biol.*, 2000, **302**, 79.
98. V.K. Misra, R. Shiman and D.E. Draper, *Biopolymers*, 2003, **69**, 118.
99. K. Chin, K.A. Sharp, B. Honig and A.M. Pyle, *Nat. Struct. Biol.*, 1999, **6**, 1055.
100. T. Hermann, P. Auffinger and E. Westhof, *Eur. Biophys. J.*, 1998, **27**, 153.
101. P. Rangan and S.A. Woodson, *J. Mol. Biol.*, 2003, **329**, 229.
102. E. Koculi, C. Hyeon, D. Thirumalai and S.A. Woodson, *J. Am. Chem. Soc.*, 2007, **129**, 2676.
103. E. Koculi, N.K. Lee, D. Thirumalai and S.A. Woodson, *J. Mol. Biol.*, 2004, **341**, 27.
104. E. Koculi, D. Thirumalai and S.A. Woodson, *J. Mol. Biol.*, 2006, **359**, 446.
105. M.S. Rook, D.K. Treiber and J.R. Williamson, *Proc. Natl. Acad. Sci. U.S.A.*, 1999, **96**, 12471.

106. C.A. Grosshans and T.R. Cech, *Biochemistry*, 1989, **28**, 6888.
107. T.S. McConnell, D. Herschlag and T.R. Cech, *Biochemistry*, 1997, **36**, 8293.
108. V.J. DeRose, *Curr. Opin. Struct. Biol.*, 2003, **13**, 317.
109. J.H. Cate, R.L. Hanna and J.A. Doudna, *Nat. Struct. Biol.*, 1997, **4**, 553.
110. R. Das, K.J. Travers, Y. Bai and D. Herschlag, *J. Am. Chem. Soc.*, 2005, **127**, 8272.
111. S. Basu, R.P. Rambo, J. Strauss-Soukup, J.H. Cate, D.A.A.R. Ferre, S.A. Strobel and J.A. Doudna, *Nat. Struct. Biol.*, 1998, **5**, 986.
112. A.M. Lambowitz and P.S. Perlman, *Trends Biochem. Sci.*, 1990, **15**, 440.
113. K.M. Weeks, *Curr. Opin. Struct. Biol.*, 1997, **7**, 336.
114. R. Schroeder, R. Grossberger, A. Pichler and C. Waldsich, *Curr. Opin. Struct. Biol.*, 2002, **12**, 296.
115. X. Chen, R.R. Gutell and A.M. Lambowitz, *J. Mol. Biol.*, 2000, **301**, 265.
116. G. Mohr, M.G. Caprara, Q. Guo and A.M. Lambowitz, *Nature*, 1994, **370**, 147.
117. G. Mohr, A. Zhang, J.A. Gianelos, M. Belfort and A.M. Lambowitz, *Cell*, 1992, **69**, 483.
118. M.G. Caprara, G. Mohr and A.M. Lambowitz, *J. Mol. Biol.*, 1996, **257**, 512.
119. X. Chen, G. Mohr and A.M. Lambowitz, *RNA*, 2004, **10**, 634.
120. A. Gampel and T.R. Cech, *Genes Dev.*, 1991, **5**, 1870.
121. A. Gampel, M. Nishikimi and A. Tzagoloff, *Mol. Cell. Biol.*, 1989, **9**, 5424.
122. K.M. Weeks and T.R. Cech, *Cell*, 1995, **82**, 221.
123. K.M. Weeks and T.R. Cech, *Science*, 1996, **271**, 345.
124. G. Bokinsky, L.G. Nivon, S. Liu, G. Chai, M. Hong, K.M. Weeks and X. Zhuang, *J. Mol. Biol.*, 2006, **361**, 771.
125. M.G. Caprara, P. Chatterjee, A. Solem, K.L. Brady-Passerini and B.J. Kaspar, *RNA*, 2007, **13**, 211.
126. I. Garcia and K.M. Weeks, *Biochemistry*, 2004, **43**, 15179.
127. A.E. Webb, M.A. Rose, E. Westhof and K.M. Weeks, *J. Mol. Biol.*, 2001, **309**, 1087.
128. S. Mohr, J.M. Stryker and A.M. Lambowitz, *Cell*, 2002, **109**, 769.
129. H.R. Huang, C.E. Rowe, S. Mohr, Y. Jiang, A.M. Lambowitz and P.S. Perlman, *Proc. Natl. Acad. Sci. U.S.A.*, 2005, **102**, 163.
130. A. Zhang, V. Derbyshire, J.L. Salvo and M. Belfort, *RNA*, 1995, **1**, 783.
131. C. Waldsich, B. Masquida, E. Westhof and R. Schroeder, *EMBO J.*, 2002, **21**, 5281.
132. C. Waldsich, R. Grossberger and R. Schroeder, *Genes Dev.*, 2002, **16**, 2300.
133. C.K. Ma, T. Kolesnikow, J.C. Rayner, E.L. Simons, H. Yim and R.W. Simons, *Mol. Microbiol.*, 1994, **14**, 1033.

134. R.A. Poot, N.V. Tsareva, I.V. Boni and J. van Duin, *Proc. Natl. Acad. Sci. U.S.A.*, 1997, **94**, 10110.
135. K. Semrad and R. Schroeder, *Genes Dev.*, 1998, **12**, 1327.
136. J.M. Burke, M. Belfort, T.R. Cech, R.W. Davies, R.J. Schweyen, D.A. Shub, J.W. Szostak and H.F. Tabak, *Nucleic Acids Res.*, 1987, **15**, 7217.
137. P.L. Adams, M.R. Stahley, A.B. Kosek, J. Wang and S.A. Strobel, *Nature*, 2004, **430**, 45.
138. F. Guo, A.R. Gooding and T.R. Cech, *Mol. Cell*, 2004, **16**, 351.

Subject Index

5′-splice site recognition in group I intron ribozyme, 185

$A^+ \cdot C$ pairing, 17
activation parameters for peptidyl transfer reaction, 278
active site,
   hammerhead ribozyme, 37
   HDV ribozyme, 97
   of the GIR1 ribozyme, 245
   of the group II intron ribozyme, 216
   of the ribosome, 279
   of the spliceosome, 255
   hammerhead ribozyme, 57
alignment and proximity, in peptidyl transfer, 279
A-minor interaction, 81, 186, 245
   in group II introns, 211
   in RNase P RNA , 162
antigenomic HDV, 93
aptamers, 124, 134
Azoarcus group I intron ribozyme,
   folding, 304
   structure, 84

bI5 ribozyme, 303
branch site reccognition by group II intron ribozyme, 214
branching reaction,
   of group II intron ribozyme, 201
   of the GIR1 ribozyme, 229
Brønsted analysis, 23, 101
Brønsted coefficient, 275
Brownian dynamics, 306

catalytic mechanism,
   hairpin ribozyme, 73
   hammerhead ribozyme, 50
   HDV ribozyme, 99
   of *glmS* ribozyme, 145
   of hammerhead ribozyme, 43
   of peptidyl transfer, 285
   of the group II intron ribozyme, 219
   RNase P RNA, 155
   of ribozymes, 7
catalytic metal ions in RNase P, 157
CBP2 in RNA folding, 308
coenzyme in *glmS*, 139
CoTC element, 123
counterion condensation, 108
CPEB3, 93
   ribozyme, 125
   structure, 127
crosslinking analysis of the hammerhead ribozyme, 42
CRS1, 221
cyclic 2′3′-phosphate, 7, 51, 165
CYT-18 in RNA folding, 308

Didymium iridis, 229
dinucleotide platform in RNase P RNA, 162
drug targetting, 148

elongation factors,
   EF-G, eEF2, 272
   EF-Tu, eEF1A, 271
exogenous base rescue, 101

folding of RNA, 5, 295
  free energy landscapes, 301
  in vivo, 302
  kinetic traps, 301
  of group I introns, 295
  of group II introns, 211
  of hairpin ribozyme, 69
  protein dependence, 307
four-way junction, 69
fragment reaction in peptide bond synthesis, 276
free energy landscapes, of folding, 301

gene expression, control, 135
general acid-base catalysis, 8, 12, 40, 74, 100, 147, 275
genomic HDV, 93
GIR1 branching ribozyme, 222, 229
  active site, 245
  biochemistry, 240
  branching reaction, 240
  origins, 247
  structural modelling, 241
GIR1 ribozyme, 229
glmS ribozyme, 134
  as drug target, 148
  catalytic mechanism, 145
  proton transfer in mechanism, 30
  structure, 141
glucosamine 6-phosphate, 135
  as *glmS* ribozyme cofactor, 30
ground state destabilization, 112
group I intron ribozyme, 178
  evolutionary origins, 197
  guanosine binding site, 180
  structure, 181
group I ribozyme,
  folding, 295
  structure, 296
group II intron ribozyme, 201
  active site, 216
  catalytic mechanism, 219
  chemical reactions, 201
  comparison with the spliceosome, 222, 253

  domain structure, 205
  folding, 211
  relationship to RNaseP, 223
  tertiary structure, 209
group II introns,
  biology, 204
  RNA folding, 211
guanosine binding pocket, 244
  in group I intron ribozyme, 180, 187
guanosine cofactor,
  in group I intron ribozyme, 180

hairpin ribozyme, 66
  catalytic mechanism, 73
  proton transfer in mechanism, 27
  similarity with VS ribozyme, 88
  structure, 67
hammerhead ribozyme, 37, 48, 129
  active site, 37
  active site structure, 57
  activity in crystal, 53
  catalytic mechanism, 43, 50
  crosslinking analysis, 42
  crystal structures, 48
  full length structure, 57
  molecular modelling structure, 43
  proton transfer in mechanism, 28
HDV ribozyme, 92, 127
  active site, 97
  catalytic mechanism, 99
  proton transfer in mechanism, 22
  structure, 95
helical junctions,
  in hairpin ribozyme, 69
  in spliceosomal RNAs, 261
  in the GIR1 ribozyme, 245
  in VS ribozyme, 77
hepatitis B, 92
hepatitis delta virus, 92, 130
  replication, 93
hierarchical folding, 297
histidine, 17
HIV, 5
homing endonuclease, 178, 231
Hoogsteen base pairing, 17
hydroxyl radical footprinting, 298

Subject Index

I-*DirI*, 230
imidazole, 101
   as pseudonucleotide, 74
in vitro selection, 124
in vivo folding of RNA, 302
induced fit,
   in peptidyl transfer center, 282
in-line nucleophilic attack, 51, 147
ISL in U6 snRNA, 258

kinetic ambiguity, 23, 103
kinetic solvent isotope effect, 23, 101
kinetic traps in RNA folding, 301

lariat, 202, 229, 255
lariat cap, 229
Lewis acid, 6, 73
LtrA, 213

mammalian ribozymes, 123
maturase, 221
metal ions, 73, 108, 136, 143, 191, 305
   and RNA folding, 5, 69
   catalytic, 111, 246, 6, 23, 40, 220
   in folding group II introns, 212
   structural, 108
microscopic reversibility, 7, 22, 74
misfolding of RNA, 300
molecular modelling of hammerhead ribozyme, 43
mRNA splicing, 253
Mss116p, 222

Naegleria, 230
non-coding RNA, 131
nucleobase participation in catalysis, 8, 41, 82, 95
nucleotide analog interference mapping (NAIM), 74, 105, 194, 212
nucleotide analog interference suppression (NAIS), 194, 210

oxyanion hole, 17

P4-P6 domain of the group I intron, 181
P5abc domain of the group I intron, 182

parallel folding pathways, 301
peptide bond formation, 270
   uncatalysed reaction, 273
peptidyl transfer reaction, 270
   active site, 279
   catalytic mechanism, 285
   enzymology, 274
   intermediates, 281
peripheral folding elements, 5, 38, 57, 71, 300
pH rate profile, 75, 102, 127, 277
pH-dependence of RNase P rates, 157
phosphorane transition state, 7, 50
phosphorothioate substitution, 25, 129, 137, 157, 164, 220, 255
   in group I intron ribozyme, 192
   in VS ribozyme, 82
   in HDV ribozyme, 106
phosphoryl transfer reactions, 1
pK_A shifting, 6, 15, 104
plant virus replication process, 49
Poisson-Boltzmann, 109
   equation, 19, 306
polyelectrolyte effect, 306
protein synthesis, 270
protein-dependent RNA folding, 308
protium, 12
proton inventory, 26, 101
proton shuttling in peptidyl transfer reaction, 286
proton transfer, 11
   in the *glmS* ribozyme, 30
   in the hairpin ribozyme, 27
   in the hammerhead ribozyme, 28
   in the HDV ribozyme, 22
   in the VS ribozyme, 29
   in the CPEB3 ribozyme, 127
Prp8, 257
pseudouridine, 262
puromycin, 272
rate enhancements by ribozymes and enzymes, 4
reverse splicing of RNA, 202
Ribonuclease P, 153
ribose zipper, 68
   in RNase P RNA, 162

ribosome, 271
riboswitch, 134
RNA chaperones, 263, 308
RNA collapse in RNA folding, 302
RNA misfolding, 300
RNA splicing by group II introns, 201
RNA world hypothesis, 4
RNase A, 17, 20
RNase P, 153
  relationship to group II intron, 223
RNase P RNA, catalytic mechanism, 155
  complex with tRNA, 162
  crystal structures, 160
  phylogenetic variation, 158

self-splicing introns, group I, 178
signal recognition particle (SRP), 153
single-molecule studies, 8
  and RNA folding, 69
  of ribozyme reactions, 71
small-angle X-ray scattering (SAXS), 8, 79, 303
S_N2 mechanism, 43, 50, 66, 93, 155, 219, 255
snRNA, 253
snRNP, 253
solvent reorganization in peptidyl transfer reaction, 287
spliceosome, 253
  relationship to group II intron, 222
splicing,
  of mRNA, 253
  regulation, 263
structure of VS ribozyme, 77
SunY (nrD) group I intron, 183
synchronous reactions, 12

TAR RNA, 5
tetrahedral intermediates in peptide bond synthesis, 273
Tetrahymena group I intron ribozyme structure, 182
tetraloop-receptor interactions in RNase P RNA, 162
therapeutic agents, 4
trans cleaving ribozymes, 131
transition state analog, 73
  for peptidyl transfer reaction, 281
trapping ribozyme reaction intermediates, 53
t-RNA, 272
  as effector in peptidyl transfer, 286
  processing by RNaseP, 153
twin-ribozyme intron, 231
  processing pathway, 234
Twort group I intron structure, 183

U2-U6 complex, 256
  in vitro crosslinking reaction, 259
U4-U6 complex in spliceosome assembly, 266

Varkud satellite (VS) ribozyme, 66
VEGFR mRNA, 5
VS ribozyme, 29, 66, 76
  A730 loop, 82
  active site, 82
  catalytic mechanism, 84
  proton transfer in mechanism, 29
  similarity with hairpin ribozyme, 88
  structure, 77
  substrate, 80

ω-G, 180, 244